Physiology

D0248742

Physiology

Series editor
Daniel Horton-Szar
BSc(Hons) MBBS (Hons), MRCGP

Northgate Medical Practice
Canterbury
Kent, UK

Faculty advisors
Gavin Donaldson
BSc(CNNA) PhD

Senior Lecturer
Barts and the London School of
Medicine and Dentistry, Queen
Mary University of London
London, UK

John C Atherton
BSc, PhD

Senior Lecturer in Medical
Education and Physiology
School of Medicine
University of Keele, UK

Mohammad Shahid
BSc, MBBS
ST2 Orthopaedics, Lister Hospital, Stevenage, Hertfordshire, UK

Ayesha Nunhuck
BSc, MBBS
GPVTS ST1, Medway Maritime Hospital, Gillingham, Kent, UK

MOSBY

ELSEVIER

Edinburgh • London • New York • Oxford • Philadelphia • St Louis • Sydney • Toronto 2008

MOSBY
ELSEVIER

Commissioning Editor	Fiona Conn/Alison Taylor
Development Editor	Ailsa Laing
Project Manager	Jess Thomson
Page Designer	Sarah Russell
Cover Design	Stewart Larking
Icon illustrations	Geo Parkin
Illustration Management	Merlyn Harvey
Illustrators	Joanna Cameron and Marion Tasker

First published 2008

ISBN: 978-0-7234-3388-0

British Library Cataloguing in Publication Data
A catalogue record for this book is available from the British Library

Library of Congress Cataloging in Publication Data
A catalog record for this book is available from the Library of Congress

Note
Knowledge and best practice in this field are constantly changing. As new research and experience broaden our knowledge, changes in practice, treatment and drug therapy may become necessary or appropriate. Readers are advised to check the most current information provided (i) on procedures featured or (ii) by the manufacturer of each product to be administered, to verify the recommended dose or formula, the method and duration of administration, and contraindications. It is the responsibility of the practitioner, relying on their own experience and knowledge of the patient, to make diagnoses, to determine dosages and the best treatment for each individual patient, and to take all appropriate safety precautions. To the fullest extent of the law, neither the Publisher nor the authors assume any liability for any injury and/or damage to persons or property arising out of or related to any use of the material contained in this book.

The Publisher

Working together to grow
libraries in developing countries

www.elsevier.com | www.bookaid.org | www.sabre.org

ELSEVIER   BOOK AID International   Sabre Foundation

ELSEVIER
your source for books, journals and multimedia in the health sciences
www.elsevierhealth.com

The publisher's policy is to use **paper manufactured from sustainable forests**

Printed in China

Physiology is often thought to be a difficult discipline to comprehend. However, knowing the basics allows one to understand the more difficult concepts. This book aims to provide the reader with a good grounding in physiology.

Furthermore, this book can be used as a revision aide when time is short before exams. It is presented in a simple logical format and then demonstrates how the physiology knowledge can be applied in a clinical setting.

To help consolidate that knowledge there are questions for each chapter which follow the typical format used by most medical schools.

This book is aimed at medical students, basic science students and doctors revising for membership exams. We hope you enjoy reading this book as much as we have done writing it.

Mohammad Shahid
Ayesha Nunhuck

More than a decade has now passed since work began on the first editions of the *Crash Course* series. Medicine never stands still, and the work of keeping this series relevant for today's students is an ongoing process. New titles build upon the success of the preceding books, keeping the series up to date with the latest medical research and developments in pharmacology and current best practice.

As always, we listen to feedback from the thousands of students who use *Crash Course* and have made further improvements to the layout and structure of the books. Each chapter now starts with a set of learning objectives, and the self-assessment sections have been enhanced and brought up-to-date with modern exam formats. We have also worked to integrate points of clinical relevance into the basic medical science material, which will not only add to the interest of the text but will reinforce the principles being described.

Despite fully revising the books, we hold fast to the principles on which we first developed the series: *Crash Course* will always bring you all the information you need to revise in compact, manageable volumes that integrate basic medical science and clinical practice. The books still maintain the balance between clarity and conciseness, and provide sufficient depth for those aiming at distinction. The authors are medical students and junior doctors who have recent experience of the exams you are now facing, and the accuracy of the material is checked by senior faculty members from across the UK.

I wish you all the best for your future careers!

Dr Dan Horton-Szar
Series Editor

Figure acknowledgements

Figures 2.7, 2.8, 8.14, 8.24, 8.27 and 8.28 from Datta S 2003 Crash Course Renal and Urinary Systems, 2nd Edn. Edinburgh, Mosby

Figures 2.10 and 6.8 from McGowan P 2003 Crash Course Respiratory System, 2nd Edn. Edinburgh, Mosby

Figures 3.6, 3.7, 3.8, 3.11, 3.12, 3.14, 3.15, 3.16 and 3.20 from Briar C et al. 2003 Crash Course Nervous System, 2nd Edn. Edinburgh, Saunders

Figures 5.6, 5.8, 5.17, 5.19 and 5.20 from Fagan T 2003 Crash Course Cardiovascular. Edinburgh, Mosby

Figures 5.7 and 5.14 from Young B et al 2006 Wheater's Functional Histology 5e. Edinburgh, Churchill Livingstone

Figure 6.4 from Boron WF, Boulpaep EL 2005 Medical and Cellular Physiology: A cellular and molecular approach, updated edition. Philadelphia, Saunders

Figures 6.20, 6.21 and 6.25 from Costanzo L 2006. Physiology, 3rd Edn. Philadelphia, Saunders

Figures 7.4, 7.13, 7.17, 7.22 and 7.23 from Long M 2002 Crash Course Gastrointestinal System. Edinburgh, Mosby

I would like to dedicate this book to my parents who have worked hard to give me the opportunities I have today.

MS

Then which of the favours of your Lord will ye deny?

AN

Part I: Physiology 1

Contents

Arterioles blood vessels that are smaller than the arteries and that branch from arteries with variable amounts of elastic and smooth tissue.

ATP (adenosine triphosphate) a general source of energy for all intracellular metabolic reactions.

Atrophy reduction in the size.

Autoregulation process by which tissue perfusion remains relatively constant despite blood pressure changes.

Baroreceptors sensitive to pressure and located in the aorta, internal carotid arteries and other large arteries in the neck and chest.

Capillaries join arterioles and venules and are present in almost every tissue in the body.

Cardiac output (CO) volume of blood ejected by one ventricle into its respective artery each minute. Calculated as the heart rate multiplied by the stroke volume.

Chemoreceptors located in the carotid sinus and aortic arches in small structures known as carotid and aortic bodies.

Diffusion the passive, simple movement of a substance down a concentration gradient.

Endocardium consists of three layers and is continuous with the endothelial lining of the large blood vessels attached to the heart.

Endoplasmic reticulum forms a network of membranes within the cell.

Facilitated diffusion this process is faster than simple diffusion with the passage of substances down their concentration gradients requiring a transporter.

Golgi apparatus membranous sacs that sort and modify proteins arriving from the granular endoplasmic reticulum, packing them into vesicles before sending them to other organelles or secreting them.

Golgi tendon organs (GTO) bundles of collagen fibres encapsulated by a connective tissue layer present at the muscle–tendon junction.

Haemopoiesis the formation of red blood cells in the bone marrow.

Haemostasis control of bleeding.

Heart rate (HR) number of ventricular contractions in one minute.

Homeostasis the maintenance of constant conditions within the body.

Hyperplasia an increase in tissue/organ size due to an increase in cell number.

Hypertrophy an increase in tissue/organ size due to an increase in cell size.

Inotropes substances that affect the force of cardiac contractility.

Lysosomes single-membraned oval organelle containing highly acidic digestive enzymes that break down bacteria, cell debris and dead organelles.

Messenger RNA (mRNA) carries the genetic code from the nucleus to the cytoplasm.

Mitochondria double-membraned, elongated, ovoid structures that function to make energy available to cells in the form of ATP.

Motor units each skeletal muscle fibre is innervated by a single motor neuron, which comprises a motor neuron and all the muscle fibres it innervates.

Muscle spindles these organs lie parallel to the skeletal muscle fibres and measure the extent of muscle stretch.

Myocardium consists of cardiac muscle cells (myocytes), which are responsible for cardiac contractility.

Myofibrils filamentous bundles on the individual muscle fibre that run along the entire length of the fibre.

Myocytes muscle cells.

Nuclear membrane/envelope two membranes surrounding the nucleus containing pores that regulate the entry and exit of molecules.

Nucleoli highly coiled structures not enveloped by a nuclear membrane and containing RNA and protein components.

Nucleus this is present in almost all cells and is its control centre.

Ossification the conversion of fibrous tissue or cartilage into bone; can either be intramembranous or endochondral.

Pericardium fibrous sac covering the whole heart.

Peroxisomes single-membraned, oval organelles that destroy the highly toxic hydrogen peroxide (H_2O_2) that is produced by certain cell reactions.

Ribosomes large particles composed of about 70 proteins and several RNA molecules. There are two subunits of different sizes, 30s and 50s, with the former being smaller.

Ribosomal RNA (rRNA) where the protein molecules are actually assembled.

Sarcomere fundamental contractile unit within the muscle, from one Z-line to the next.

Sarcoplasm muscle fibre matrix where the myofibrils are suspended.

Sarcoplasmic reticulum (SR) an endoplasmic reticulum equivalent in the muscle fibre. Runs longitudinally along the myofibrils and wraps around groups of myofibrils.

Sinoatrial (SA) node part of the heart that causes the pacemaker potential.

SOB shortness of breath

Stroke volume (SV) volume of blood ejected in one ventricular contraction.

Total peripheral resistance (TPR) resistance to blood flow in the circulatory system.

Transcription mRNA synthesis.

Transfer RNA (tRNA) transfers amino acids to the ribosomes to manufacture proteins.

Translation formation of proteins from the mRNA.

Transmural pressure the pressure across the wall of the vessel; can be affected by external and internal pressures.

Venules collect the blood from the capillaries and transport it to the veins.

PHYSIOLOGY

Objectives

In this chapter, you will learn to:
- Describe in detail the structure and function of the mammalian cell and the different cell organelles
- Relate the structure of the above to their function
- Discuss how energy is formed within the cell, e.g. Krebs cycle
- Outline the structure and function of DNA and RNA
- Describe in detail the sequence of events involved in protein synthesis
- Describe the different components that make up the cell membrane
- Discuss the different forms of transport involved in the movement of molecules across cell membranes

OVERVIEW OF BODY SYSTEMS

Circulation and transport

About 50-65% of the body consists of liquids, termed 'fluid'. Two-thirds of this fluid is contained within cells (i.e. it is intracellular) and one-third is outside the cells (extracellular). The different types of fluid have different constituents (Fig. 1.1).

Extracellular fluid is a medium from which cells take up nutrients and into which they discharge metabolic waste products. About 75% of the extracellular fluid is found in between the cells, where it is called interstitial fluid; once it passes into the lymphatic vessels its name changes to lymph. Other extracellular fluids include cerebrospinal fluid and pericardial fluid; plasma, which is found in the blood vessels, comprises the remainder.

As blood flows through the capillaries, some of the plasma passes through the capillary pores to become the interstitial fluid that bathes the cells. The continual bidirectional movement of extracellular fluid from plasma into the interstitial fluid permits the exchange of nutrients and waste products.

- Main components of the intracellular fluid: K^+, Mg^{2+}, PO_4^{3-}
- Main components of the extracellular fluid: Na^+, Cl^-, HCO_3^-, oxygen, glucose, amino acids, fatty acids

Nutrient intake

The nutrients in the extracellular system originate from the:

- Respiratory system: blood flowing through the lung picks up oxygen via diffusion from the alveoli.
- Gastrointestinal system: nutrients from the gut, such as sugars, amino acids and fatty acids, are absorbed into the blood.
- Liver: the liver chemically modifies substances absorbed from the gut to facilitate their use by the body. Other body tissues or organs store some of these nutrients (e.g. fat cells, endocrine glands) until they are required.
- Musculoskeletal system: moves the body to enable it to physically obtain food and avoid adverse surroundings.

Removal of metabolic end products

- Lungs: carbon dioxide diffuses into the alveoli and is expelled into the atmosphere via exhalation.
- Kidneys: the kidneys remove the majority of waste products not needed by cells (e.g. urea, uric acid, excess ions and water), but not CO_2. At the same time, the kidneys reabsorb those products needed by the body (glucose, electrolytes, amino acids and water).

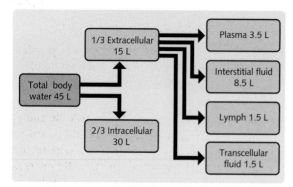

Fig. 1.1 Distribution of water in a 70-kg male.

Regulation of body functions

The nervous and hormonal systems work together to regulate muscular, secretory and metabolic activity and so achieve stability within the body.

Nervous system

The nervous system comprises three parts.

1. Central nervous system

The central nervous system (CNS) consists of the brain and spinal cord. It stores information and processes thoughts and emotions. Signals generated by the CNS are transferred via the motor output of the peripheral nervous system to initiate an appropriate response.

2. Peripheral nervous system

The peripheral nervous system (PNS) has two main functions:

1. Sensory: receptors detect the state of the surroundings and relay information to the CNS.
2. Motor: after the sensory information from the PNS receptors has been processed by the CNS, signals from the CNS instruct the motor component of the PNS to carry out the desired response.

3. Autonomic nervous system

The autonomic nervous system (ANS) innervates the internal organs. It controls – at a subconscious level – functions such as gastrointestinal motility, heart rate, blood vessel diameter and endocrine secretion.

Hormonal secretion

Eight primary endocrine glands secrete substances called hormones (Fig. 1.2) into the extracellular fluid, which transports them to their site of action.

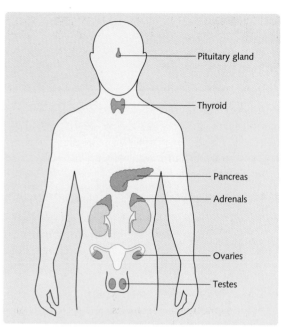

Fig. 1.2 Endocrine glands of the body.

Hormones regulate metabolic processes, e.g. the thyroid gland secretes the hormone thyroxine, which increases the rate of many reactions in the body.

Homeostasis

This term refers to the maintenance of constant conditions within the body; all the tissues and organs of the body work together to achieve homeostasis. In general, it depends on a negative feedback loop so that, when a change occurs in the body, the body aims to restore the status quo by reversing the direction of change. For example, if you are suddenly moved from a cold environment to somewhere hot, your body temperature will increase. As the body must keep its temperature within certain limits to optimize function, it sets in motion a number of actions to reduce its temperature. These actions include:

- Vasodilatation of the surface vessels: so that heat is lost by convection.
- Sweat production: this will increase to cool the body.
- Reduced metabolic heat production.

Such responses tend to keep the body's internal environment constant i.e. they sub-serve homeostasis.

THE CELL

Structure and function

Microstructure of the cell

Protoplasm is the collective name given to the basic components of the cell, i.e. the material from which the cell is made. The protoplasm therefore includes the cell nucleus and cytoplasm (Fig. 1.3). The protoplasm is made up of:

- Water: comprises the majority of the cell (around 80%).
- Proteins: form 15% of the cell. There are two types:
  - structural proteins: form the body of the cell
  - globular proteins: these are mainly enzymes.
- Lipids: there are three main types within the cell:
  - phospholipids and cholesterol: form the outer and intracellular membranous barriers
  - triglycerides: provide the cell with energy.
- Carbohydrates: stored as glycogen and provide the cell with energy.
- Electrolytes, which are used for cellular reactions (e.g. action potential): potassium ions (K^+), magnesium ions (Mg^{2+}), phosphate ions (PO_4^{3-}), sulphate ions (SO_4^{2-}), bicarbonate ions (HCO_3^-).

> Water is the major component of all cells.

Organelles and their function

Within the cell are highly organized physical structures – organelles – which carry out specific functions.

Nucleus

This organelle is present in almost all cells and is the 'control centre' (Fig. 1.4). It stores and transmits genetic information (in the form of DNA) to the next generation. The genes within the cell's nucleus determine the types of protein made by the cell.

Nuclear membrane/envelope

These two membranes, which surround the nucleus, contain pores that regulate the entry and exit of molecules.

Nucleoli

The nucleoli are highly coiled structures that are not enveloped by a nuclear membrane. They contain RNA and protein components

Ribosomes

These large particles are composed of about 70 proteins and several RNA molecules (Fig. 1.5). There are two subunits of different sizes – 30s and 50s – the former being smaller.

Protein molecules are synthesized by ribosomes from amino acids, using genetic information derived from DNA in the nucleus and carried to the ribosomes by RNA messenger molecules. The proteins formed are either released in the cytosol

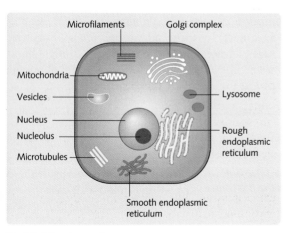

Fig. 1.3 Cell and organelles.

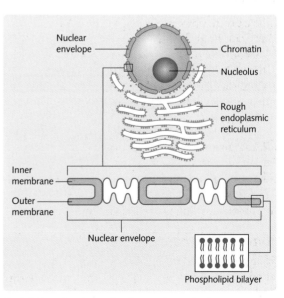

Fig. 1.4 The cell nucleus and nuclear membrane structure.

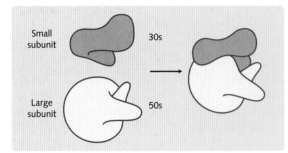

Fig. 1.5 Ribosomal structure.

(the fluid part of the cell's cytoplasm) or transferred to other organelles via the Golgi apparatus (see below).

Endoplasmic reticulum

The endoplasmic reticulum (ER) forms a network of membranes within the cell. There are two types:

1. Granular (or rough) ER: carries surface ribosomes. Proteins are synthesized on the attached ribosomes, enter the lumen of the ER and are either distributed within the cell or secreted to other cells.
2. Agranular (or smooth) ER: no ribosomes on the surface. This type of ER synthesizes fatty acids and regulates cellular levels of calcium (Ca^{2+}), which controls many of the cell's activities.

Golgi apparatus

These membranous sacs (Fig. 1.6) sort and modify proteins arriving from the granular ER, packaging them into vesicles before sending them to other organelles within the cell or secreting them.

Mitochondria

These are double-membraned, elongated, ovoid structures. The outer membrane is smooth and the inner is folded into tubes (cristae), thereby increasing

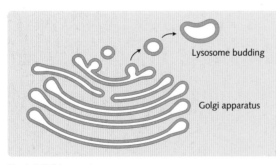

Fig. 1.6 Golgi apparatus.

the surface area. The cristae extend into the inside of the mitochondria (matrix) and contain DNA for mitochondrial protein synthesis.

The mitochondria function to make energy available to cells in the form of adenosine triphosphate (ATP).

Lysosomes

These single-membraned, oval organelles contain highly acidic digestive enzymes that break down bacteria, cell debris and dead organelles.

Peroxisomes

These single-membraned, oval organelles destroy the highly toxic compound hydrogen peroxide (H_2O_2), which is a by-product of certain cell reactions. Additionally, the perixosomes in the liver and kidney cells manufacture H_2O_2 and use it to detoxify various ingested molecules.

Filaments

The cytoplasm of most cells contains protein filaments – the cytoskeleton – which maintain the cell's shape and produce cell movements. The four main types, and their functions, are outlined in Figure 1.7.

Energy production in the cell

Cells use oxygen to break down foodstuffs (carbohydrates → glucose, proteins → amino acids, fats → fatty acids) and form the compound adenosine triphosphate (ATP). ATP is a general source of energy for all intracellular metabolic reactions.

Structure of ATP

ATP is a nucleotide containing the base adenine, the pentose sugar ribose and three phosphate molecules (Fig. 1.8). The end two phosphate molecules are connected by a high-energy bond. When this bond is hydrolysed to adenosine diphosphate (ADP), large amounts of energy are liberated. Loss of another phosphate to form adenosine monophosphate (AMP) releases yet more energy. Each covalent bond broken liberates about 30.6 kJ/mol. This is about the same as the energy in a single peanut.

Formation of ATP

The majority (95%) of ATP is formed in the mitochondrial matrix (Fig. 1.9). There are four stages of production:

1. Glycolysis Glycolysis is the formation of pyruvic acid (PA). Glucose, fatty acids and amino acids enter the cell cytoplasm and are converted to pyruvic acid, releasing energy for the formation of two molecules of ATP.

Fig. 1.7 Cell filaments: their proteins and functions.

Fig. 1.7 Cell filaments: their proteins and functions

Filament type	Protein subunit	Location	Function
Microfilament	Actin	All cells	Movement of organelles in cytoplasm
Intermediate filament	Several proteins	Cells exposed to frequent mechanical stress	Muscle contraction
Muscle: thick filament	Myosin	All cells – single filaments Muscle cells – bundles of filaments	
Microtubule	Tubulin	Nerve cells	Maintain cylindrical shape
		All cells	Movement of organelles in cytoplasm
		All cells	Chromosome separation during cell division
		Ciliated epithelium	Movement

2. Conversion of PA to acetyl-CoA PA enters the mitochondrial matrix, where enzymatic changes transform it into acetyl-CoA.

3. Krebs cycle Krebs cycle is sometimes called the citric acid or tricarboxylic acid cycle. Acetyl-CoA enters a series of chemical reactions in the mitochondrial matrix in which it is split into acetyl and CoA. The acetyl portion enters the cycle and the CoA is used in the formation of more acetyl-CoA. During the Krebs cycle, for every glucose molecule:

- Seven molecules of water are added.
- Sixteen hydrogen atoms and four molecules of carbon dioxide (CO_2) are released.
- Two molecules of ATP are formed.

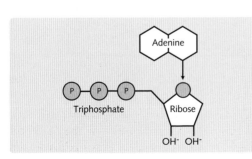

Fig. 1.8 Structure of adenosine triphosphate (ATP).

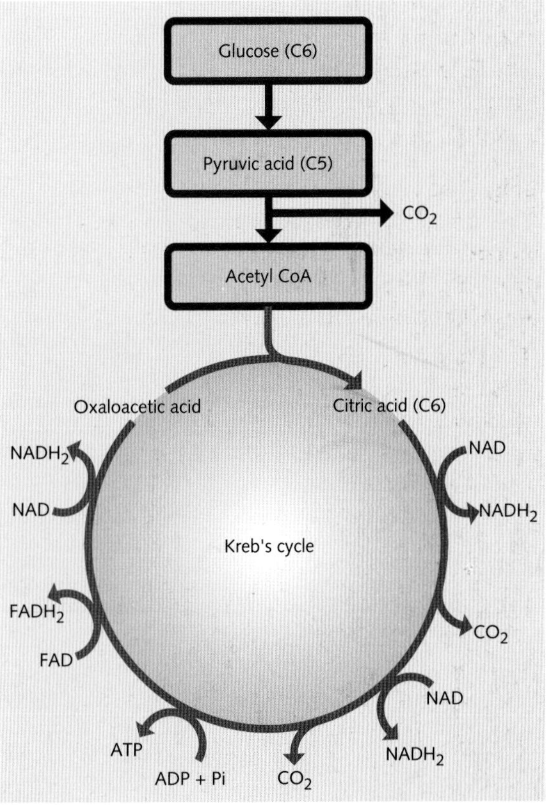

Glucose (C6)

Pyruvic acid (C5)

CO_2

Acetyl CoA

Oxaloacetic acid

Citric acid (C6)

$NADH_2$

NAD

NAD

$NADH_2$

Kreb's cycle

$FADH_2$

CO_2

FAD

NAD

ATP

$NADH_2$

ADP + Pi

CO_2

Fig. 1.9 Krebs cycle.

The CO_2 diffuses out of the mitochondria and is eventually expired as a waste gas from the lungs.

4. Oxidative phosphorylation   The formation of large quantities of ATP (Fig 1.10). Only a small net quantity of ATP (two molecules) is formed during Krebs cycle; more is produced during oxidative phosphorylation. There are seven main steps:

1. Hydrogen is split into a hydrogen ion (H^+) and an electron.
2. The electron enters the electron transport chain on the inner mitochondrial membrane.
3. The electron is transported from electron acceptor to electron acceptor (i.e. flavoproteins and cytochromes B, C and A) until it reaches cytochrome A3 (cytochrome oxidase).
4. Cytochrome oxidase helps form ionic oxygen, which combines with the H^+ to form water.
5. The large amounts of energy that are released during the transport of electrons pumps the H^+ from the inner matrix of the mitochondrion to between the inner and outer membrane (outer chamber).
6. The high concentration of H^+ in the outer chamber flows over the enzyme ATPase, which is attached to the inner mitochondrial membrane. This energy from the H^+ flow is used by ATPase to convert ADP to ATP.
7. ATP is transferred – by facilitated diffusion – from the mitochondrion to the cell cytoplasm.

How much ATP is produced?

In total, 38 molecules of ATP are produced from one molecule of glucose:

- Two molecules during glycolysis.
- Two molecules during Krebs cycle: for each cycle, one ATP is formed. As glucose splits into two pyruvic acid molecules, two molecules of ATP are produced.
- 34 molecules during oxidative phosphorylation.

One molecule of glucose produces a net of 38 molecules of ATP.

Other substrates (e.g. fats) can produce more ATP than carbohydrates. In fact, 1 g of fat can produce twice as much ATP as 1 g of carbohydrate. Protein is used as a substrate only when all the carbohydrate and fat reserves have been utilized, i.e. in a starvation state.

Functions of ATP

- Membrane transport of ions.
- Synthesis of chemical compounds, e.g. protein synthesis on ribosomes, formation of cholesterol, phospholipids, etc.
- Mechanical work, e.g. muscle contraction, ciliary function.

Fig. 1.10 Oxidative phosphorylation, the electron transfer chain and the citric acid chain.

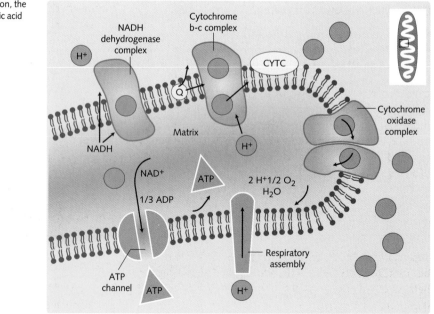

Genetic control of the cell

Genes control cell function. They determine what proteins – in the form of enzymes and structural proteins – are produced within the cell. Every gene is made of deoxyribonucleic acid (DNA), which regulates ribonucleic acid (RNA) to dictate the formation of the particular proteins.

DNA

DNA is a double helical chain composed of:

- Phosphoric acid, part of the backbone of the DNA molecule.
- A deoxyribose sugar, part of the backbone of the DNA molecule.
- Four nitrogenous bases:
  - two purines: adenine and guanine
  - two pyrimidines: thymine and cytosine.

The bases connect the two DNA strands via hydrogen bonds:

- Adenine can bind only to thymine.
- Guanine can bind only to cytosine.

This is known as complementary base pairing.

One amino acid is coded for by a sequence of three bases (a triplet).

When DNA is split into its individual strands, the exposed bases provide the genetic code. Every three successive bases (triplet) codes for one amino acid. Chains of amino acids form proteins. The four bases can be arranged in 64 different three-letter combinations ($4 \times 4 \times 4 = 64$). As only 21 different amino acids are synthesized in the body, there are more than enough combinations and some amino acids are represented by more than one code. There are also codes for stop signals, which indicate that the end of a genetic message has been reached (Fig 1.11).

The remainder of the DNA has been classed as redundant DNA. However, new research is challenging this idea and hinting at the possibility of this 'junk DNA' having important functions.

Super-coiling   DNA is a very long molecule and must be packaged very well to fit into a cell. The double helix gives it a natural twist but further twisting packs it even tighter; this is super-coiling.

Fig. 1.11 Examples of codons.

	U	C	A	G	
U	Phe	Ser	Tyr	Cys	U
	Phe	Ser	Tyr	Cys	C
	Leu	Ser	Stop	Stop	A
	Leu	Ser	Stop	Stop	G
C	Leu	Pro	His	Arg	U
	Leu	Pro	His	Arg	C
	Leu	Pro	Gin	Arg	A
	Leu	Pro	Gin	Arg	G
A	Ile	Thr	Asn	Ser	U
	Ile	Thr	Asn	Ser	C
	Ile	Thr	Lys	Arg	A
	Met	Thr	Lys	Arg	G
G	Val	Ala	Asp	Gly	U
	Val	Ala	Asp	Gly	C
	Val	Ala	Glu	Gly	A
	Val	Ala	Glu	Gly	G

Fig. 1.11 Examples of codons.

For proteins to form, the DNA code must be transferred to an RNA code. As DNA is in the cell nucleus and the majority of cell function occurs in the cytoplasm, the RNA acts as an intermediary to these processes.

RNA

RNA is identical to DNA with two exceptions:

1. Deoxyribose is replaced with ribose.
2. Thymine is replaced by the pyrimidine uracil.

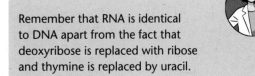

Remember that RNA is identical to DNA apart from the fact that deoxyribose is replaced with ribose and thymine is replaced by uracil.

There are three types of RNA:

1. Messenger RNA (mRNA): carries the genetic code from the nucleus to the cytoplasm.

2. Transfer RNA (tRNA): transfers amino acids to the ribosomes to manufacture proteins.
3. Ribosomal RNA (rRNA): where the proteins molecules are actually assembled.

Protein synthesis

Transcription: mRNA synthesis

1. RNA polymerase attaches to the DNA promoter and moves along the helix (Fig. 1.12).
2. The DNA double helix unwinds and the strands separate.
3. RNA nucleotides attach – via hydrogen bonds – to the exposed bases on one DNA strand.
4. Covalent bonds are formed between the RNA nucleotide phosphate and ribose, producing mRNA.
5. When RNA polymerase comes across a DNA stop signal it breaks away from the DNA strand. The stop signal indicates that the mRNA is complete.
6. The hydrogen bonds holding the mRNA to the DNA strand break and the mRNA enters the nucleoplasm.

A three-base sequence in the mRNA transcript is known as a codon and is complementary to the three-base sequence in DNA (triplet).

Splicing

The primary mRNA transcript contains specific segments (exons) that code for amino acids; the remainder of the transcript does not code for any protein (introns). Splicing removes the introns and combines the remaining exons. This results in the final mRNA.

Translation

- The mRNA passes from the cell nucleus to the cytoplasm.
- rRNA binds to one end of the mRNA.
- The three-base anticodon in an amino-acid–tRNA complex pairs with its corresponding codon on the mRNA.
- The amino acid on the tRNA forms a covalent bond with the adjacent amino acid to elongate the polypeptide chain.
- The tRNA is liberated from the bound amino acid and released.
- The ribosome travels one codon along the mRNA and the procedure repeats until a termination sequence is reached.

Post-translational modifications

When the polypeptide chain has been assembled, various chemical groups might be attached and/or the protein might be split into several smaller side chains.

There are four stages of protein synthesis: transcription, splicing, translation and post-translational modifications.

Cell movement

There are two types of cell movement (Figs 1.13 and 1.14):

Fig. 1.12 DNA synthesis.

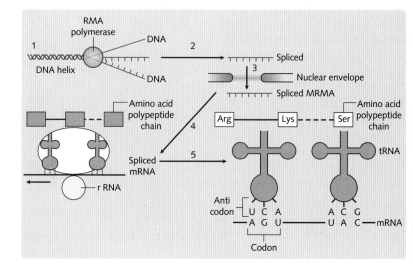

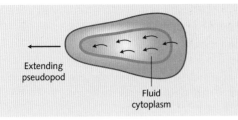

Fig. 1.13 Amoeboid movement.

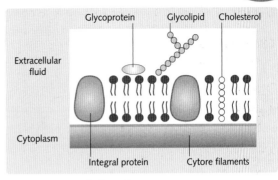

Fig. 1.15 Structure of cell membranes.

1. Amoeboid movement: a protrusion from the cell – a pseudopodium – attaches to the substrate across which the cell is moving. The remainder of the cell body drags itself towards the pseudopodium (e.g. white blood cells through tissues). Amoeboid movement is initiated by chemical substances produced by the tissues (chemotaxis). The cells can move towards (positive chemotaxis) or away from (negative chemotaxis) an area of chemotactic substances.
2. Ciliary movement: whip-like movements of cilia propel the cell. Such movement occurs on the inside surfaces of the respiratory airways and fallopian tubes. In the respiratory airways, ciliary movements propel mucus towards the trachea; in the fallopian tubes they propel the ovum towards the uterine cavity.

- Cilia are composed of nine double and two single microtubules, which are enveloped by a cell membrane.
- Flagella (e.g. sperm tail) are structurally similar to cilia but are longer and move in a different pattern to cilia.

Membrane physiology
Cell membranes
The cell membrane consists of three components: lipids, proteins and carbohydrates (Fig. 1.15).

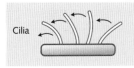

Move in whiplash motion

Fig. 1.14 Movement by cilia and flagella.

Lipids
Phospholipid molecules form a lipid bilayer, which envelops the entire cell and organelle. One part of the phospholipid molecule is hydrophilic (water loving) and the other is hydrophobic (water hating). This results in the molecules lining up with the hydrophilic portions back to back in contact with water. Because there is no chemical bond linking the phospholipids to each other they are free to move independently. This is sometimes termed 'the fluid mosaic model' and permits cells to change considerably without disruption to their structure.

The phospholipid bilayer regulates the movement of certain substances into and out of the cell. It prevents the entry of water-soluble substances (such as urea, glucose and ions) and allows the passage of fat-soluble substances (alcohol, oxygen and carbon dioxide).

Cholesterol, a steroid, inserts itself in the membrane with the same orientation as the phospholipid molecules. It functions to immobilize the first few hydrocarbon groups of the phospholipid molecules. If cholesterol was absent (as in a bacterium) a cell would need a cell wall. It also prevents crystallization of hydrocarbons and the membrane from shifting.

Cell membranes regulate the entry and exit of substances into and out of the cell.

Proteins
There are two types of protein in the cell membrane:

1. Integral proteins: extend all the way through the bilayer (i.e. are transmembranous) and form

structural channels through which ions and other water-soluble substances can permeate. They also form receptors for enzyme binding.

2. Peripheral proteins: located on the inside of the membrane and form enzymes.

Carbohydrates (glycocalyx)

The carbohydrates in the cell membrane combine with proteins (to form glycoproteins) and lipids (to form glycolipids) that protrude outside the cell. They function to:

- Repel other negative objects (because of their negative charge).
- Attach to other cells so can identify/be identified.
- Provide binding sites for hormones.
- Cells involved in immunity use the glycocalyx for recognition of host/foreign cells.

Membranous junctions

Adjacent cells can be joined together by different types of junction: tight junctions, gap junctions and spot desmosomes (Fig. 1.16).

Tight junctions

These junctions form an impermeable bond between adjacent cells and direct the passage of substances through the cells by preventing passage between them. They are found in the epithelial cell sheet lining the small intestine.

Gap junctions

These are protein tunnels, which form between adjacent cells. They allow for the passage of small molecules and/or ions between the cells. Cells connected by gap junctions are capable of working synchronously as a unit instead of individually. In the heart, gap junctions allow the atria and the ventricles to contract in sequence.

Spot desmosomes

These filamentous adhesions between nearby cells serve as mechanical reinforcements. Tissues under stress, e.g. cardiac muscle, are connected by spot desmosomes.

Transport

The movement of molecules through the cell membrane occurs by diffusion/passive and active transport.

Diffusion

Diffusion is the movement of molecules from an area of higher concentration to an area of lower concentration as the result of random thermal molecular motion. There are several different types:

> It is important to differentiate between the different types and subtypes of diffusion. This is a commonly asked question in exams.

Simple diffusion Movement of molecules occurs through a membrane opening (Fig. 1.17). This opening either passes through the lipid bilayer (allowing the passage of lipid-soluble substances) or takes the form of protein channels, e.g. sodium and potassium (Na^+ and K^+) channels, through which water- and lipid-insoluble molecules pass. The protein channels regulate their permeability via a gating system.

Voltage gating The electrical potential across the cell membrane influences the entry of certain substances. For example, during an action potential an impulse passes down a neuron; the resulting reduction in the voltage causes sodium channels in the adjacent portion of the membrane to open.

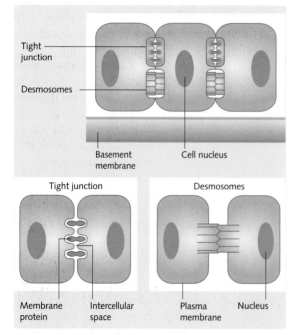

Fig. 1.16 Membranous junctions, showing a tight junction and desmosomes.

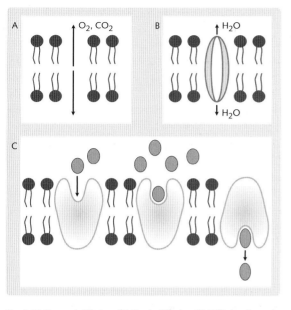

Fig. 1.17 Types of diffusion. (A) Simple diffusion. (B) Diffusion through channels: proteins and charged molecules are usually pulled through channels by water. (C) Facilitated diffusion: molecules (e.g. glucose) bind to protein, triggering a change in protein shape. This transports the glucose molecules across the membrane.

This allows the influx of sodium ions (Na^+) into the neuron and hence the continuation of the nerve impulse

Chemical gating Binding of another molecule permits entry. For example, acetylcholine opens the acetycholine channel, which allows the nerve signals to propagate.

Facilitated diffusion A carrier protein is required to transport a substance across the membrane. A good example is found in the liver cells, which control the concentration of glucose in the blood. Liver cells store excess glucose as glycogen when blood sugar levels are high. The breakdown of glycogen is closely controlled by hormones. The breakdown products of glycogen are impermeable to the liver cell membrane and so a transport protein, which functions by facilitated diffusion, allows the movement of glucose.

Osmosis This is the diffusion of water, across a semipermeable membrane (i.e. a membrane that allows the passage of certain, small molecules but which prevents the passage of large molecules). The water molecules diffuse from a region of higher water concentration to lower water concentration. *Note*: a low water concentration implies a high solute concentration:

- The solutions inside and outside cells contain water; the cell wall is a semipermeable membrane. Thus osmosis applies to the movement of water molecules into and out of the cell.
- If the solution outside the cell contains a *higher* concentration of water molecules than the interior of the cell (i.e. the external solution is *hypertonic* for water), then water molecules will diffuse into the cell. This will cause the cell to swell and lyse (break).
- If the solution outside the cell contains a *lower* concentration of water molecules than the inside (i.e. the external solution is *hypotonic* for water), then water will diffuse out of the cell. This will cause the cell to shrink (crenate).
- If the concentrations of water molecules in the intracellular and extracellular concentrations are identical (*isotonic*), there is no net movement of water.

Active transport

This energy-dependent system transports ions or molecules across a membrane against an electrochemical gradient. It can thus maintain the concentration of a substance against a concentration gradient, e.g. the K^+ concentration is high intracellularly and low extracellularly.

Active transport depends on two types of transmembrane carrier protein, which derive their energy sources in different ways:

- Primary transmembrane carrier proteins: receive their energy from ATP. They transport many different ions (e.g. Na^+, K^+, Ca^{2+} into and out of muscle cells; H^+ and K^+ into and out of the gastric glands). The most common type is the Na^+/K^+ pump system hereafter referred to as Na^+/K^+-ATPase , which is present in all cells of the body. This pumps Na^+ out of the cell, K^+ into the cell and maintains the electrical potential across the cell membrane (see Chapter 3).
- Secondary transmembrane carrier proteins: energy is generated from differences in ionic concentration between two sides of a membrane.

As the result of the Na^+/K^+-ATPase, K^+ is predominantly intracellular and Na^+ is predominantly extracellular.

Cotransport

This term refers to transport across a cell membrane when a carrier is occupied by two substances simultaneously. There are two types:

- Antiport: a membrane carrier creates a gradient for the movement of one substance in one direction and another substance in the opposite direction.

- Symport: two substances move in the same direction by means of a common carrier.

Physiology of the blood and body fluids

Objectives

In this chapter, you will learn to:
- Explain the different body fluid compartments and outline the methods and underlying principles by which these compartments can be measured
- Define osmolarity and osmolality, the effects of osmotic pressure, osmosis and the pressure of a gas in a solution
- Discuss how ions move between biological membranes and between the body and the external environment
- Describe the functions and components of blood
- Explain the S-shaped oxygen dissociation curve and how it is affected by carbon monoxide
- Discuss the Bohr and Haldane effects and how they affect carbon dioxide transport
- Describe the coagulation pathway, including the role of thrombin and platelets, the events involved in haemostasis, and the resolution of a clot

OVERVIEW OF BODY FLUIDS AND FLUID COMPARTMENTS

The major component of the human body is water, which accounts for 63% of an adult male. An increased body fat content is associated with ageing, obesity and being female. Consequently, the percentage of water in women falls to 52%. Dissolved within this water are carbon dioxide (CO_2), nutrients, proteins and charged particles (ions).

Fluid in the body is distributed into different compartments (Fig. 2.1):

- Intracellular fluid (ICF): the fluid inside the cells.
- Extracellular fluid (ECF): all fluids outside cells, comprising:
  - 75% interstitial fluid (ISF): the ECF that bathes the cells and lies outside the vascular system
  - 25% plasma: the non-cellular part of the blood (within the vascular system).

ISF and plasma are in a state of continual exchange via pores in the highly permeable capillary membrane. The two fluids therefore have a similar composition, with the exception of large proteins, which are trapped within the capillaries in the vascular system.

Transcellular fluid is another small compartment of body fluid. Although it can be viewed as a specialized type of ECF, their compositions vary greatly.

Osmolarity and osmolality

Adding solute dilutes the concentration of pure water, and so increasing solute concentration decreases water concentration (Fig. 2.2).

The tendency of water (solvent) to diffuse from a region of higher concentration to an area of lower water concentration until equilibrium is reached underlies the principle of osmosis. As explained in Chapter 1, osmosis is the diffusion of solvent through a selectively (semi-) permeable membrane from a less concentrated solution to a solution with a higher concentration of solute. Note that the membrane allows the passage of solvent but not solute. In biological systems, this solvent is water (Fig. 2.3).

The pressure at which water is drawn from the weak solution (on the left side of Fig. 2.3A) into the more concentrated solution (on the right side of Fig. 2.3A) is known as the osmotic pressure; the higher the solute concentration, the higher the osmotic pressure. Water will be drawn into the right side until there is osmotic equilibrium (Fig. 2.3B). If a pressure is applied on the more concentrated side (Fig. 2.3A), net movement of water can be prevented.

Fig. 2.1 Body composition and fluid components.

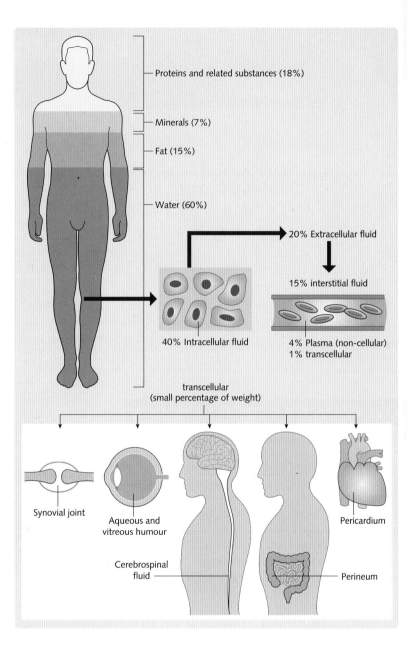

Proteins and related substances (18%)

Minerals (7%)

Fat (15%)

Water (60%)

20% Extracellular fluid

15% interstitial fluid

40% Intracellular fluid

4% Plasma (non-cellular)
1% transcellular

transcellular
(small percentage of weight)

Synovial joint

Aqueous and
vitreous humour

Cerebrospinal
fluid

Pericardium

Perineum

In Figure 2.3, osmotic pressure exerted by the solution forces water to move from left to right, as hydrostatic pressure forces movement in the opposite direction

In an ideal solution, osmotic pressure is similar to the pressure of a gas, with respect to temperature and volume:

$$P = \frac{n\text{RT}}{V}$$

Where P = pressure of a gas, n = number of particles, R = the gas constant, T = absolute temperature and V = the volume of the gas.

Therefore, at a given temperature, osmotic pressure will be proportional to the number of particles per unit volume. Solute particles are seen as the osmotically active particles, the total concentration of which, regardless of exact composition, is referred to as *osmolarity*, which is expressed in osmoles (Osm):

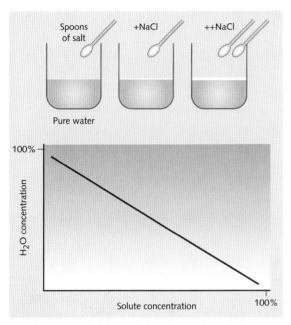

Fig. 2.2 The effect of adding solute to water on water and solute concentration.

- 1 Osm = 1 mole (6.02×10^{23}) solute particles.
- 1 mOsm = 1/1000 mole of solute particles.

The nature of the solute will dictate the osmolarity. Dissolving 1 mole of a non-ionizing compound, such as glucose, in water will give a solution of 1 Osm. However, dissolving 1 mole of sodium chloride (NaCl), which dissociates into Na^+ and Cl^-, will give a solution of 2 Osm.

Although the above is true for an ideal solution, it is not the case for bodily solutions, in which the number of osmotically active particles is less than the dissociated particles as a result of ion combinations. This effect is more pronounced with increasing concentration: the more NaCl dissolved per unit volume, the less it behaves like an ideal solution.

Osmolarity

- Osmolarity (mOsm/L) is defined as the number of osmoles per unit volume.

This can be illustrated by putting x Osm of solute into a beaker then adding water to make up 1 L of solution. Clearly, the water added would be less than 1 L.

Osmolality

- Osmolality (mOsm/kg) is defined as the number of osmoles per unit weight.

Normally, osmolality is about the same as osmolarity. The normal value for plasma is 280–295 mOsm/kg. At these normal plasma values, there is no net water movement. Lower or higher values will cause cell swelling (with danger of lysis/bursting) or shrinking, respectively.

Osmolarity = osmoles per volume.
Osmolality = osmoles per weight.

THE DIFFUSION OF IONS ACROSS BIOLOGICAL MEMBRANES

The selective permeability properties of the cell membrane play an important role in cell function by controlling the entry into cells of small molecules and ions. Diffusion across biological membranes can be divided broadly into passive and active forms of transport (see also Chapter 1).

Passive transport

The movement of small molecules and ions is dictated by their electrochemical concentration gradient. They move from high to low concentrations, or to neutralize a charge imbalance between two zones.

Size, electrical charge, shape and weight affect the rate of diffusion. If a substance is lipid soluble, diffusion across the cell membrane (which is formed

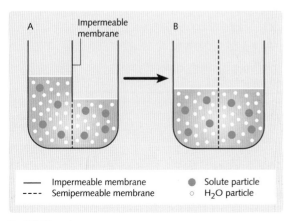

Impermeable membrane	● Solute particle
---- Semipermeable membrane	○ H_2O particle

Fig. 2.3 Osmosis.

from a lipid bilayer; see Chapter 1) will occur more readily. However, with polar substances, diffusion rates through water-filled ion channels are greater.

If the solutions either side of a membrane comprise only diffusible ions, diffusion occurs until equilibrium is reached and the ion distribution on each side is the same. At this point, this is the value of diffusible anions × diffusible cations.

Non-ionic diffusion

Although weak acids and bases cross cell membranes with difficulty in their ionic and dissociated forms, some have increased solubility in their undissociated form. Diffusion of such undissociated substances is called non-ionic diffusion; it occurs in the kidneys and gastrointestinal tract.

The Gibbs–Donnan effect

Non-diffusible ions trapped on one side of a membrane affect the passage of other ions. Negatively charged proteins (anions) will attract positive ions (cations) but repel other anions. This effect is described by the Gibbs–Donnan equation. The concentration of diffusible ions on side A of the membrane is the same as that on side B:

$$\frac{\text{cation A}}{\text{cation B}} = \frac{\text{anion B}}{\text{anion A}}$$

and:

cation A x anion A = cation B x anion B

Active transport

Carrier proteins can transport substances against their electrical and chemical gradients by using energy from ATP hydrolysis. A typical example of active transport is the Na^+/K^+-ATPase, which moves Na^+ out of and K^+ into a cell.

FLUID MOVEMENT BETWEEN BODY COMPARTMENTS

The body's fluid compartments are normally in osmotic equilibrium, although they contain different amounts of various ions (Fig. 2.4):

- ICF: K^+ contributes $\approx 50\%$ of osmolality.
- ISF and plasma: Na^+ and Cl^- are responsible for $\approx 80\%$ of osmolality.

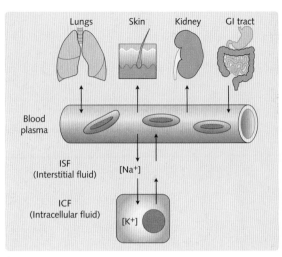

Fig. 2.4 Intercompartmental fluid exchange.

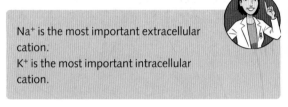

Na^+ is the most important extracellular cation.
K^+ is the most important intracellular cation.

Plasma and interstitial fluid exchange

In the capillaries, the osmolality of plasma is approximately 1 mOsm/L greater than the osmolality of ICF and of ISF. Much of this is due to plasma proteins. This osmotic pressure draws fluid into the capillaries and is counterbalanced by the capillary hydrostatic pressure, which is 20 mmHg greater than that of the ISF.

The exchange of water and ions occurs across the thin capillary wall, which is composed of endothelial cells. Substances can pass via:

- Vesicular transport (not discussed further but requires energy expenditure).
- Junctions between endothelial cells.
- Fenestrations (when present).

As well as this vesicular transport, simple diffusion and filtration are responsible for transport.

Simple diffusion

This is responsible for 90% of exchange – relating mainly to net efflux of O_2 and glucose from plasma and influx of CO_2 into plasma.

Filtration

This is responsible for 10% of exchange. The rate of filtration relies on Starling forces, which are derived from two aspects of the ISF and capillary fluid:

1. Oncotic pressure (π), which resists filtration.
2. Hydrostatic pressure (P), which favours filtration.

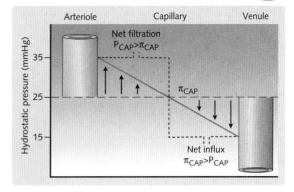

Fig. 2.5 Fluid movement across the capillary wall, as dictated by Starling forces.

Starling's hypothesis

Fluid movement = $K_f[(P_{CAP} - \pi_{CAP}) - (P_{ISF} - \pi_{ISF})]$
Where P_{CAP} = capillary hydrostatic pressure, P_{ISF} = ISF hydrostatic pressure, π_{CAP} = capillary oncotic pressure, π_{ISF} = ISF oncotic pressure and k_f = filtration coefficient which takes account of the capillary surface area and permeability per unit surface area. The oncotic pressure of the capillary is 25 mmHg throughout. Hydrostatic pressure of the capillary, however, decreases progressively from the arteriolar end (32 mmHg) to the venous end (12 mmHg).

Oncotic pressure

Oncotic pressure is the osmotic pressure produced by the proteins that are confined to a compartment/space by their relatively large size. Its value remains constant throughout the capillary. The effect of these proteins is to draw fluid into the compartment/space in which they are confined. In the plasma, oncotic pressure is 17 mmHg but, owing to imbalance of ions as the result of the Donnan effect, there are more Na^+ ions within the capillary and therefore a higher osmotic pressure of 25 mmHg.

Hydrostatic pressure

The following determine vessel hydrostatic pressure:

- Arteriolar blood pressure.
- Venous blood pressure.
- Arteriolar resistance, on which depends the extent to which blood pressure is transferred along the capillary.

Hydrostatic pressure is maximal at the arteriolar end of the capillary (32 mmHg), where it exceeds oncotic pressure (which is 25 mmHg) and thus favours filtration. At the venous end, fluid re-entry into the capillary is favoured; hydrostatic pressure (12 mmHg) is lower than oncotic pressure (25 mmHg) (Fig. 2.5).

Exchange between interstitial fluid and the lymphatic vessels

The overall efflux of fluid from the capillaries would be expected to cause an increase in ISF hydrostatic pressure. However, this fluid, along with plasma proteins lost from the vascular space, enters a network of lymphatic channels, which is present in all organs and tissues. The fluid is returned to the circulatory system when the lymphatic system empties into the venous system via the thoracic duct in the neck. Normal lymph flow is 2–4 L per day.

FLUID AND ION MOVEMENT BETWEEN THE BODY AND THE EXTERNAL ENVIRONMENT

A careful balancing of fluid intake against output maintains the composition of body fluids (Fig. 2.6). Daily water intake can be from two sources:

- Ingestion of fluids as liquids and as water in food: 2000 mL.
- Oxidative metabolism of food: 400 mL.

Daily water loss occurs by several different mechanisms:

- Lung: water evaporates continuously from the respiratory tract. The amount varies with climate and humidity.
- Skin: diffusion through the skin occurs independent of sweating and is minimized by the cornified cholesterol-filled skin layer. Sweating can account for variable water loss. Values are normally 100 mL/day, although it can increase to 1–2 L/h, depending on weather or exercise.

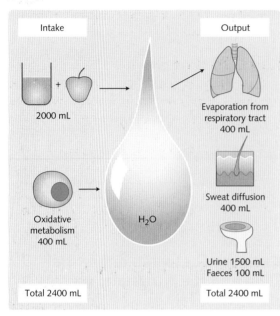

Fig. 2.6 Water balance.

- Faeces: water loss can be excessive with diarrhoea compared to the small amount normally lost.
- Urine/kidney: the excretion of electrolytes and water by the kidney is the most important mechanism the body has to regulate fluid balance. The rate of water excretion is adjusted according to the body's needs and water intake. This occurs by filtration and reabsorption and can maintain body fluid volumes despite change in fluid intake or loss elsewhere, e.g. haemorrhage.

There is a minimal necessary water loss of 1200 mL, which needs to be balanced by intake, and explains how untreated severe diarrhoea can kill in a few days.

MEASURING BODY FLUID COMPARTMENTS

It is not always possible to remove fluid for direct measurement and, instead, in-situ methods are used. A fixed amount of a substance is injected and the concentration then measured, to give its volume of distribution:

$$\text{Volume of distribution} = \frac{\text{Amount of substance injected}}{\text{Concentration of substance}}$$

This calculation relies on the principle of conservation of mass. However, in real life, excretion and metabolism of the indicator substance takes place. Hence, to measure concentration x hours after administering the injection:

$$\text{Volume of distribution} = \frac{\begin{array}{c}\text{Amount of substance injected -}\\ \text{Amount metabolized after x hours}\end{array}}{\begin{array}{c}\text{Concentration of substance}\\ \text{after x hours}\end{array}}$$

Dye can be injected to assist with measurement. The dye must be confined to the compartment being measured; it also needs to be distributed equally, as well as having no influence on water or other solute distribution. Two methods of measurement are used, depending on the rate of excretion from that compartment:

1. Single injection method: a test substance with a slow rate of excretion is used.
2. Constant infusion method: a test substance with rapid excretion is used.

Slow injection method

After introduction of the dye, the concentration is measured at intervals. The logarithm of concentration is plotted against time with the concentration of the substance (denoted by using square brackets: [substance]) determined by extrapolating the straight portion of the graph back to time = 0 (Fig. 2.7):

$$\text{Compartment volume} = \frac{\text{Amount of substance injected}}{\text{[substance] at time} = 0}$$

Constant infusion method

After an initial loading dose, the substance is infused at rate equal to the rate at which it is being excreted, so that plasma concentrations remain constant (they are checked at intervals). The infusion is then stopped and urine collected until all the substance is excreted. The amount present in the body when the infusion is stopped = amount excreted (Fig. 2.8):

$$\text{Compartment volume} = \frac{\text{Amount of substance excreted}}{\text{Plasma [substance]}}$$

Plasma volume

This is measured by the dilution principle using a high-molecular-weight substance that persists in

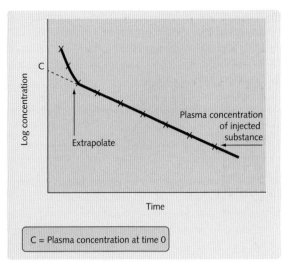

C = Plasma concentration at time 0

Fig. 2.7 The dilution principle: single injection method, showing change in plasma concentration of injected substance with time.

the vascular compartment. Examples include radio-iodinated human serum albumin or the diazo compound Evans blue dye (T-1824).

Blood volume

This can be derived from the haematocrit (fraction of total blood volume) and the plasma volume:

$$\text{Total blood volume (TBV)} = \frac{\text{Plasma volume}}{1 - \text{haematocrit}}$$

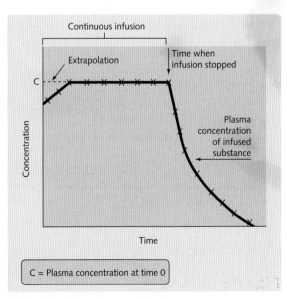

C = Plasma concentration at time 0

Fig. 2.8 The dilution principle: constant infusion method, showing change in plasma concentration of infused substance with time.

So, if plasma volume = 3.5 L and haematocrit = 0.3, then TBV = 5 L.

Red cell volume

Red cell volume (RCV) can be measured in a variety of ways, depending on what information is known:

1. RCV = TBV – plasma volume

2. $\text{RCV} = \dfrac{\text{Haematocrit} \times \text{plasma volume}}{1 - \text{haematocrit}}$

3. Direct dilutional method: radioactive chromium (^{51}Cr) is used to label or tag red blood cells. The fraction of red blood cells tagged is measured.

Extracellular fluid volume

The substances that are used to measure ECF by the dilutional principle also disperse in both plasma and ISF but are cell-membrane impermeable and so do not enter cells. Examples are:

- Thiosulphate.
- Inulin (can be excluded from bone and cartilage).
- Mannitol.

Small amounts of radiochloride or radiosodium diffuse intracellularly. The lymphatic system cannot be separated from the ECF and is therefore measured with it. Normal ECF volume is 15 L (20% body weight).

Interstitial fluid volume

Direct measurements of ISF cannot be made and it is therefore calculated as:

ISF = ECF – Plasma volume

Although absolute ECF volume is less in infants and children, the ECF:ICF ratio is larger. Hence, dehydration is frequently more severe and develops more rapidly in children than in adults.

Total body water

Radioactive water – tritium (^{3}H$_2$O) or deuterium (^{2}H$_2$O) – is used by the dilutional principle. Normal total body water values are:

- 63% in males = 45 L of 70 kg body weight.
- 52% in females = 36 L of 70 kg body weight owing to greater proportion of fat; fat cells have a lower water content than muscle.

Transcellular fluid

Transcellular fluid (TCF) is a small compartment usually considered as part of the ECF representing ~5% of ECF (~1L) and includes a number of small volumes such as cerebrospinal fluid, intraocular, pleural, peritoneal and synovial fluids. To include digestive secretions (which can be excessive and are, strictly speaking, outside the body) as transcellular may be debated.

BLOOD PHYSIOLOGY

Functions and components of the blood

Blood is the only liquid connective tissue. It comprises 8% of total body weight (5 L in a normal adult). The functions of blood relate to its composition:

- Transport: of gases, nutrients, waste products and hormones.
- Immunological: defence against bacteria, viruses and foreign bodies by leukocytes.
- Homeostatic: temperature, pH, haemostasis and fluid exchange.

There are two main components in blood (Fig. 2.9):

1. Plasma (55%): a watery substance containing dissolved solutes and proteins in suspension.
2. Cells and cellular fragments (45%):
   - erythrocytes (99%), also know as red blood cells (RBCs)
   - platelets (<1%)
   - leukocytes (<1%) or white blood cells (WBCs).

Plasma

Plasma comprises:

- Water: forms a medium for the suspension and transport of proteins, solutes and gases and so influences partial pressures and gas exchange. Water is important in temperature regulation because it releases heat. It also removes waste and breakdown products.
- Solutes: electrolytes in particular create osmotic pressure. Ions, e.g. HCO_3^-, are important in buffering pH change.
- Proteins: are important buffers and exert oncotic pressure:
  - Albumin: particularly important for vascular oncotic pressure and fluid exchange. it also transports fatty acids, lipid-soluble hormones and some drugs
  - Globulins: α and β globulins transport hormones and iron, respectively. γ Globulins

Fig. 2.9 Blood composition.

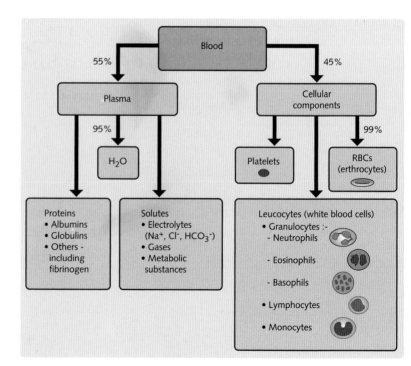

(antibodies) defend against viruses and bacteria.

- Other components: including fibrinogen, which is important in the process of blood clotting.

Cells

- RBCs: small, flexible, anucleate cells containing haemoglobin (~8 μm diameter). The biconcave shape suits their function in gas exchange and their transport.
- Platelets: small cell fragments derived from megakaryocytes in the bone marrow. Initiate haemostasis and thrombus formation at injury sites.
- WBCs: numbers increase during infection, surgery or strenuous exercise:
  - neutrophils (60%): engulf and phagocytose bacteria; also involved in inflammation. Numbers increase with bacterial infection, inflammation, burns and stress
  - lymphocytes (20%): there are type B and type T cells, the immunological roles of which include generation of the specific immune response, including antigen–antibody reactions. Numbers increase in viral infections and some leukaemias
  - monocytes (5%): phagocytose after transforming into macrophages. Numbers increase in viral or fungal, infections, tuberculosis and some chronic diseases
  - eosinophils (3%): destroy worm parasites. In allergic reactions, they combat histamine. Numbers increase in parasitic infections, allergic reactions and autoimmune diseases
  - basophils (< 1%): amplify the inflammatory response via the release of heparin and vasoactive substances. Numbers increase in allergic reactions, cancers and leukaemias.

Gas transport in the blood

After gas exchange at the lungs, gas transport completes the O_2 and CO_2 trade between cells and the external environment.

Oxygen transport

The factors affecting O_2 delivery to tissues are:

- The speed of delivery or blood flow: this depends on vascular constriction/dilatation.

- The amount of O_2 that can be carried by the blood: this depends on the:
  - amount of dissolved O_2
  - proportion of haemoglobin in the blood
  - affinity of haemoglobin for O_2.

Hence, O_2 is transported in the blood in two forms:

- Dissolved in plasma: 2%.
- In chemical combination with haemoglobin in RBCs: 98%.

O_2 dissolves in plasma up to about 3 mL/L and, in accordance with Henry's law, is proportional to its partial pressure. This small amount reflects the low solubility of O_2 in plasma. Clearly, this is not sufficient to meet even the body's resting metabolic needs of 250 mL O_2/min, as a normal cardiac output of 5 L/min would supply only 15 mL/min of O_2 in this form. Hence, a carrier molecule is needed to transport the required O_2. This molecule is haemoglobin.

Henry's law

The weight of a gas absorbed by a liquid with which it does not combine chemically is proportional to the partial pressure of the gas to which the liquid is exposed.
Note: both partial pressure and solubility determine the amount of gas absorbed.

Haemoglobin

The molecular structure of this conjugate protein molecule is largely responsible for its O_2-carrying properties (Figs 2.10 and 2.11). Each of the four subunits consists of a polypeptide chain attached to a haem group, which can reversibly bind with one O_2 molecule. Hence, up to four molecules of O_2 can combine with each molecule of haemoglobin (Hb); a combination that exhibits cooperativity among the four binding sites. Haemoglobin has a greater affinity for binding the second and third O_2 molecules than the first and fourth, and these differences in affinity give rise to the characteristic 'S' shape of the O_2 dissociation curve (see later).

- Men have 150 g Hb/L of blood (15 g/dL).
- Females have 130 g Hb/L of blood (13 g/dL).

In adults, normal Hb is known as HbA, and consists of two α and two β chains. Any change to

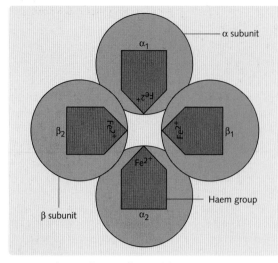

Fig. 2.10 Schematic diagram of haemoglobin.

the sequence of the polypeptide chains results in structural differences, with consequent alteration to the properties of the Hb (see later).

Haemoglobin binding

The rapid reversible combination of O_2 with Hb to form oxyhaemoglobin (HbO_2) is written as:

$$Hb + O_2 \leftrightarrow HbO_2$$

(DeoxyHb)   (OxyHb)

(Purplish)   (Bright red)

Depending on the number of binding sites occupied by O_2, Hb can be considered as either saturated (carrying all four molecules of O_2) or partially saturated (carrying less than four molecules of O_2).

The O_2-carrying capacity of Hb is 1.34 mL O_2/g Hb. This figure takes into account the small amounts of methaemoglobin (ferric iron core) in the O_2-carrying state compared to the normal ferrous Fe^{2+} state, and also allows for Hb that has combined with carbon

monoxide (CO): in neither state can Hb transport O_2. Therefore, in an adult man with 15 g of Hb/dL blood, the O_2-carrying capacity will be:

$$15 \times 1.34 = 20.1 \text{ mL } O_2/\text{dL blood}$$

The actual O_2-carrying capacity will depend on the amount of Hb in the blood, with the pressure of O_2 (PO_2) driving the O_2 into the blood.

O_2 saturation of Hb refers to the proportion of Hb bound to O_2. When the amount of reduced (deoxygenated) Hb is >5 mg, the lips, tongue and oral mucosa in particular take on a bluish pallor, known as central cyanosis. This reflects an O_2 loading problem through either reduced gaseous exchange or pulmonary blood flow. Peripheral cyanosis-blueness of the nail beds and other extremities-results from secondary local causes such as sluggish blood flow, vasoconstriction, cold temperatures; that is it can occur independently of, but always in the presence of, central cyanosis.

The oxygen dissociation curve

Blood PO_2 not only determines the amount of dissolved O_2 in plasma, but also how much O_2 binds to Hb, i.e. the O_2 saturation. This is represented by the O_2 dissociation curve (Fig. 2.12), which relates the PO_2 to the O_2-carrying capacity of Hb under the following conditions:

- pH 7.4.
- Temperature 37 °C.
- Normal PCO_2 (40 mmHg).

The graph is 'S'-shaped because Hb undergoes four sequential reactions to enable each subunit to bind with O_2. These reactions become progressively easier, except for the binding of the fourth oxygen molecule. The S shape of the curve is explained as follows:

- When PO_2 is low: the curve is steep and small changes in PO_2 lead to large changes in Hb

Fig. 2.11 Components of haemoglobin.

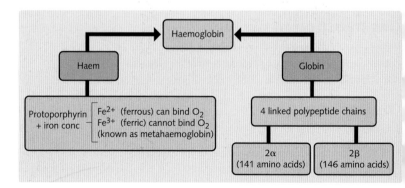

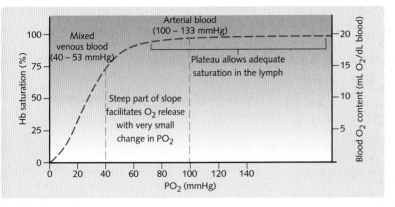

Fig. 2.12 Oxygen dissociation curve.

saturation. This corresponds to gas exchange between the systemic capillaries and tissues. The result is that, for a small drop in PO_2, the tissues can take large amounts of O_2.

- When $PO_2 > 70$ mmHg: the curve is almost flat. This confers the advantage that when alveolar PO_2 drops, Hb can still become fully saturated with O_2. Importantly, this means that someone with moderate lung disease or who is hypoventilating can still easily load blood with O_2.

Various influences alter the behaviour of O_2 dissociation by shifting the position of the O_2 dissociation curve to the right or the left (Fig. 2.13).

Right shift   This equates to reduced O_2 affinity of Hb, which results in:

- Easier unloading of O_2 at tissues at a given PO_2.
- Lower O_2 saturation.

A shift to the right occurs under the following conditions:

- pH < 7.4.
- Temperature $> 37\,°C$.
- $\uparrow PCO_2$: known as the Bohr effect. This is attributed to its action on $[H^+]$ and is discussed later in the chapter.
- $\uparrow 2,3$ diphosphoglycerate (2,3-DPG), which binds to the β chains of Hb.

Left shift   This equates to increased affinity of Hb for O_2 resulting in:

- Easier O_2 binding and uptake at any PO_2.
- Higher O_2 saturation.

A shift to the left occurs when conditions causing a right shift are reversed, such as:

- pH > 7.4.
- Cooler temperatures: note that O_2 is more soluble in plasma at lower temperatures.

Effects of 2,3-diphosphoglycerate   2,3-DPG is a normal metabolic by-product of RBC metabolism

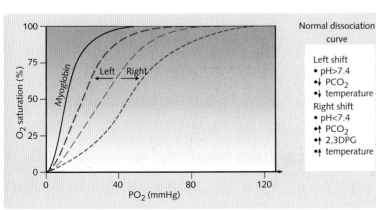

Fig. 2.13 Factors affecting the oxygen dissociation curve.

Normal dissociation curve

Left shift
- pH>7.4
- ↓ PCO$_2$
- ↓ temperature

Right shift
- pH<7.4
- ↑ PCO$_2$
- ↑ 2,3DPG
- ↑ temperature

and large increases in levels of 2,3-DPG are seen in chronic hypoxia (chronic lung disease or at high altitudes). Its function is to facilitate the unloading of O_2 for use by peripheral tissues by allowing greater O_2 release at any PO_2. It does this by binding to the β chains of Hb, thereby reducing its affinity for O_2.

2,3-DPG is lost from stored blood, so that 1-week-old blood has very low levels. This means that unless 2,3-DPG levels are corrected, the blood will have very little tissue unloading.

Other forms of haemoglobin

Myoglobin This haem protein is found within muscle. Like the Hb subunit, myoglobin consists of one haem group with a polypeptide chain. It has the following features:

- Higher affinity for O_2 than normal adult Hb (HbA): the O_2 dissociation curve is relatively to the left. Therefore, myoglobin takes up O_2 from Hb in the capillaries.
- Acts to transport and temporarily store O_2 in skeletal muscle and has limited availability during anaerobic conditions, lasting only a few seconds.
- Binds O_2 at PO_2 values greater than those in venous blood PO_2. Hence it would be unsuitable as a blood carrier for O_2 (because it would not surrender O_2 to the tissues).

Fetal haemoglobin The fetal circulation has a lower PO_2 (30 mmHg after deoxygenation) than the maternal uterine arterial blood from which O_2 diffuses. Fetal blood contains mainly fetal Hb (HbF), the structure of which differs from HbA in that γ chains replace the β chains of HbA. HbF is crucial in directing O_2 transfer from maternal circulation to the fetus:

- HbF has a higher affinity for O_2 and its dissociation curve is positioned to the left of that of HbA. Therefore, HbF is 20–50% more saturated with O_2 than HbA at low PO_2 levels.
- The concentration of Hb in fetal blood (the [HbF]) > the concentration in maternal blood ([HbA]). This results in greater carriage of O_2 in fetal circulation.
- CO_2 unloading from fetal to maternal circulation: the acidic affects of increased CO_2 levels promote HbA dissociation of O_2, shifting its curve to the right. As CO_2 diffuses from fetal blood into maternal, this Bohr effect works in

both fetal and maternal directions: double Bohr shift. O_2 release from maternal to fetal circulation and increasing fetal blood O_2 capacity are thereby promoted.

Sickle haemoglobin The Hb variant HbS is found in sickle-cell disease. It consists of a single amino acid substitution at position 6 in the β-globin chain of HbA. This has the following consequences:

- The O_2 dissociation curve is shifted to the right.
- HbS is a poorly soluble deoxygenated form that polymerizes, rendering the RBC fragile and crescent- or sickle-shaped: this further predisposes to thrombus formation, and hence vaso-occlusive crises caused by blockage of blood vessels can arise.

Sickle-cell disease is an autosomal recessive disorder with heterozygotes symptomatic only when hypoxic. However, homozygotes experience manifestations of infarction and ischaemia from vaso-occlusive crises (such as leg ulceration and bone pain, among others).

Thalassaemias

This is a group of disorders where there is an imbalance in the 1:1 ratio of α- and β-globin chains, secondary to defective synthesis in either of these chains. As a consequence, the globin chains precipitate, causing:

- Inhibition of erythropoeisis in RBC precursors.
- Haemolysis of mature RBCs.

Each chain is coded by four genes, with symptoms varying from a mild microcytic anaemia with two gene deletions, to the death of either the fetus or the newborn if all four genes coding a chain are deleted. Disease is described as minor, intermediate or major, depending on the number of normal genes.

Carbon monoxide poisoning and carboxyhaemoglobin

Carboxyhaemoglobin (carboxyHb) is formed when carbon monoxide (CO) binds to Hb. CO interferes with O_2 transport in a number of ways:

- CO competes with O_2 for Hb-binding sites and renders the iron atom to which it is bound unable to bind any O_2 molecules.
- The affinity of CO for Hb is significantly higher (250 times greater) than that of O_2. It also takes a long time to clear.

- Exposure to very low levels of CO can seriously compromise the O_2-carrying capacity of blood: inspiration of 0.1% CO halves the O_2-carrying capacity because 50% of Hb would be in the form of carboxyHb.
- CO shifts the O_2 dissociation curve to the left, even further to the left than the curve for myoglobin (Fig. 2.13), decreasing both O_2 loading at the lungs and O_2 unloading at the tissues.
- CO poisoning is detected by blood carboxyHb levels and cherry-red colouring of the patient (from its pigment colour, not cyanosis). There would also be few if any symptoms.

Carbon dioxide transport

CO_2 in the blood is found in the following forms:

- Dissolved in plasma.
- As bicarbonate ions (HCO_3^-) in plasma.
- As carbamino compounds (chemical combination with proteins) in whole blood.

The relative proportions of the above are subject to arteriovenous differences.

Dissolved CO_2

Between 5 and 10% of CO_2 is carried dissolved in the plasma. As with dissolved O_2, the actual amount dissolved depends on the partial pressure of CO_2. Henry's law applies and CO_2 is approximately 20 times more soluble than O_2:

CO_2 solubility at 37°C is:
0.6×10^{-3} mL/mmHg CO_2/1 mL of plasma

Therefore, in 100 mL mixed venous blood PCO_2 = 45 mmHg, the amount of dissolved CO_2 is:

$0.6 \times 10^{-3} \times 100 \times 45 = 2.7$ mL CO_2/dL blood or 27 mL CO_2/L blood

Bicarbonate ions

The majority of CO_2 is transported as HCO_3^-: 90% of CO_2 in arterial blood and ≈60% of CO_2 in venous blood is in this form:

$$CO_2 + H_2O \leftrightarrow H_2CO_3 \leftrightarrow H^+ + HCO_3^-$$

H_2CO_3 = carbonic acid, which is a weak acid.
   The formation of HCO_3^- ions from CO_2 consists of two reactions:

1. $CO_2 + H_2O \leftrightarrow H_2CO_3$: the formation of carbonic acid. This reaction requires the enzyme carbonic anhydrase, which is present in RBCs and which speeds up the reaction so that it is thousands of times faster than in the plasma.

2. $H_2CO_3 \leftrightarrow H^+ + HCO_3^-$: the dissociation of carbonic acid into H^+ and HCO_3^- is normally rapid. However, it is limited by accumulation of the products, which need to be removed if the reaction is to proceed. This can be achieved by:
   - Hb buffering of H^+: the reduced form of Hb is less acidic than the non-reduced form and so accepts H^+ more readily:

     i $H^+ + Hb^- \rightarrow HHb$
     ii $H^+ + HbO_2^- \rightarrow HHb + O_2$

   - The chloride shift: HCO_3^- movement from RBC to plasma. A specific anion exchange (AE) is responsible for the movement of HCO_3^- out of and Cl^- into the RBC. This Cl^- movement is known as the chloride shift (Hamburger phenomenon) and is important in maintaining electrical neutrality.

Carbamino compounds

In the tissues, CO_2 reacts readily with the terminal amine (NH_2) group on blood proteins (i.e. Hb) to form carbamino compounds – including carbaminoHb – that contribute up to 30% of CO_2 transport in venous blood. This reaction does not require an enzyme and occurs more readily with reduced (less acidic) Hb. At the lungs, CO_2 is easily released by carbamino compounds.

The Haldane effect and the Bohr effect

The carriage of CO_2 in the blood is increased by deoxygenation. This is a result of reduced Hb being a weaker acid than its oxygenated form. Hence reduced Hb has the following actions:

- Promotes HCO_3^- formation and carriage by more readily accepting H^+ in RBCs.
- Reacts more readily with CO_2 to produce carbaminoHb.

At the tissues, both Bohr and Haldane effects facilitate peripheral gas exchange (Fig. 2.14):

- ↑PCO_2 and ↓pH promote O_2 unloading from HbO_2 (the Bohr effect), shifting the O_2 dissociation curve to the right.
- ↓PO_2 encourages CO_2 transport by the mechanisms described above (the Haldane effect).

At the lungs, again these effects are important (Fig. 2.15):

Fig. 2.14 Red blood cell carbon dioxide uptake and oxygen release at the peripheral tissues.

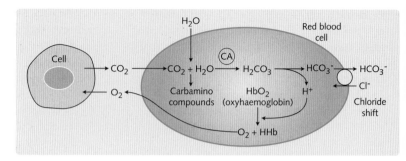

- $\uparrow PO_2$ and $\downarrow PCO_2$ facilitate CO_2 release from carbamino compounds by forming oxyHb.
- Release of H^+, which was buffered by Hb in its reduced form, is crucial for the reverse reaction that forms HCO_3^- and H^+ and produces CO_2, which is then exchanged at the lungs and exhaled.

Importantly, pH remains stable at both sites of gas exchange:

- At the tissues, reduced Hb gains acid.
- At the lungs loss of acid from reduced Hb results in acidic oxygenated Hb formation.

Carbon dioxide dissociation curve

PCO_2 dictates the total blood concentration of CO_2 in all its different forms of transport (Fig. 2.16). Importantly, the CO_2 dissociation curve displays an absence of flat portions and therefore does not demonstrate any saturation. The variability of PCO_2 is therefore greater than that of PO_2.

Haemostasis

Haemostasis refers to the control of bleeding. Following injury to a blood vessel, responses occur in two phases:

1. Rapid: reactions of blood vessel and platelets:
   - slowing of blood flow

- formation of a platelet mass at the site of injury
- diffusion of tissue factors from the extravascular compartment, initiating the extrinsic coagulation pathway.
2. Slow: intrinsic coagulation pathway: formation of insoluble fibrin threads that stabilize the platelet mass.

Three mechanisms are involved with haemostasis:

1. Vasoconstriction.
2. Formation of a platelet plug.
3. Coagulation (formation of a blood clot) with eventual clot retraction.

Vasoconstriction

Immediately after an injury to a blood vessel, the smooth muscle in the vascular wall contracts, decreasing diameter and blood loss. This occurs in response to:

- Nervous reflexes: as a result of pain and traumatized vessels.
- Local myogenic spasm.
- Platelet thromboxane A_2: this is more important in smaller than in larger vessels.

Formation of a platelet plug

Platelet adhesion, activation and aggregation occur (Fig. 2.17).

Fig. 2.15 Red blood cell carbon dioxide unloading at the lungs.

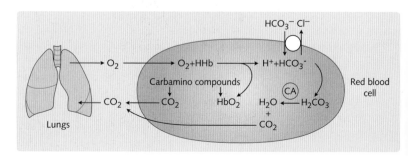

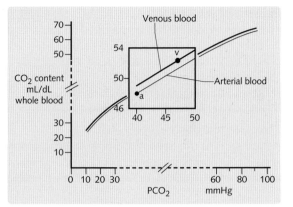

Fig. 2.16 Carbon dioxide dissociation curve. For any PCO$_2$, total CO$_2$ content increase as PO$_2$ decreases (Haldane effect). For a change in PCO$_2$ in the physiological range (a, 40mmHg to v, 47mmHg), total CO$_2$ content changes by ~4mL/dL whole blood.

Adhesion
Damage to the vascular endothelium exposes collagen and other connective tissue to which platelets adhere.

Activation
During activation, platelets undergo:

- Shape change: they swell then extend many projections that facilitate greater contact with other platelets.

- Granule release: contractile proteins contract, causing release of the following from granules:
  - ADP and thromboxane A$_2$: which activate and make platelets sticky
  - serotonin and thromboxane A$_2$: which cause vasoconstriction, so reducing blood loss.
- Expression of new receptors: IIb and III.

Activated platelets recruit nearby platelets so that the effects are amplified.

Aggregation
The sticky, activated platelets attract one another and adhere, forming a loose clump. This platelet plug is effective at blocking small holes and coagulation will not be necessary. However, injury to a blood vessel often requires coagulation.

Coagulation
Coagulation comprises a sequence of enzyme-catalysed conversions of inactive factors to more active forms, culminating in the conversion of fluid blood into a solid gel or clot.

Blood clot formation begins a matter of seconds after severe trauma to the vascular wall, and in minutes after injury to other areas of the body.

Fig. 2.17 Platelet structure.

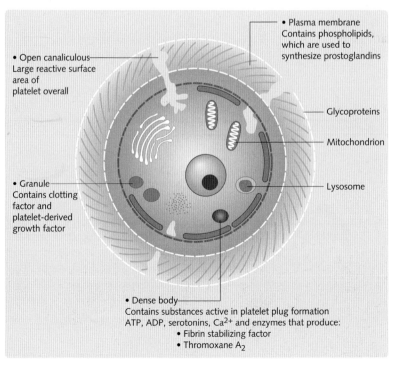

- Plasma membrane
Contains phospholipids, which are used to synthesize prostoglandins

- Open canaliculous
Large reactive surface area of platelet overall

Glycoproteins

Mitochondrion

- Granule
Contains clotting factor and platelet-derived growth factor

Lysosome

- Dense body
Contains substances active in platelet plug formation
ATP, ADP, serotonins, Ca^{2+} and enzymes that produce:
- Fibrin stabilizing factor
- Thromoxane A$_2$

Two pathways are involved (Fig. 2.18):

1. The intrinsic pathway:
   - is triggered by trauma to blood and exposure to collagen
   - involves many enzyme-catalysed steps
   - is slow (minutes).
2. The extrinsic pathway:
   - requires factors external to blood vessels, e.g. tissue factor (tissue thromboplastin)
   - few enzymes and steps are involved
   - is limited by the amount of tissue factor
   - is fast (seconds).

Both pathways produce prothrombinase (pro-thrombin activator), the formation of which appears to be the rate-limiting step in haemostasis. After this stage, pathways follow a common set of reactions, which is known as the final common pathway (see Fig. 2.18):

1. Prothrombinase and ionized Ca^{2+} cause prothrombin → thrombin.
2a. Thrombin and ionized Ca^{2+} cause fibrinogen (soluble) → fibrin fibres (insoluble).
2b. Thrombin activates fibrin-stabilizing factor (XIII), which cross-links the fibrin fibres.

The cross-linked fibrin strands mesh plasma, the platelet plug, plasmin and RBCs to form a blood clot. In clinical practice:

- The intrinsic pathway is monitored by partial or accelerated thromboplastin time (APTT).
- The extrinsic pathway is monitored using the prothrombin time (PT).

Role of thrombin

Thrombin formed in the first stage on the final common pathway exerts positive feedback effects on the coagulation cascade (Fig. 2.19):

- Acceleration of the formation of prothrombinase.
- Platelet activation.

Actions of thrombin

- Polymerization of fibrinogen.
- Activation of factor XIII, which cross-links fibrin strands.
- Positive feedback effects on the coagulation cascade.

Role of platelets

- Platelet phospholipids are required for the assembly of prothrombinase: these interact with activated factors X and V, and with Ca^{2+}, to produce prothrombinase.
- IIb and IIIa receptors bind to fibrin causing platelet aggregation with the fibrin glue.

Thrombus formation

A thrombus is a clot that forms within an intact blood vessel. This results from inappropriate activation of haemostasis with one of the following consequences:

- The thrombus dissolves spontaneously.
- The thrombus remains intact, with the risk of embolization.

Blood clots, fat from a fracture and air bubbles transported by the blood can all form emboli. They can lodge in a blood vessel, blocking it and interrupting the blood supply.

Venous and arterial thrombi differ:

- Arterial thrombi:
  - contain large platelet and small fibrin components
  - formation is associated with atheroma formation and turbulent blood flow
  - have a rough endothelial surface that attracts the platelets.
- Venous thrombi:
  - contain large fibrin and small fibrin components
  - formation is associated with slow blood flow, which causes a large increase in procoagulant factor concentration.

Clot retraction

Within a few minutes of its formation, the clot begins to contract due to platelets applying tension to the fibrin fibres that are attached to damaged blood vessels, the ends of which are therefore brought closer together. In addition, fluid is squeezed out of the clot. Platelets are suited to clot retraction because they:

- Release factor XIII (fibrin-stabilizing factor): causing more cross-linking of fibrin and permitting further compression of the clot.

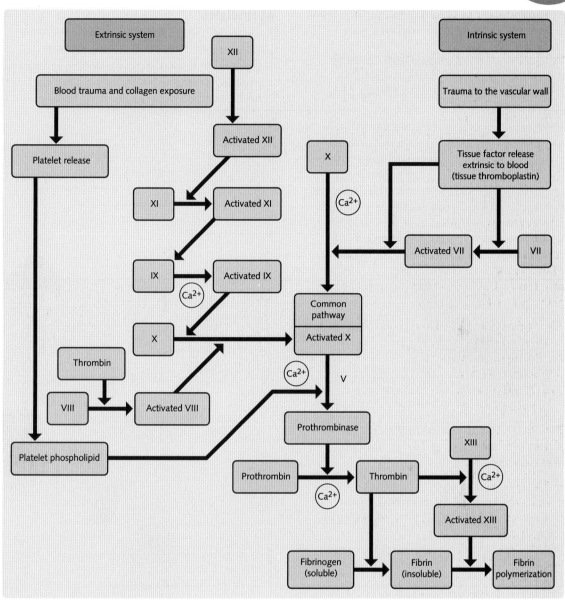

Fig. 2.18 Coagulation pathways.

- Activate the self-contractile proteins thrombosthenin, actin and myosin: these allow the platelet to pull harder on the fibrin fibres.

Normal prevention of coagulation

The normal vascular system employs a number of mechanisms to keep haemostasis in check:

Prevention of activation of a haemostatic plug

- Smooth endothelial cells discourage any activation of the intrinsic pathway.

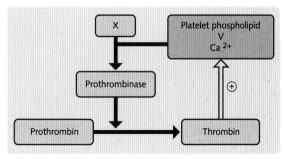

Fig. 2.19 Thrombin positive feedback loop.

- The glycocalyx layer on the endothelium repels both platelets and clotting factors.

Inhibition of the coagulation cascade

- Endothelial-bound thrombomodulin: binds thrombin and activates protein C, which is a plasma protein.
- Protein C: inactivates factors V and VIII.
- Antithrombin III (α-globulin and most important circulating anticoagulant): combines with thrombin, inactivating it for up to 20 minutes and blocking its effects on fibrinogen.
- Heparin molecule: is not active by itself but potentiates antithrombin III 100–1000-fold, with the added effect of removing activated factors XII, XI, X and IX.

Fibrinolysis

Plasmin is particularly important for removing inappropriately formed small blood clots; it acts to digest fibrin fibres and inactivate clotting substances: fibrinogen, prothrombin, factors V, VIII and XII. Plasmin is formed from the inactive plasma enzyme plasminogen by:

- Tissue plasminogen activator (t-PA): released by damaged endothelial cells at the periphery. It activates plasminogen in the presence of fibrin.
- Clotting factors: thrombin and activated factor XIII can activate plasminogen.

Physiology of the nervous system

Objectives

In this chapter, you will learn to:
- Describe the divisions of the nervous system
- State the different cell types and their arrangement within the nervous system
- Describe the action potential and the different ions involved
- Explain the different types of synaptic transmission
- Describe in detail the process of chemical transmission
- Describe the types of neurotransmitter
- Discuss the different sensory receptors in the skin and mucous membranes
- Explain how pain is regulated peripherally and centrally
- Relate the peripheral and central control of pain to how the most common forms of analgesia work
- Describe the different divisions of the brainstem and their functions
- Discuss the structure and function of the visual, auditory, olfactory and gustatory systems

This chapter provides a general overview of neurophysiology and of the sensory systems. The neurophysiology of motor control and higher functions is covered in *Crash Course: Nervous System*.

OVERVIEW OF THE NERVOUS SYSTEM

The nervous system receives information about the internal and external environments from the different sensory organs. In turn, it influences how the body responds. The nervous system functions at three main levels:

1. Sensory division: sensory/afferent receptors (e.g. pain receptors, receptors for vision and hearing) provide information to the brain about the environment.
2. Information processing: the brain collates this information and decides what action to take via its motor system.
3. Motor division: the motor/efferent system responds appropriately via contraction of muscle and/or gland secretion in the body.

Anatomy of the nervous system

The nervous system is divided into two broad parts (Fig. 3.1):

- Central nervous system (CNS).
- Peripheral nervous system (PNS).

Central nervous system

The CNS comprises the brain and spinal cord.

Peripheral nervous system

The PNS has two main divisions: the somatic and autonomic nervous systems.

Somatic nervous system

The somatic nervous system contains the sensory (afferent) and motor (efferent) nerves that supply the skin, muscles and joints. Single neurons relay information between the CNS and the target cells. For example, touching something painful stimulates a sensory (afferent) nerve, which carries the information that the hand is touching something painful to the CNS to be processed. A signal from the CNS then travels down the motor (efferent) pathway to the muscle, which responds, e.g. it contracts and moves the hand out the way.

Autonomic nervous system

The autonomic system (ANS) regulates the majority of the visceral organs and blood vessels. Two-neuron chains connected by a synapse relay information between the CNS and effector organs. The ANS is divided into the parasympathetic, sympathetic and enteric systems:

Fig. 3.1 Division of the central and peripheral nervous systems.

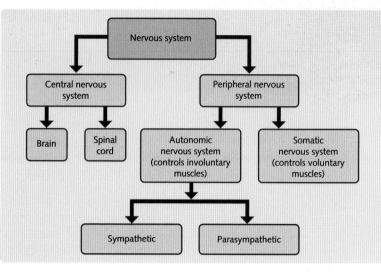

- Parasympathetic: involved in secretion from glands and in gastrointestinal motility.
- Sympathetic: prepares the body for a flight/fight/frolic response.
- Enteric: nerves found in the gastrointestinal tract, regulating glands and smooth muscle.

2. Dendrites: a branching network of soma that communicates with nearby cells.
3. Axon: a long extension from the soma that transmits the action potential to the terminal boutons.
4. Terminal boutons: swellings that form the presynaptic membrane; they contain the neurotransmitters that propagate the action potential to other excitable cells.

CELLULAR PHYSIOLOGY OF THE NERVOUS SYSTEM

Neurons: structure and function

The basic cell of the nervous system is the neuron. These are excitable cells, which transmit information in the form of action potentials (see later) to other excitable cells. The neuron has four main regions (Fig. 3.2):

1. Soma: the main body of the neuron that contains the organelles (see Chapter 1).

The four main regions of the neuron are the: soma, dendrites, axon and terminal boutons.

Arrangement of neurons

Neurons can be arranged in different patterns depending on their location:

- Layers, e.g. the cerebral and cerebellar cortices.
- Rods, e.g. motor neurons in the spinal cord.
- Clumps (nuclei), e.g. cranial nerve nuclei in the brainstem.

Classification of neurons

Projection neurons: Golgi type 1

These neurons influence cells located in different parts of the nervous system. This function is helped by the fact that the cells have long neurons and sometimes produce collateral branches.

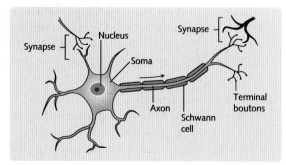

Fig. 3.2 Structure of a neuron. The arrow indicates the direction of impulse transmission.

Interneurons: Golgi type 2

These neurons influence cells located in close proximity. They have shorter axons than the projection neurons, although they too sometimes produce collateral branches, which increase their ability to relay information.

Neuronal excitation and inhibition

An electrical potential gradient exists across the nerve cell membrane. This gradient is caused by the relative permeability of the cell membrane to sodium ions (Na^+) in the extracellular fluid and potassium ions (K^+) in the intracellular fluid. The value of this potential determines whether an action potential is fired.

Resting potential

There is a potential difference of $-70\,mV$ across the membrane of a resting neuron, the inside being negative relative to the outside. This is due to a greater concentration of K^+ ions inside the cell, which results in a negative inside and a positive outside and hence a potential difference (diffusion potential).

Diffusion takes place until an equilibrium state, in which the electrical force attracting the positive K^+ into the cells is equal to the chemical force of the concentration gradient, is reached. The electrical potential (E_K) at this equilibrium point can be calculated using the Nernst equation:

$$E_K = (RT/zF)\log_e [K^+]_o/[K^+]_i$$

Where R = the ideal gas constant [$8.314510\,J\,K^{-1}\,mol^{-1}$]; T = thermodynamic temperature ($+273\,K$); F = Faraday's constant [$96485.3\,C\,mol^{-1}$]; Z = ionic valency ($+1$ for K^+); $[K^+]_o$ = concentration of K^+ outside cell; and $[K^+]i$ = concentration of K^+ inside cell.

In fact, this equation suggests that the value of E_K is not $-70\,mV$ but $-90\,mV$. The explanation for the difference is that other ions, such as Na^+, diffuse across the cell membrane in the opposite direction to the K^+. The Goldman–Hodgkin–Katz equation is similar to the Nernst equation but, as it takes ions other than K^+ into consideration, is far more accurate.

There is always some unregulated leakage of Na^+ into the cell and K^+ out of the cell along the concentration gradient, but this is counteracted by an energy-dependent Na^+/K^+-ATPase exchange pump, which stabilizes the resting membrane potential.

This pump ejects three Na^+ in exchange for every two K^+ transported into the cell.

Action potentials

An action potential (AP) is the feature of muscle and nerve cells that results in self-propagating membrane depolarization (Fig. 3.3). When a nerve cell is stimulated the electrical potential across the membrane changes. If the stimulus is strong enough and reaches a certain value (threshold potential), an AP is generated. APs have three distinct properties:

1. They are all or nothing: this means that an AP will occur when the threshold potential has been exceeded. Increasing the stimulus further does not change the shape or size of the AP.
2. They have a refractory period: during an AP it is either impossible or very difficult to stimulate a second AP. There are two types of refractory period:
   - absolute: in the early part of the AP no further stimulus is possible
   - relative: a further AP is possible but a larger-than-normal stimulus is required.
3. They are self-propagating: once the AP has been initiated it spreads throughout the excitable tissue.

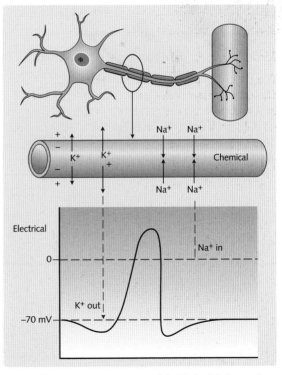

Fig. 3.3 Changes in ionic movement and electrical activity in an action potential.

Ionic mechanism of an action potential

Action potentials have three properties: they are all or nothing, have a refractory period and are self-propagating.

In 1952, Hodgkin and Huxley published a series of papers in which they investigated membrane currents through the membrane of the squid giant axon. Theirs was the first ever recording of the AP and forms the basis of neuroscience today.

The propagation of a nerve impulse along an axon begins when synapses on a neuron receive neurotransmitters from adjacent nerve endings (Fig. 3.4). This causes the electrical potential across the cell membrane to change, setting off a chain of events that results in the action potential.

Make sure you can describe and draw a diagram of an action potential. It is a favourite with examiners.

There are three stages to the generation of an AP:

1. The threshold: to initiate an AP, the stimulus must increase the membrane potential to about 20 mV above the resting potential of –70 mV.

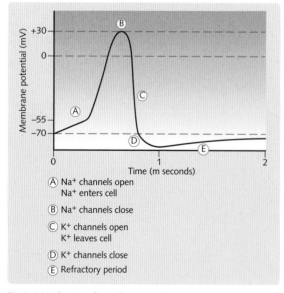

Fig. 3.4 Mechanism of an action potential.

When this happens, Na$^+$ enters the neuron via voltage-gated Na$^+$ channels.

2. Upstroke (depolarization): if the threshold is reached then more voltage-gated Na$^+$ channels open and further repolarization takes place (positive feedback). The membrane potential rises towards 0 mV and then overshoots to +30 mV.

3. Downstroke (repolarization): once 0 mV is passed, the positive intracellular charge resists further Na$^+$ entry and the slow-inactivation gates begin to close. The slow voltage-gated K$^+$ channels are fully open at +30 mV and K$^+$ leaves the cell and the internal negativity of the membrane is returned. There is a period of hyperpolarization when the voltage-gated K$^+$ channels are not fully closed permitting additional K$^+$ to leave - the absolute refractory period. Restoration of the resting membrane potential occurs when voltaged-gated Na$^+$ and K$^+$ channels are closed. The Na$^+$/K$^+$-ATPase continues to actively pump Na out of and K into the cell throughout the AP.

After potentials

Sometimes the Na$^+$ and K$^+$ channels do not return to their previous states. This causes either an over- or an undershoot, leading to hyper- or hypopolarization, respectively.

Myelinated and unmyelinated fibres

Myelin

Myelin comprises proteins and lipids (i.e. fatty substances) that form a sheath around some nerves (Fig. 3.5) and increases the speed of transmission of impulses. There are roughly twice as many myelinated fibres as unmyelinated.

Myelin sheaths in the PNS are formed by Schwann cells, which envelop the naked axon and produce a cellular membrane containing the lipid substance

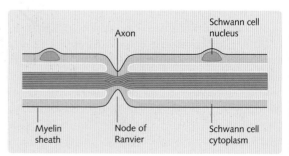

Fig. 3.5 Myelinated fibre.

sphingomyelin. This insulates the axon, decreases the ion flow through the membrane and consequently reduces its capacitance.

Between every two Schwann cells there is a small, exposed area of axon where ions can still flow between the axon and extracellular fluid; these are the nodes of Ranvier. APs jump from node to node by saltatory conduction (Fig. 3.6). The importance of saltatory conduction is twofold:

1. Nerve transmission speed is increased as the depolarization jumps from node to node.
2. As only the nodes depolarize, less ATP is required and energy is conserved.

Demyelination

Lack of myelin can cause a number of diseases, of which multiple sclerosis and Guillain–Barré syndrome are the most important:

- Multiple sclerosis: this disease of unknown aetiology causes multiple plaques of demyelination within the brain and spinal cord. Clinical features

include optic neuropathy, diplopia (double vision), vertigo (dizziness) and spastic paraparesis.

- Guillain–Barré syndrome: this is thought to have an autoallergic basis and follows 1–3 weeks after a viral infection. It is a demyelinating neuropathy that results in the patient complaining of weak distal muscles and/or distal weakness, which progress more proximally with time. It can sometimes lead to respiratory failure.

Guillain–Barré and multiple sclerosis are demyelinating diseases.

Speed of conduction

The speed of conduction of an AP depends on:

- Myelination versus unmyelinated fibres: see later.
- Fibre diameter: the thicker the fibre, the faster the speed of conductance.

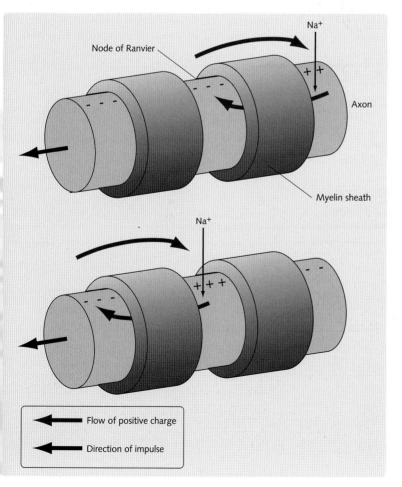

Fig. 3.6 Saltatory conduction.

Na⁺

Node of Ranvier

++

Axon

Myelin sheath

Na⁺

+ + +

Flow of positive charge

Direction of impulse

Synaptic transmission

A synapse is the gap between the end of a nerve fibre and the target cell, across which nerve impulses pass. There are two types (Fig. 3.7):

1. Chemical synapse: the action potential travels along the first (presynaptic) neuron and causes release of a neurotransmitter substance at the synapse. This diffuses across the synapse to the other (postsynaptic) neuron. It acts on the postsynaptic membrane receptors, causing either their excitation or inhibition.
2. Electrical synapse: these have direct channels (usually gap junctions) that conduct ions (electricity) from one cell to another.

THE PROCESS OF NERVOUS TRANSMISSION

There are two types of transmission:

1. Chemical.
2. Via a second messenger.

Chemical transmission (Fig. 3.8)

1. The AP spreads over the presynaptic terminal. Depolarization causes opening of voltage-gated calcium channels in the presynaptic membrane.
2. Large numbers of Ca^{2+} ions enter the terminal boutons of the presynaptic membrane.
3. As the result of the influx of Ca^{2+} ions, neurotransmitter-containing vesicles migrate towards the end of the presynaptic terminal, using an actin cytoskeleton. When they reach the synapse, they exocytose and release their contents in the synaptic cleft.

4. The postsynaptic neuronal membrane contains large numbers of receptor proteins. The neurotransmitter diffuses across the cleft and binds to these receptors. Once bound, the neurotransmitter causes a change in the postsynaptic membrane potential and can activate either iontrophic receptors or second-messenger systems.
5. Enzymes in the synaptic cleft inactivate the neurotransmitter.

Ionotrophic receptors are coupled with an ion channel can either inhibit or stimulate the postsynaptic neuron. There are two types:

1. Cation/excitatory postsynaptic potential (EPSP) ionotrophic receptors: these allow Na^+ to pass into the postsynaptic neuron, causing excitation and propagation of the AP down the postsynaptic neuron.
2. Anion/inhibitory postsynaptic potential (IPSP) ionotrophic receptors: these allow Cl^- to enter, which inhibits the postsynaptic neuron.

Second-messenger transmission

Ion channels close within milliseconds and so do not permit prolonged postsynaptic neural changes as is required in memory formation. The second-messenger system allows a prolonged response via second messengers such as cAMP or calmodulin.

Removal of the transmitter substance

To prevent continued action, the neurotransmitter must be removed from the synapse. This is done by:

- Active reuptake of the neurotransmitter into the presynaptic terminal.
- Diffusion of transmitter out of synaptic cleft into the extracellular fluid.
- Enzyme degradation, e.g. acetylcholine (ACh) is broken down by acetylcholinesterase.

Fig. 3.7 Comparison of electrical and chemical synapses.

Fig. 3.7 Comparison of electrical and chemical synapses		
Feature	Electrical	Chemical
Cytoplasmic continuity	Yes	No
Delay	None	0.8–1.5 ms
Agent	Ion	Neurotransmitter
Space between cells	2 nm	30–50 nm
Direction of signal	One way or both ways	One way
Variation in function	Either on or off	Modifiable activity levels

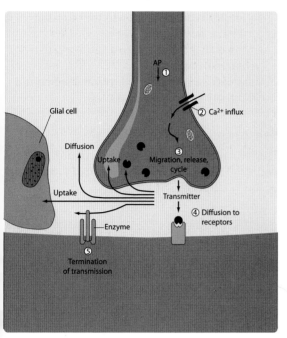

Fig. 3.8 Synaptic transmission.

Summation

The excitation of a single presynaptic terminal on the surface of a neuron will rarely cause an AP. This is because even though an EPSP might be generated, on its own it is insufficient to pass the threshold. Many EPSPs need to summate to reach the threshold and produce an AP. There are two mechanisms for this:

1. Spatial: activation of many terminals in close proximity to each other.
2. Temporal: the same channel reopens continuously so more EPSPs can occur sequentially.

Facilitation of neurons

Facilitation is when the summated effects of the EPSPs are insufficient to exceed the threshold for an AP to occur. When this happens, signals arising from other neurons can also excite the facilitated neuron to help it reach threshold and fire an AP.

Neurotransmitters
Rapidly acting neurotransmitters

Small, rapidly acting neurotransmitters are involved in an acute response, such as the transmission of signals to the brain and motor signals back to the muscles. These neurotransmitters are synthesized in the cytosol of presynaptic terminals.

Acetylcholine

ACh is secreted in the motor cortex, basal ganglia, motor neurons innervating skeletal muscle, preganglionic neurons of the autonomic nervous system and postganglionic neurons of both the sympathetic and parasympathetic nervous systems. It can have both inhibitory and excitatory effects. Reduced ACh concentrations in certain parts of the brain are associated with Alzheimer's disease.

Norepinephrine (noradrenaline)

This is secreted by the locus caeruleus in the pons, where it is involved in the control of sleep. It is also the main neurotransmitter of ganglion cells in the sympathetic nervous system, which excite some organs (e.g. myocardium to contract stronger) and inhibit others (e.g. lung bronchi to dilate). Norepinephrine also causes constriction of peripheral blood vessels. Blockage of its removal at synapse is the mode of action of tricyclic antidepressants.

Dopamine is secreted by neurons that originate in the substantia nigra, the area of the brain that initiates movement. A lack of dopamine is responsible for Parkinson's disease.

Amino acids

There are two types:

1. Excitatory, e.g. glutamate and aspartate: glutamate is the most common amino acid and stimulates two receptors; NMDA and AMPA. The former is thought to be involved in the formation of memory.
2. Inhibitory, e.g. GABA and glycine: GABA is secreted by nerve terminals present in the basal ganglia, cerebellum and the spinal cord. Glycine is mainly present in the spinal cord, particularly in the interneurons.

Slow-acting neuropeptide transmitters

These are larger in size than the rapidly acting transmitters and cause more prolonged responses, e.g. long-term changes in the numbers of receptors and synapses. They are manufactured in the neuronal cell body and have a broad range of functions, e.g. pain modulation by opioids.

SENSATION AND PAIN

Be aware of the difference between sensation and perception:
- Sensation: the conscious or subconscious awareness of external/internal stimuli. The reaction to the sensation depends on the CNS target of the impulses. For example, touching a hot oven will elicit the spinal reflex arc, which results in the rapid removal of the hand from the heat.
- Perception: the conscious awareness and interpretation of the sensations. One example is the awareness of extremes of temperature.

Sensation

Sensory modalities

The term 'modalities' describes the different types of sensation, of which there are two:

1. General:
   - Somatic: provide information from sensory receptors in the skin and mucous membranes and include:
   - tactile, e.g. touch, vibration and pressure
   - pain
   - thermal

   - proprioception, which tells the brain the relative positions of the body in space.
   - Visceral: provides information about the internal organs.
2. Special: these include smell, vision, hearing and balance. These are discussed later.

Sensory receptors (Fig. 3.9)

The different types of sensory receptor monitor particular stimuli and can be classified functionally or structurally (Fig. 3.10).

Commit the different types of sensory receptor to memory. They are sometimes asked about in histology exams.

The mechanism of sensation

For example, as demonstrated by an increase in external temperature:

1. The receptor is stimulated, e.g. the increase in temperature stimulates the skin thermoreceptors.
2. The receptor converts this into a graded potential in a neuron.
3. When this reaches threshold it triggers an action potential.

Fig. 3.9 Dermis with hair, Pacinian corpuscle, magnified with pain receptors.

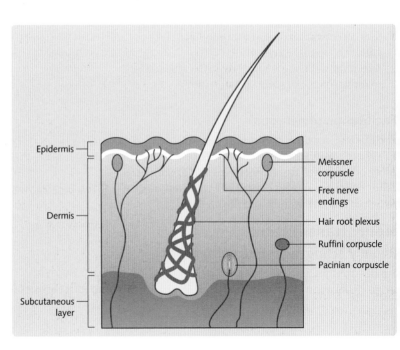

Fig. 3.10 Types of sensory receptor

Functional attributes of receptors		
Type	**Stimulus**	**Where present**
Mechanoreceptor	Mechanical pressure	Skin
Thermoreceptor	Temperature changes	Skin
Nociceptor	Pain	Skin, eyes, muscles
Osmoreceptor	Osmotic pressure of body fluids	Hypothalamus – detect osmolality of body fluids and result in ADH regulation
Chemoreceptor	Chemical changes	Mouth (taste), nose (smell)
Photoreceptor	Light changes	Retina
Structural attributes of receptors		
Type	**Description**	**Associated with**
Free nerve endings	The dendrites are not covered inmyelin	Pain, tickling, touch
Encapsulated nerve endings	These dendrites are surrounded by a connective tissue substance	Touch, e.g. Pacinian corpuscles
Separate cells	These synapse with other neurons to detect sensation	Taste, e.g. gustatory receptors in the taste buds

ADH, antidiuretic hormone.

Fig. 3.10 Types of sensory receptor.

4. The action potential propagates towards the CNS via afferent fibres, e.g. temperature signals travel through the spinothalamic tract.
5. Certain regions of the cerebral cortex, e.g. the hypothalamus, process the information.
6. The cerebral cortex sends signals via efferent fibres to particular regions of the body to respond to the stimulus, e.g. if the body is too hot then sympathetic tone is reduced and the blood vessels dilate passively.

Adaptation

During a continued stimulus, the generator potential decreases in amplitude causing fewer and fewer action potentials. Consequently, the perception of sensation declines. An example is when you go into your flatmate's room and it has that pungent smell(!). After a while you can't smell it any more. This is not because the smell has gone but because your olfactory system has adapted.

Somatic sensations

These arise from stimulation of sensory receptors in the skin and mucous membranes:

Tactile receptors

Tactile sensation includes touch, pressure, vibration, itching and tickling. It is sensed by a combination of the following receptors (the first three activate encapsulated large-diameter myelinated A fibres):

- Meissner corpuscles: comprising free nerve endings. These are egg-shaped masses in the skin dermal papillae. They sense fine touch and predominate in the finger tips, clitoris, lips, nipples and tip of the penis.
- Mechanoreceptors: e.g. Merkel discs, flattened free nerve endings in the stratum basale of skin, for fine touch sensation.
- Ruffini corpuscles: stretch receptors are present deep in dermis, ligaments and tendons.
- Hair root plexus: found in hairy skin, detect movements of hair.
- Pacinian corpuscles: large, multilayered connective tissue capsules. They are found in the dermis and subcutaneous layer, where they detect pressure.
- Free nerve endings: small-diameter, unmyelinated C fibres stimulated by chemicals such as bradykinin, for tickle and itch sensation.

Thermal receptors
- Free nerve endings.
- Cold receptors: found in the stratum basale where they mainly attach to medium-diameter myelinated A fibres.
- Warmth receptors: present in dermis. They are attached to small-diameter, unmyelinated C fibres.

Photoreceptors
Photoreceptors are light-sensitive proteins involved in the function of photoreceptor cells. An example is the rhodopsin in the retina.

Chemoreceptors
A chemoreceptor is a cell, or a group of cells, that transduces a chemical signal into an action potential. There are two types: distance and direct. An example of distance chemoreceptors are olfactory receptor neurons in the olfactory system. Examples of direct chemoreceptors include taste buds in the gustatory system and carotid bodies that detect changes in pH inside the body.

Pain

Pain serves as a protective function by alerting the body to tissue damage. Nociceptors are the receptors for pain and found almost everywhere. They are activated by thermal, mechanical or chemical stimuli.

Understanding how pain is regulated will allow you to prescribe appropriate analgesia on the wards.

There are two types of pain:

1. Fast: occurs within 0.1 s of application of the stimulus:
   - conducted along medium-diameter myelinated A delta fibres
   - very well localized
   - an example would be a hand on a hot stove.
2. Slow: occurs 1–2 s after application of the stimulus and persists:
   - conducted along small-diameter, unmyelinated C fibres
   - localized and diffuse
   - an example would be biliary colic.

Pain can be superficial somatic (in the skin), deep somatic (in deeper structures like muscles, joints and tendons) and visceral organ pain, which can occur away from the organ that is damaged, i.e. referred pain. This is due to the damaged visceral organ and the referred region being innervated by the same spinal cord segment, e.g. liver damage refers to shoulder tip. In this example, the damaged liver presses on the diaphragm, which is innervated by the C3, C4 and C5 nerve branches, which also supply sensation to the shoulder tip.

Regulation of pain
- Peripheral regulation: activity in the low-threshold mechanoreceptors can inhibit the spinothalamic nerve impulses, e.g. rubbing the area of pain.
- Central regulation (Fig. 3.11): the CNS has its own pain control areas:
  - periaqueductal grey areas of the midbrain and pons
  - raphe magnus nucleus: located in the lower region of the pons and superior medulla
  - reticular formation regions in the dorsal horn: nucleus reticularis paragigantocellularis and locus coeruleus – stimulation of opiate and 5-HT receptors in these areas can inhibit the pain pathway and cause analgesia.

Analgesia
Three classes of opioid receptor are found throughout the CNS:

1. Mu (μ).
2. Delta (δ).
3. Kappa (κ).

Opiate-like peptides are derived from three large molecules: proenkephalin, propiomelanocortin and prodynorphin. They act on the μ and δ-opioid receptors in the spinal cord, brainstem and hypothalamus to block pain signals. This is endogenous analgesia.

The opioid morphine mimics endogenous opioid peptides and stimulates the μ-receptors, resulting in non-endogenous analgesia, euphoria, sedation and respiratory depression. All opioids have certain side-effects (Fig. 3.12):

- Respiratory depression: reducing the sensitivity of the brainstem to arterial PCO_2 ($PaCO_2$).
- Nausea/vomiting.

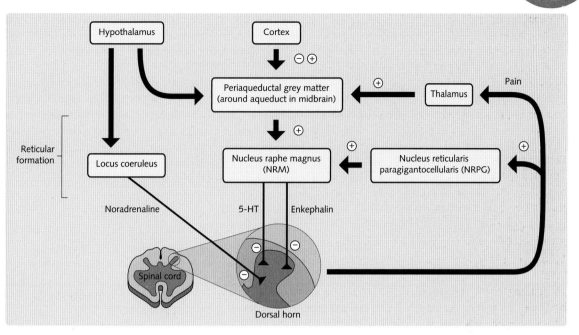

Fig. 3.11 Central regulation of pain.

- Constipation: due to reduced motility of the gastrointestinal tract.

Repeated administration of morphine causes:

- Tolerance: decreased responsiveness to the drug so that more must be taken to achieve the same effect.
- Dependence: both physical (where a withdrawal causes physical symptoms, e.g. influenza) and psychological (drug-seeking behaviour).

Overdose with opiates can result in a comatose patient, respiratory depression and pin-point pupils. Treatment is with intravenous μ-antagonists such as naloxone.

Non-steroidal anti-inflammatory drugs (NSAIDs)
These decrease the sensitization of nociceptors, which occurs in inflammation, by inhibiting the formation of inflammatory mediators.

The brainstem

The brainstem is sandwiched between the spinal cord and the diencephalon. It consists of the medulla oblongata, pons and midbrain. It is here that the cranial nerves, and their respective nuclei, originate.

Medulla oblongata
The medulla oblongata contains three important functional regions:

1. White matter: contains all the sensory (ascending) and motor (descending) tracts:
   - the pyramids are bulges of white matter consisting of large motor tracts; around 90% of these tracts decussate (cross over) to the opposite side of the body, which is the reason why the left brain controls the right side of the body and vice versa.
2. Olives: small olive-shaped structures, just lateral to the pyramids; proprioceptive impulses enter the medulla oblongata here and are then conducted to the cerebellum for processing.
3. Nuclei: regions of grey matter where neurons synapse with each other. There are many different types:
   - gracile: both the gracile and cuneate nuclei are implicated in sensations of touch and in consciousness
   - cuneate: is also involved in proprioception, pressure and vibration
   - cranial nerve nuclei: there are between 8 and 12
   - vital body function nuclei: regulate the cardiovascular, respiratory, vomiting and coughing centres.

Fig. 3.12 Opioid drugs

μ Agonist	Bioavailability and administraion	Metabolism	Potency and length of action	Clinical use	Notes
Morphine	Poor availability when given orally due to high rate of first-pass metabolism. Intravenous administration gives reliable dosing	Active metabolite morphine-6-glucuronide	$t_{1/2}$ 3 h	Acute and chronic pain	Cannot be given in labour as fetal liver cannot conjugate
Diamorphine (heroin)	More lipid soluble. Given orally or by intramuscular, intravenous or subcutaneous injection	Partly to morphine	Very potent, rapid onset, $t_{1/2}$ 2 h	Acute and chronic pain	
Codeine	High oral bioavailability	To other opioids including morphine	One-sixth potency of morphine	Mild pain, headache, dental pain	Potent antitussive, low side effect profile
Pethidine	High lipid solubility. Given orally and by intramuscular injection	Metabolite norpethidine interacts with MAOIs	One-tenth potency of morphine	Acute pain, labour	Does not cause miosis
Fentanyl	High lipid solubility. Given intravenously, epidurally, transdermally		Very potent, short acting	Intraoperative pain	Intraoperative analgesia
Buprenorphine	Increased first-pass metabolism. Given sublingually, intrathecally		$t_{1/2}$ 12 h, slow onset	Acute and chronic pain	Partial agonist and difficult to reverse effects in overdose
Methadone	Given orally or by injection		$t_{1/2}$ >24 h, very slow onset	Maintenance of drug addicts	Does not produce euphoria

Fig. 3.12 Opioid drugs and their pharmacology.

The pons

The word 'pons' means bridge. This 2.5-cm structure rests superior to the medulla and connects different parts of the brain via bundles of axons, which can be part of either the ascending or the descending tract. The nuclei for cranial nerves (CN) V, VI, VII and VIII, and the nerves that relay information from the cerebellum to the cerebral cortex, are also present in the pons.

Midbrain

This extends from the pons to the diencephalon and contains:

- Cerebral peduncles: contain motor axons from the cerebrum to the spinal cord (corticospinal/pontine/bulbar) and sensory axons from the medulla to the thalamus.

- Colliculi: four elevations in the posterior part of the midbrain:
  - superior × 2: some of the optic tracts enter here and regulate reflex eye movements, e.g. scanning, papillary reflex, accommodation reflex and the reflexes that occur with the eyes, head and neck with regard to visual stimuli
  - inferior × 2: conduct information from the auditory pathway to the thalamus.
- Nuclei substantia nigra: darkly pigmented, neurons extend from here to the basal ganglia, where they release dopamine and regulate subconscious muscle activities. The absence of these neurons causes Parkinson's disease.
- Red nuclei: cerebellar and cerebral neurons synapse here and regulate muscular movements.

- Cranial nerve nuclei 3 (oculomotor) and 4 (trochlear) are present here.

Reticular formation

The reticular formation (Fig. 3.13) is a loosely arranged network of white and grey matter in the brainstem. It has the following functions:

- Motor control: via modulation of spinal interneurons and the transmission of information to the cerebellum.
- Sensory control: exerts some control over activity in spinal reflex arcs. It is also important in the regulation of pain perception.
- Visceral control: involved in the regulation of the respiratory and cardiovascular systems.
- Control of consciousness: a certain region of sensory axons, which project to the cerebral cortex (reticular activating system), maintains consciousness. Damage to this region results in prolonged coma.

The brainstem and autonomic function

The autonomic nervous system (ANS) regulates most of the visceral body functions, e.g. vessel tone, gastrointestinal motility, sweating. The ANS is activated by centres located in the spinal cord, brainstem and hypothalamus. These structures relay signals to the particular visceral organ, eliciting a response via the two major subdivisions of the ANS:

1. Parasympathetic
2. Sympathetic.

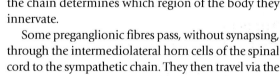
The sympathetic nervous system prepares the body for a flight/fight/frolic situation.

Sympathetic nervous system (Fig. 3.14)

The sympathetic fibres originate in the spinal cord between cord segments T1 and L2. Each sympathetic pathway from the spinal cord to the tissues comprises two neurons, one pre- and one postganglionic. The cell body of each preganglionic neuron lies in the intermediolateral horn of the spinal cord and the fibres pass from here to the anterior cord root in the corresponding spinal nerve.

When the spinal nerve leaves the spinal canal, the sympathetic fibres pass through the white ramus into the sympathetic chain ganglia. The course of the fibre can then:

- Communicate with postganglionic neurons in the ganglion it enters.
- Move up or down in the chain and synapse with other ganglia in the chain.
- Migrate through the chain and radiate out from the chain to synapse in a peripheral sympathetic ganglion.

The level at which the sympathetic fibres exit from the chain determines which region of the body they innervate.

Some preganglionic fibres pass, without synapsing, through the intermediolateral horn cells of the spinal cord to the sympathetic chain. They then travel via the splanchnic nerves to terminate in the adrenal medullae. These fibres influence the secretion of epinephrine (adrenaline) and norepinephrine (noradrenaline).

Parasympathetic nervous system (Fig. 3.15)

The parasympathetic fibres leave the CNS via cranial nerves III, VII, IX and X and spinal nerves S2–S4. Around 75% of the fibres exit via cranial nerve X (the vagus nerve) and distribute to the entire thorax and abdominal regions.

As with the sympathetic system, the parasympathetic system has both pre- and postganglionic neurons. However, the majority of preganglionic fibres reach their target organ uninterrupted and the parasympathetic

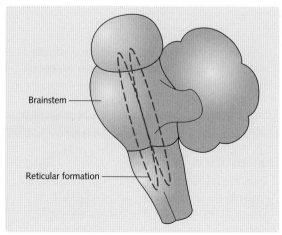

Brainstem

Reticular formation

Fig. 3.13 Reticular formation.

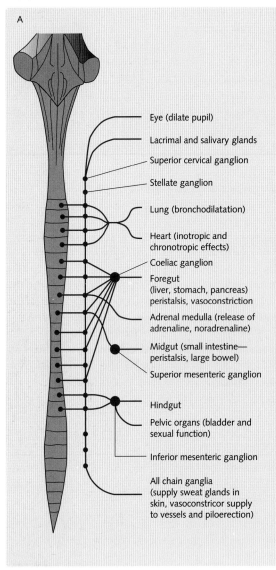

Cord segment	Organ	Effect	Origin of postganglionic neuron
Th1–2	Eye Lacrimal gland Salivary glands	Pupillary dilatation, vasoconstriction	Sup. cervical ganglion
Th1–4	Heart	Heart rate, stroke volume, coronary artery dilatation	Sup., middle and inf. cervical and upper thoracic ganglia
Th1–4	Skin & muscles of head & neck	Sweat, vasoconstriction and piloerection	Sup. and middle cervical ganglia
Th2–7	Bronchi Lungs	Dilatation, vasodilatation	Inf. cervical and upper thoracic ganglia
Th3–6	Skin & muscles of upper extremity	Sweat, vasoconstriction and piloerection	Stellate and upper thoracic ganglia
Th6–10	Stomach	Decreased peristalsis and secretion,	Coeliac ganglion
	Pancreas Small intestine, ascending and transverse large bowel	vasoconstriction, vasoconstriction, decreased peristalsis and secretion	Coeliac ganglion Coeliac, sup. and inf. mesenteric ganglia
Th10–L2	Skin & muscles of lower extremity	Sweat, vasoconstriction and piloerection	Lower lumbar and upper sacral ganglia
Th11–L2			Inf. mesenteric, hypogastric ganglia and pelivc plexus
	Descending large bowel & rectum	Decreased peristalsis and secretion	
	Ureter & bladder	Relax detrusor, contract internal sphincter	Hypogastric ganglion and pelivic plexus

Inf., inferior; sup., superior.

Fig. 3.14 Organization of the sympathetic nervous system.

system thus has longer preganglionic fibres. The postganglionic neurons tend to be located in the wall of the target organ and, after synapsing with the preganglionic fibres, innervate the organ tissues.

Enteric nervous system
The neurons for this third division of the ANS are located in the walls of the gastrointestinal tract, pancreas and gall bladder. The enteric nervous system comprises two plexi:

1. Meissner's/the submucosal plexus: which lies between the mucous membrane and circular muscle layer: This regulates gastrointestinal secretion and blood flow.

2. Auerbach's/the myenteric plexus: which is present between the longitudinal and circular muscle layers: This controls the gastrointestinal movements.

THE VISUAL SYSTEM (FIG. 3.16)

Definitions
- Anterior chamber: the space between the cornea and the iris is filled with aqueous humor.
- Choroid: the vascular portion of the posterior segment of the eye; it provides nutrition to the retina.

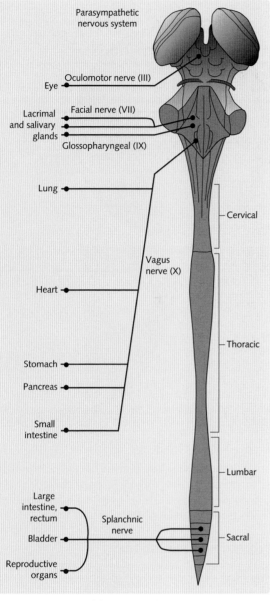

Fig. 3.15 Organization of the parasympathetic nervous system.

- **Lacrimal apparatus:** group of structures that produces and drains lacrimal fluid (tears). The lacrimal glands secrete lacrimal fluid, which drains into the lacrimal ducts. It then enters the nasolacrimal duct and finally drains away into the nasal cavity.
- **Lens capsule:** a fibrous capsule over the lens.
- **Lens:** ball-like structure that changes shape to focus the image on the retina.
- **Lids:** provide protection for the eye and the glands within the eyelid.
- **Limbus:** where the cornea meets the sclera.
- **Meibomian glands:** secrete an oily substance that helps stabilize the tear film.
- **Optic disc:** that part of the fundus where all the nerve fibres from the retina meet to form the optic nerve.
- **Optic nerve (cranial nerve II):** sensory nerve that runs from the retina to various parts of the brain.
- **Posterior chamber:** the space between the iris and the lens; filled with aqueous fluid.
- **Retina:** nervous tissue that lines the posterior three-quarters of the eyeball. Photoreceptive cells in the retina convert the light image into a bioelectrical signal for the brain to understand.
- **Sclera:** the fibrous shell of the eye.
- **Vitreous:** a jelly-like material that fills the posterior segment of the eye.
- **Zonular ligaments:** the ligaments connecting the ciliary muscle to the perimeter of the lens, stabilizing the position of the lens.

Anatomy

Make sure you can draw the layers of the retina and know the role of each of the cells.

- **Ciliary body:** aqueous fluid is produced here.
- **Ciliary muscles:** these smooth muscles attach the ciliary body to the lens via the zonular ligaments.
- **Conjunctiva:** lines the inner surface of the eyelids.
- **Cornea:** clear tissue in front of the eye. It has an epithelial surface layer; a stroma, which comprises most of the corneal tissue; a tough layer called Desçemet's membrane and an endothelium on the inner surface.
- **Iris:** the coloured part of the eye. Consists of dilator and constrictor muscles and vascular tissue. The pupil is the opening in the centre of the iris.

The retina forms the inner surface of the eyeball. It has two layers:

1. **Deep – pigmented – layer:** this melanin-coated epithelial sheet prevents the scattering of light rays and focuses them on the retina.
2. **Superficial – neural – layer:** processes the incoming light source and sends information to the optic nerve. This information is analysed in the brain and an image is formed. The neural layer is divided into three cell levels:

Fig. 3.16 Cross-section through the eye showing the main structures.

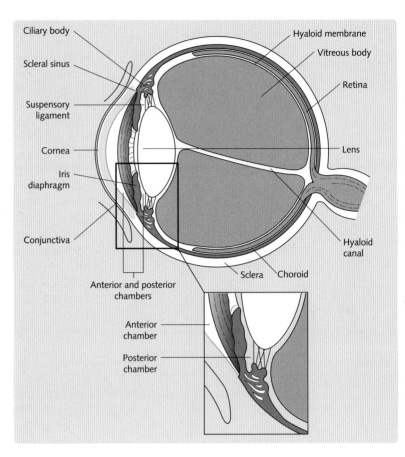

- Photoreceptive cells.
- Bipolar cells.
- Ganglion cells.

These levels (Fig. 3.17) are separated by two synaptic cell layers where synaptic contacts are made. Two other types of cell are present: horizontal and amacrine cells. These alter the signals from the photoreceptive cells, bipolar cells and ganglion cells.

Photoreceptive cells

Two cell types transduce light rays into receptor potentials:

1. Rods: contain the blue–green light-absorbing photopigment rhodopsin. Dim light stimulates the rods, enabling to see with minimal light, but not in colour.
2. Cones: react to bright light and produce colour vision. The cones contain three different types of photopigment, which absorb blue, green and yellow–orange light. These photopigments are composed of two parts: opsin (glycoprotein) and retinol (a vitamin A derivative). Each photopigment contains a different variant of opsin, which permits the absorption of different wavelengths of light and hence the different colours.

Processing visual stimuli

1. Light enters the eye through the pupil and strikes the photopigment in the retina.
2. This results in:
   - isomerization: a change in conformation of retinol from the *cis* to the *trans* form
   - bleaching: transretinol separates from opsin. It looks colourless.

Seeing in the dark

Sodium ions enter the photoreceptor via ligand (cGMP)-gated Na^+ channels. This depolarizes the rods and cones to about –30 mV. (Remember, to elicit an action potential the resting membrane potential must be at least –70 mV.) The –30 mV partial depolarization causes release of glutamate, which triggers IPSPs that hyperpolarize bipolar cells and prevent transmission to the ganglion cells.

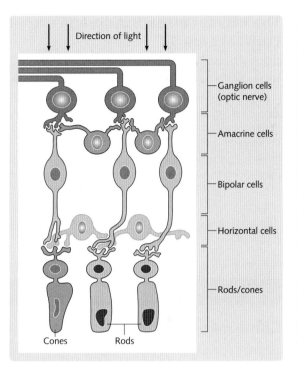

Fig. 3.17 Layers of retina.

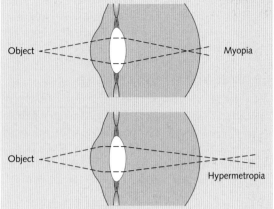

Fig. 3.18 Myopia and hypomyopia.

Seeing when light is present and isomerization occurs

The ospin undergoes a conformational change, which reduces cGMP levels. This causes the ligand-gated Na⁺ channels to close and the membrane potential becomes closer to −70 mV. This hyperpolarization decreases glutamate and prevents the inhibition of the ganglion cell, which transmits a signal to the optic nerve and to the optic region of the brain where it is analysed.

Reconversion

When no more light is present, the enzyme retinol isomerase converts *trans*-retinol back to *cis*-retinol.

Regeneration

The *cis*-retinol then binds to opsin and the pigment resynthesizes.

Optics

Light that enters the eye is refracted by the cornea and – to a lesser degree – by the lens so that it is focused exactly on the retina (convergence). If the rays do not meet together at the same retinal point then the image produced is unfocused (Fig 3.18). If images of near objects fall on the retina but images of far objects do not, then the person is nearsighted (myopic). Conversely, a person is farsighted (hypermetropic) if images of far objects focus on the retina but images of near images do not.

Accommodation

Objects that are very close to the eye hit the cornea at a greater angle and so must be refracted more to meet at the retina. The lens is responsible for this extra refraction. The ciliary muscles contract, causing the zonular fibres to relax. This eases the tension on the lens and it becomes shorter and broader in shape, and refracts the rays so they meet the same point on the retina.

Visual acuity

This is how precisely an image is seen: the greater the precision the higher the acuity. The area of greatest acuity in the retina is the central fovea, which contains only cones.

Clinically, visual acuity is tested by using a Snellen chart. This contains letters in decreasing size. Patients stand 20 feet away and, if they can see the letters one would normally see at 20 feet, then they are said to have 20/20 vision.

Dark and light adaptation

After a person has been sitting in a bright room for a while, much of the photochemical in the rods and cones will have been converted to retinol (which is subsequently converted to vitamin A) and opsins. This reduction in the amount of photosensitive chemical reduces the sensitivity of the eye to light even further.

Conversely, sitting in a dark room for a while results in the photochemical being converted back into light-sensitive pigments, so much dimmer levels of light can be detected.

Central visual processing

After they have left the eyes, the optic nerves pass through the optic chiasma, where some nerve fibres cross and travel to the opposite side of the brain from which the visual stimuli came (Fig. 3.19). After the chiasma, the axons become the optic tract, which ends in one of the following locations:

- Lateral geniculate nucleus (LGN): the retinal fibres terminate in six discrete layers of the LGN: ipsilateral fibres in layers 2, 3 and 5 and contralateral fibres in layers 1, 4 and 6. A given area of the retina projects to a certain level in the LGN. This organization is preserved all the way to the visual cortex.
- Pretectal nuclei and superior colliculus in the midbrain: activate the pupillary light reflex and govern rapid directional eye movements.
- Suprachiasmatic nucleus in the hypothalamus: involved with the control of circadian rhythms.

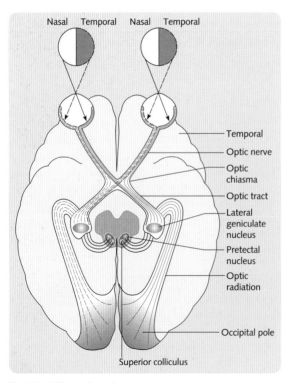

Nasal   Temporal   Nasal   Temporal

Temporal
Optic nerve
Optic chiasma
Optic tract
Lateral geniculate nucleus
Pretectal nucleus
Optic radiation
Occipital pole
Superior colliculus

Fig. 3.19 Pathway of visual processing.

The visual cortex/Brodmann's area cortical area 17

The primary visual cortex lies in the medial occipital cortex and contains six layers of cells arranged in millions of vertical hypercolumns, each representing a different function, e.g. orientation columns enable lines to be discriminated; ocular dominance columns enable the perception of the depth of an object; there are also colour-detecting regions. All this information is processed individually and then brought together to form an image. Although it is unclear precisely how these cells interact to form an image, the process no doubt involves excitatory and inhibitory interplay between the cells.

THE AUDITORY SYSTEM

Conduction of sound

Sound waves are changes in pressure that are transmitted through the air; sound is defined by the amplitude (loudness) and frequency (pitch) of the sound waves. Amplitude is measured in decibels (dB), which is a logarithmic scale; frequency is measured in hertz (Hz), with the normal range being between 20 and 20 000 Hz.

The auditory system, which converts sound waves into electrical signals, consists of two main areas:

1. The ear, which is divided into the external, middle and inner regions.
2. The neural pathway to the brainstem and auditory cortex.

The ear

> Thinking of the ear as being divided into three regions will help you to understand what each section does and how they relate to each other.

The ear is divided into three regions (Fig. 3.20).

The external ear

This consists of the pinna, external auditory canal and eardrum. It collects the sounds and transmits them to

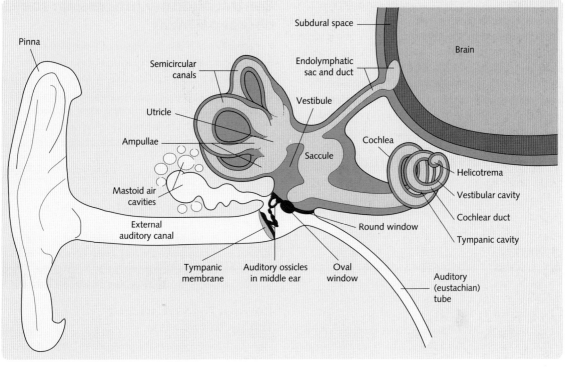

Fig. 3.20 Components and relations of the outer, middle and inner ear.

the external auditory meatus, from where they strike the eardrum (tympanic membrane).

The middle ear

This small, epithelium-lined, air-filled cavity contains the auditory ossicles. These are three small bones called the malleus, incus and stapes (or hammer, anvil and stirrup, respectively). The malleus is attached to the inside surface of the eardrum and to the incus, which articulates with the stapes. They transmit sound to the inner ear.

The muscles in the middle ear include the tensor tympani, which limits extreme movements of the eardrum to prevent excessive (and damaging) vibrations in the inner ear, and the stapedius, which attenuates large vibrations on the stapes to protect the oval window where the sound will travel.

The eustachian/pharyngotympanic/auditory tube connects the middle ear and the oropharynx. It allows pressure in the middle ear to equal atmospheric pressure. This permits sound vibrations to travel freely.

The round and oval windows are openings that separate the middle from the inner ear.

The inner ear

This contains two distinct functional divisions, the cochlea (for hearing) and the vestibule and semicircular canals (for balance); these resemble tunnels within the temporal bone. Within the canals is a series of membranous sacs (labyrinths) containing sensory epithelium and endolymph (a fluid). The canals are further surrounded by another fluid (perilymph). The content of endolymph is similar to that of intracellular fluid, whereas perilymph resembles extracellular fluid.

The cochlea resembles a snail's shell. It is divided into three channels, the scala vestibuli, scala tympani and scala media/cochlear duct. The scala media contains endolymph; the others are bathed in perilymph. The floor of the scala media is formed by the basilar membrane, which contains hair cells covered with stereocilia on their surfaces. Attached to the scala media is a gelatinous structure (the tectorial membrane) which lies above the inner and outer hair cells and comes into contact with their stereocilia.

The organ of Corti is the functional unit of hearing and contains the sound receptors that convert mechanical vibrations into nerve impulses.

The epithelium of the organ of Corti consists of supporting cells and hair cells. The bases of the hairs are surrounded by the afferent fibres of the cochlear nerve. The tips of the hairs protrude into the endolymph, the longest of them being embedded in the overlying, gel-like tectorial membrane.

Axons from the auditory nerve synapse at the base of the hair cells and leave the cochlear and temporal bone through the internal auditory canal to the brainstem.

Events involved in hearing
(Fig. 3.21)

1. The pinna channels sound waves to the external auditory meatus.
2. The sound hits the eardrum, causing it to vibrate.
3. The vibration is transmitted from the malleus to the incus and finally to the stapes.
4. The vibrating stapes pushes the oval window membrane in and out.
5. The moving oval window causes the perilymph in the cochlea to form a wave, which travels through the cochlea.
6. The fluid thrill passes the basilar membrane, causing it to move in a wave-like pattern that travels the length of the cochlea and dissipates at the round window.
7. As the basilar membrane is displaced by the perilymph wave, the stereocilia at the apex of each inner and outer hair cell move.
8. Mechanical movement can open or close cation channels in the stereocilia. Opening results in an inward K^+ current, graded depolarization and the release of a neurotransmitter (glutamate) that causes the cochlear fibres to transmit a faster stream of impulses to the brain.

Central auditory processing: the neural pathway

- Nerve fibres from the organ of Corti enter the cochlear nuclei in the medulla oblongata (Fig. 3.22).
- Here they synapse with second-order neurons and the majority travel to the opposite side of the brain (a minority stay ipsilateral), where they terminate in the superior olivary nucleus (SON).
- The fibres then pass to the lateral lemniscus. Some terminate in this nucleus, but many pass through and synapse in the inferior colliculus.
- They then travel to and synapse in the medial geniculate nucleus. Finally, the pathway continues as the auditory radiation, finally ending in the auditory cortex.

The auditory cortex

The auditory cortex is located in the superior temporal gyrus of the temporal lobe. It is divided into two separate areas: primary auditory cortex (PAC) and secondary auditory cortex (SAC). The PAC is stimulated by impulses form the medial geniculate body; the SAC receives impulses from the PAC.

Approximately six tonotopic maps have been found in both association cortices. An association cortex is part of the cerebral cortex involved with advanced steps of sensory information processing. Tonotopic maps are associative areas of the brain that performs a topology preserving mapping of acoustic frequencies on the auditory cortex. Each map discriminates certain characteristics of sounds. For instance, one distinguishes sound frequencies, another tells you where the sound is coming from and so on.

Fig. 3.21 Events in the stimulation of auditory reception in the right ear.

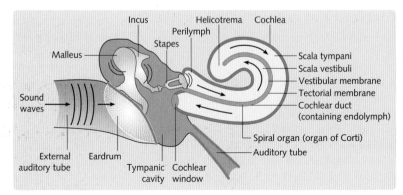

Fig. 3.22 Central auditory processing.

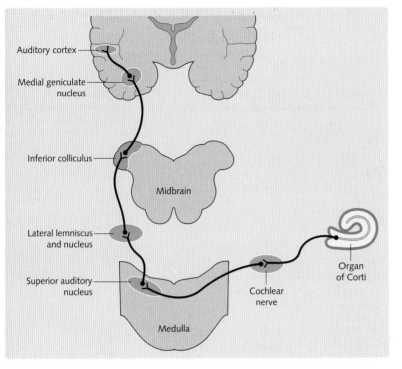

THE OLFACTORY SYSTEM

The sensations of smell and taste arise from interactions between certain molecules and specific receptors. Signals are then sent to the limbic and other higher cortical areas, which analyse the incoming information.

> Olfactory neurons are one of the few neurons that can regenerate.

Anatomy of the olfactory system

The epithelial lining of the nasal cavity contains three types of cell:

1. Olfactory receptors: these number 10–100 million and are first-order neurons. The dendrites of these are covered with cilia.
2. Supporting cells: these provide the olfactory receptors with nutrition, physical support and remove toxins.
3. Basal stem cells: produce new olfactory receptor cells.

Physiology of the olfactory system

Odours enter the nasal cavity and stimulate the nasal cilia on the olfactory receptor. This is linked to a second-messenger system that opens Na^+ channels. The influx of Na^+ results in depolarization and a subsequent action potential. This propagates along the olfactory nerve (CN I), which terminates below the frontal lobes of the cerebrum and lateral to the crista galli in olfactory bulbs. Within these bulbs, the olfactory receptors synapse with second-order neurons to form the olfactory tract. This projects to the lateral olfactory area in the temporal lobe. Further connections exist between this lateral olfactory area to the hypothalamus, thalamus, limbic system and frontal lobe. Thus smell excites many areas of the brain, which explains why some smells can evoke memories and others can cause nausea.

THE GUSTATORY SYSTEM

Anatomy of the gustatory system

The receptors for this system are the taste buds, of which there are approximately 10 000. They predominate on the epithelial surface of the tongue, soft palate, pharynx and epiglottis. On the tongue they are found in three types of papillae: circumvallate, fungiform and foliate. They detect four modalities; salt, sour, bitter and sweet. Each taste bud contains three types of cell:

1. Gustatory receptor cells (GRC): project a single microvillus.
2. Supporting cells: provide nutrition and physical support to the gustatory receptor cells.
3. Basal cell: develop into supporting cells and then gustatory receptor cells.

Physiology of the gustatory system

Chemicals (e.g. food) make contact with the microvillus on the GRC and result in depolarization with the release of a neurotransmitter substance (Fig. 3.23). The signal for depolarization varies for each taste modality:

- Saltiness: Na^+ from salty food enters the GRC via Na^+ channels. This causes the influx of Ca^{2+}, which results in the release of neurotransmitter.
- Sourness: H^+ from acid in the sour substance either enters Na^+ channels or blocks K^+ channels, causing depolarization.
- Sweetness: molecules bind to a receptor site coupled with a G protein. This causes an increase in cAMP, which activates protein kinase A. This blocks K^+ channels, causing depolarization.
- Bitterness: molecules either block K^+ channels directly or act via second messengers to cause depolarization.

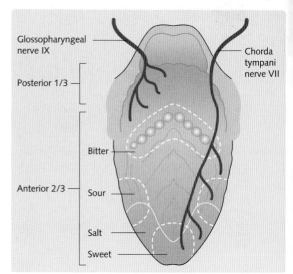

Fig. 3.23 Anatomy of the tongue.

Once the neurotransmitter is released it stimulates the first-order neurons in the gustatory pathway. These include CN VII (which serves the anterior two-thirds of the tongue), CN IX (serving the posterior third of the tongue) and CN X (the pharynx and epiglottis).

The afferent neurons enter the medulla, where they synapse in the gustatory nucleus. They then propagate to the thalamus, hypothalamus and limbic system. From the thalamus, some project to the parietal lobe, which gives the conscious perception of taste.

Physiology of the musculoskeletal system

4

Objectives

In this chapter, you will learn to:
- Describe the basic structure of the skeleton
- Discuss the structure and function of the different types of muscles
- Explain the mechanism of muscle contraction
- Discuss the different sources of energy used in muscle contraction
- Outline the central and peripheral control of the musculoskeletal system
- Describe the structure of long bones and the different types of cells involved in bone remodelling
- Describe the different functions of bone

OVERVIEW OF THE MUSCULOSKELETAL SYSTEM

The musculoskeletal system has many important functions, which are facilitated by its structure.

Function

- Support and protection of organs.
- Maintain body posture.
- Locomotion and work.
- Storage of minerals, e.g. calcium, phosphate.
- Haemopoiesis.

Structure

The musculoskeletal system consists of muscles, skeleton, joints and connective tissue.

Muscle types

There are three types of muscle: skeletal, cardiac and smooth.

The three different types of muscle have the following features (Fig. 4.1):

1. Skeletal (striated) muscle:
   - attached to the skeleton
   - maintains posture and move limbs
   - under voluntary control.
2. Cardiac (striated muscle of the heart):
   - contraction of the heart
   - under involuntary control.
3. Smooth (non-striated) muscle:
   - lines blood vessels and hollow organs of the body
   - maintains blood vessel tone
   - under involuntary control.

The skeleton

The skeleton has four components:

1. Bone.
2. Cartilage.
3. Ligaments and tendons.
4. Joints.

Bone

Bone is a highly specialized, hard form of connective tissue. It has a variety of functions:

- Structural support.
- Attachment for muscles, tendons and ligaments.
- Locomotion via joints.

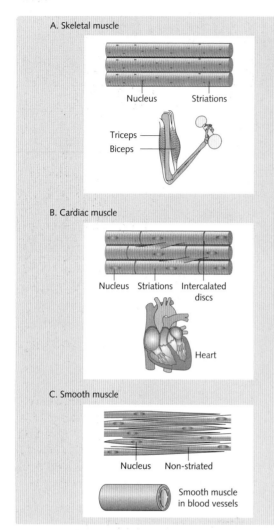

A. Skeletal muscle

Nucleus   Striations

Triceps
Biceps

B. Cardiac muscle

Nucleus Striations Intercalated
                         discs

Heart

C. Smooth muscle

Nucleus   Non-striated

Smooth muscle
in blood vessels

Fig. 4.1 Muscle types.

- Mineral homoeostasis.
- Haemopoiesis.

Cartilage

This is a semi-rigid form of connective tissue and is present at points of mobility, e.g. costal cartilage, which attaches the ribs to the sternum.

Fig. 4.2 Types of joint (A) synovial, (B) fibrous and (C) cartilaginous.

Ligaments and tendons

Both are composed of fibrous tissue.

- Ligaments: connect bones or cartilage and strengthen joints.
- Tendons: connect muscle and bone so that when the muscle changes length the attached skeleton moves with it.

Joints

A joint is the point of union between two or more bones. The three main types (Fig. 4.2) of joint are classified according to the material used to adjoin the bones:

1. Synovial: the most common type of joint, which permits free movement between the adjoining bones. Synovial joints are united by an articular capsule, e.g. knee, elbow, hip, shoulder.
2. Fibrous joints: these are united by fibrous tissue. Movement is usually restricted, e.g. skull sutures, interosseous membrane between radius/ulna and tibia/fibula.
3. Cartilaginous joints: these are united by hyaline or fibrocartilage, e.g. intervertebral discs.

SKELETAL MUSCLE: STRUCTURE AND FUNCTION

Three types of muscle are present in the body: skeletal, cardiac and smooth. Cardiac and smooth muscle are discussed in Chapter 5. The following text relates to skeletal muscle.

Be able to distinguish between myofibrils, muscle fibres, sarcomeres and sarcoplasm.

A
Synovial joint
(ball and socket)
e.g. hip/glenohumeral
joints

B
Fibrous joint
e.g. skull sutures

C
Cartilagenous joint
e.g. intervetebral discs

Microstructure

Skeletal muscle consists of a number of individual muscle fibres grouped together in bundles called fasciculi (Fig. 4.3). These are bounded by a coat of connective tissue called the epimysium. Connective tissue is also present in the muscle body which transfers the mechanical force generated by the muscle to the skeleton.

Skeletal muscle fibres

Long, thin cylindrical cells containing many nuclei. They are up to 30 cm in length and 10–100 μm in diameter.

Myofibrils

Filamentous bundles found in individual muscle fibres; they run along the entire length of the fibre. Myofibrils consist of a system of longitudinal thin and thick filaments arranged in a regular pattern. The thin filaments consist mainly of the protein actin, which is arranged spirally along the filamentous protein tropomyosin. At regular intervals (40 nm) there is a regulatory protein complex made up of troponins C, I and T. Thick filaments consist mainly of the protein myosin. These too are arranged in a regular manner, with each thick filament surrounded by six thin filaments.

The overlap of thick and thin filaments gives the muscle its striated appearance. The A band (dark in colour) contains myosin filaments overlapping actin filaments. It has a high reflective index under light microscopy and can refract polarized light (i.e. it is anisotropic). The I band (light in colour) contains only actin filaments and has a lower refractive index (isotropic). In the centre of each I band is a Z line, a disc of material running across the muscle fibre and joining one myofibril to another.

Sarcomere

This is the fundamental contractile unit within the muscle, from one Z line to the next.

Sarcoplasm

Muscle fibre matrix in which the myofibrils are suspended. The sarcoplasm contains:

- Ions: potassium, magnesium, phosphate.
- Proteins: enzymes.
- Mitochondria: a lot of energy is required for muscle contraction.

Sarcoplasmic reticulum

This is the muscle fibre's equivalent of endoplasmic reticulum. The sarcoplasmic reticulum (SR) runs

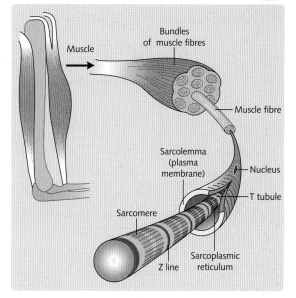

Fig. 4.3 Microstructure of skeletal muscle.

longitudinally along the myofibrils and wraps around groups of myofibrils.

Sarcolemma

The cell membrane surrounding the muscle cell. At each Z line, the sarcolemma invaginates to form a T tubule. Where the T tubule and SR meet, the SR enlarges to form the terminal cisternae. A triad is a T tubule sandwiched between two SRs. This plays an important role in excitation–contraction coupling.

Ion balance and the resting membrane potential

The ionic balance between the intracellular fluid (ICF) and extracellular fluid (ECF) (Fig. 4.4) of a muscle cell is maintained by:

- The selective permeability of the cell membrane to K^+ and Cl^-.
- Relative cell membrane impermeability to Na^+.

At rest, the ionic difference across the sarcolemma creates a resting potential of −90 mV. This difference in charge is maintained by a Na^+/K^+-ATPase, which moves two K^+ into the cell and three Na^+ ions out, thus creating a net negative charge inside the cell.

Neuromuscular junction

The skeletal muscle fibres are innervated by large, myelinated nerve fibres that arise from

Fig. 4.4 Distribution of ions in the ICF and ECF of muscle cells

Ion	ICF (mmol/L)	ECF (mmol/L)
Na^+	12	145
K^+	155	4
H^+	13×10^{-5}	3.8×10^{-5}
Ca^{2+}	8	1.5
Cl^-	3.8	12
HCO_3^-	8	27
A^-	155	0

Fig. 4.4 Distribution of ions in the intracellular (ICF) and extracellular fluid (ECF) of muscle cells.

the anterior horns of the spinal cord. The nerve fibres branch and, when they reach the muscle fibre, lose their myelin sheath and end in a small swelling (terminal bouton). This nerve terminal together with the underlying membrane on the target cell (i.e. skeletal muscle cell) constitute the neuromuscular junction (NMJ; Fig. 4.5). The space between the two is the synaptic cleft. Each muscle fibre has only one NMJ.

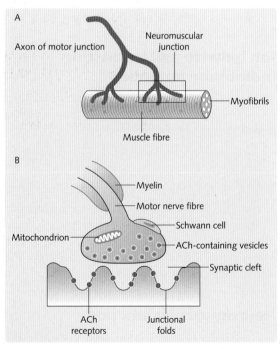

Fig. 4.5 Neuromuscular junction (A) gross, (B) more detailed.

Presynaptic nerve terminal/terminal bouton

The terminal bouton contains:

- Synaptic vesicles: these contain the neurotransmitter acetylcholine (ACh).
- Mitochondria: ATP is required to power the action potential and to synthesize ACh.
- Active zones: these are the sites of neurotransmitter release. They lie opposite a junctional fold in the postsynaptic membrane.
- Voltage-activated Ca^{2+} channels: these are adjacent to active zones and open in response to an AP. Ca^{2+} then enters and causes the vesicles to migrate to the active zones and release the neurotransmitter.

Postsynaptic membrane

The membrane on the other side of the synapse – the postsynaptic membrane – contains:

- Acetylcholine receptors (AChR): the binding of two ACh molecules to an AChR induces a conformational change in the receptor, which results in the opening of the voltage-gated Na^+ channels.
- Voltage-gated Na^+ channels: located throughout the junctional folds and in membranes adjacent to the NMJ. They open when the neurotransmitter ACh binds to AChR.
- Acetylcholinesterase: the enzyme that breaks down ACh into acetate and choline. It is found in the junctional folds.

Transmission across the synapse
Secretion of acetylcholine

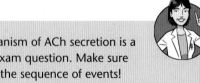

The mechanism of ACh secretion is a frequent exam question. Make sure you know the sequence of events!

1. Nerve impulse reaches the NMJ.
2. Voltage-gated Ca^{2+} channels open and Ca^{2+} diffuses into the terminal bouton.
3. The Ca^{2+} attracts the ACh vesicles to the neural membrane.

4. The vesicles fuse with the neural membrane and empty their ACh into the synaptic cleft.
5. The binding of two molecules of ACh to an AChR causes the channel in the receptor to open.
6. Na^+ enters the channel and creates a potential change at the muscle membrane, called the end-plate potential (EPP).
7. This EPP initiates an AP at the muscle membrane and causes muscle contraction.
8. The ACh in the synaptic space continues to activate the AChR, so when muscular contraction is no longer required the ACh must be removed. This is done by two means:
   - acetylcholinesterase is attached to the basal lamina; this enzyme breaks ACh into acetic acid and choline
   - a small amount of ACh diffuses out of the synaptic space and is broken down by plasma cholinesterases.
9. Recycled choline reacts with acetyl CoA from the mitochondria and, after catalysis with acetyltransferase, is reformed as ACh and packaged into vesicles in the terminal bouton.

Excitation–contraction coupling

The skeletal muscle fibre is so large that an AP spreading along its surface membrane causes almost no current flow deep within the fibre. To overcome this, transverse tubules (T tubules) penetrate the muscle fibre and transmit the AP. In response to an AP, the sarcoplasmic reticulum in the myofibrils releases Ca^{2+}, which causes the muscle fibre to contract. This is known as excitation–contraction coupling.

SKELETAL MUSCLE: CONTRACTION

Contractile filaments

Myosin

Myosin comprises six polypeptide chains, two of which are heavy and the remaining four light. The heavy chains intertwine to form a double helix, one end of which is folded into a globular structure known as a head; the other end is the tail (Fig. 4.6).

The four light chains attach to the myosin head and control function during muscle contraction. The head also has ATPase activity, which is required to cleave ATP to provide the energy required in contraction.

Each myosin filament is made up of 200 or more individual myosin molecules, which are bundled together. The bundled tails form the body, part of the helix forms the arm, which has the head attached to it. The protruding tail and head form the cross-bridges with the actin filament.

Actin filament

Each actin filament consists of three protein components: actin, tropomyosin and troponin.

Actin

Double-stranded F-actin comprises the backbone of the actin filament. Each F strand is made up of

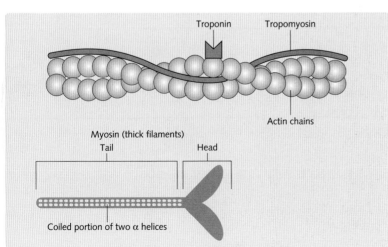

Fig. 4.6 Structures of actin and myosin.

G-actin molecules. Attached to these is one molecule of ADP, which is the attachment site for the myosin cross-bridges.

Tropomyosin

Attached loosely to the F-actin strands, the tropomyosin molecules cover the myosin-binding site on the actin filament so that no interaction can occur and there is no muscular contraction in the resting state.

Troponin

This is a complex of three subunits: I, T and C. Its main function is to attach tropomyosin to actin. It also binds to Ca^{2+}, thus initiating the contraction process.

Mechanism of contraction
(Fig. 4.7)

1. Ca^{2+} ions combine with troponin C.
2. The combination of Ca^{2+} and troponin C causes a conformational change in the troponin complex.
3. This displaces the tropomyosin from its resting position over the actin active sites.
4. The myosin heads from the cross-bridges attach to the actin active site.
5. ATP is hydrolysed by the ATPase on the myosin head to ADP. The energy released causes the head to tilt (power stroke) and drag the actin filament with it.
6. The cycle then repeats, with the head breaking away from the active actin site and attaching to a new site further along.
7. The cycles continue until the actin filament pulls the Z membrane up against the ends of the myosin filaments or until the load on the muscle becomes too great for further pulling to occur.

Relaxation

During AP repolarization, Ca^{2+} is actively pumped back into the SR. This causes the tropomyosin to obstruct the myosin binding sites on the actin myosin heads, so the actin and myosin filaments cannot attach. The filaments then slide back to their original position, passively and without the expenditure of energy.

Bioenergetics of muscle contraction

The mechanism of muscle contraction requires energy for:

- The power stroke, which is the actual muscle contraction.

- Pumping Ca^{2+} from the sarcoplasm into the SR when contraction is over.
- Pumping Na^+ and K^+ ions through the muscle fibre membrane to maintain an appropriate ionic environment for the propagation of action potentials.

> The mechanism of muscle contraction is a favourite of the examiners. Try not to get confused with troponin and tropomyosin.

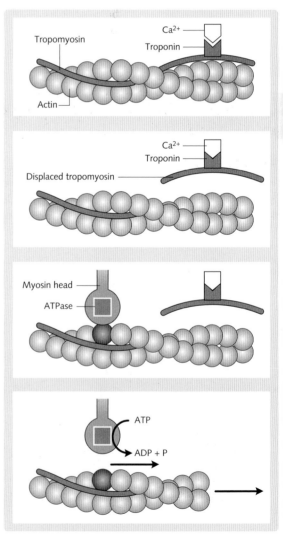

Fig. 4.7 Mechanism of muscle contraction.

Sources of energy

- Phosphocreatine: carries a high-energy phosphate bond similar to that of ATP. It can be used to phosphorylate ADP to ATP by the enzyme creatine kinase.
- Glycogen: is stored in muscle cells. Rapid enzymatic breakdown to pyruvic acid and lactic acid liberates energy that converts ADP to ATP. There are two advantages to this mechanism:
  - it can occur even in the absence of oxygen, although only for about 1 minute, as there would be too much of a build-up of lactic acid to go on longer than this
  - ATP formation is 2.5 times greater via glycolysis than when foodstuffs react with oxygen.
- Oxidative metabolism: combining oxygen with various foodstuffs (fats, carbohydrates and proteins) liberates ATP. Around 95% of all energy derived is from this source.

MUSCLE FUNCTION AND MOTOR CONTROL

Motor units

A motor unit consists of a group of muscle fibres and the motor neuron that innervates them. Each muscle fibre is innervated by a single motor neuron (Fig. 4.8).

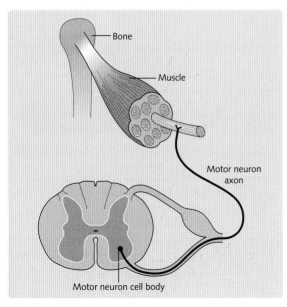

Fig. 4.8 Motor unit.

However, the axons of each motor neuron can branch and innervate many muscle fibres. The number of muscle fibres innervated by a single motor neuron axon is functionally important. All the muscle fibres in a single motor unit contract and relax together. Muscles requiring fine control, such as the laryngeal, therefore have fewer muscle fibres per motor neuron. The opposite is true for muscle involved in gross movements. This motor unit : muscle fibre ratio also dictates the strength of contraction.

Motor unit = single motor neuron and all muscle fibres it innervates.

Muscle mechanics

There are two types of muscle contraction (Fig. 4.9):

1. Isometric: the muscle does not shorten during contraction. An example would be if you held a dumb-bell steady with your arm outstretched. These movements are important in maintaining posture.
2. Isotonic: the muscle shortens during contraction, while the muscle tension remains constant. This is divided into:
   - concentric: the muscle shortens and pulls on another structure such as the tendon, e.g. repeatedly lifting a dumb-bell using the biceps brachii
   - eccentric: the muscle lengthens in a controlled manner, e.g. when you lower the dumb-bell gently as oppose to dropping it.

Maintaining muscle contraction

Twitch contraction

This refers to a single contraction from a single AP. As this is insufficient to cause any useful movement, summation occurs to add together the twitches and cause contraction. There are two types of summation (Fig. 4.10):

1. Multiple fibre summation: this is the increase in the number of motor units contracting simultaneously to produce sustained contraction.
2. Frequency summation/tetanization: initially, twitches follow each other sequentially at low frequency. The frequency of stimulation increases until the twitches overlap and contraction occurs. However, there is still time between contractions for

Fig. 4.9 Types of muscle contraction (A) isotonic (B) isometric. In isotonic contraction the muscle contracts and either shortens or lengthens. In isometric contraction, the muscle neither shortens or lengthens, but it does contact.

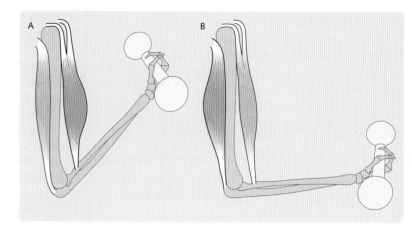

partial relaxation (unfused tetanization). When a critical period is reached the repeated contractions merge with each other appearing smooth and continuous with no relaxation in between.

Length–tension relationship

The force, or tension, of muscle contraction depends on the length of the sarcomeres before contraction. There is an optimal range of sarcomere length at which the force of contraction is at its maximum. To understand this, think about the structure of muscle: if muscle is overstretched, the sarcomere will overstretch. This will minimize the overlap between the actin and myosin heads, thus reducing tension. Similarly, if the sarcomere is shortened, the actin–myosin interactions decline and, again, this reduces contraction. This principle is important in heart failure, in which the chambers of the heart (i.e. the myocardium) enlarge.

Fig. 4.10 Maintaining muscle contractions. (From Sherwood L, Human Physiology: From Cells to Systems, Brooks/Cole Pub Co., 2001).

The majority of resting muscle fibres are held very close to their optimum length via attachments to bones and tendons. This ensures no overstretching occurs.

Types of muscle fibre

Most mammals have three types of muscle – fast (fast is divided into fast and very fast) and slow – which relate to their speed of contraction. The extraocular muscles are the fastest, followed by the jaw (40 ms) and those in the hand and feet (50 ms). The slowest are the gastrocnemius (100 ms) and soleus (120 ms).

Slow muscle has a greater density of capillaries, higher resting blood flow, is red in colour because of a high myoglobin content, and is more resistant to fatigue than fast muscle. Fast muscle is predominately white in colour, with less myoglobin but more of the glycogen and enzymes responsible for glycolysis and anaerobic metabolism. Fast muscle is capable of intense activity but fatigues rapidly (Fig. 4.11).

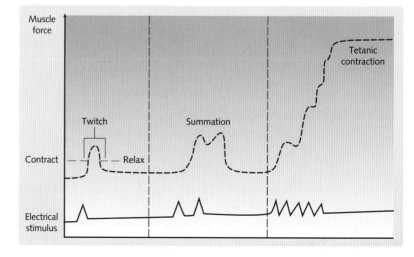

Fig. 4.11 Characteristics of muscle types

Fig. 4.11 Characteristics of muscle fibre types.

Property	Type I fibres	Type II A fibres	Type II B fibres
Contraction time	Slow	Fast	Very fast
Size of motor neuron	Small	Large	Very large
Resistance to fatigue	High	Intermediate	Low
Activity used for	Aerobic	Long-term anaerobic	Short-term anaerobic
Force production	Low	High	Very high
Mitochondrial density	High	High	Low
Capillary density	High	Intermediate	Low
Oxidative capacity	High	High	Low
Glycolytic capacity	Low	High	High
Where found	Postural muscles	Leg muscles	Bicep muscles
Colour	Red	Red pink	White/pale
Myoglobin content	High	High	Low

Know the three types of muscle fibre and be able to name a few examples of muscles relating to these.

Muscle plasticity

This refers to the continuous remodelling of muscles in the body to match function. Many different factors can be adjusted, such as the muscle length, diameter, vascular supply and strength. To a small degree, fibre types can even be altered:

- Hypertrophy: when the total mass of the muscle increases through the enlargement of pre-existing cells. The main cause of this is continuous maximal contraction of the muscle, as in weight training.
- Hyperplasia: an increase in muscle mass through an increase in the actual number of muscle cells. The main cause is the splitting of the muscle fibres. It is much less common than hypertrophy.

- Hypertrophy: increase in mass by enlargement of pre-existing cells.
- Hyperplasia: increase in mass by an increase in cell numbers.

- Atrophy: a lack of muscle stimulation, e.g. nerve denervation or muscle disuse causes a reduction in the fibre size.

Central control of movement

Central control of movement is achieved by three key regions in the brain: motor cortex, basal ganglia and cerebellum.

Motor cortex

Anatomically, the motor cortex is divided into three main areas (Fig. 4.12):

Primary motor cortex

This lies in the frontal lobe (Fig. 4.13) and is involved in the planning and ongoing control of voluntary movements requiring the coordination of several muscles. It projects to all contralateral body motor neurons and its innervation is represented by a motor homunculus, which demonstrates how some regions

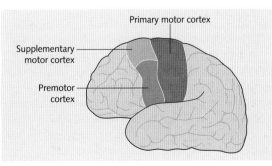

Fig. 4.12 Divisions of the motor cortex.

Fig. 4.13 Sensory and motor homunculi.

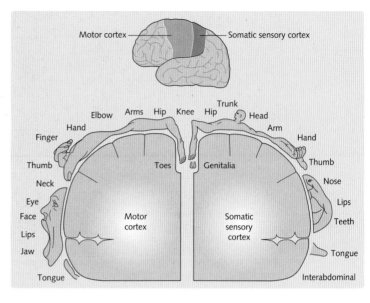

are generously supplied by motor neurons (e.g. lips, finger tips) whereas others (e.g. knee) are not. The primary motor cortex receives inputs from the supplementary cortex, premotor area, cerebellum and somatosensory cortex.

Supplementary cortex

This lies anterior to the primary motor cortex and projects to distal muscles. It is involved in programming motor sequences and bimanual coordination. Input is received from the basal ganglia via the thalamus, and also from the contralateral supplementary motor area.

Premotor cortex

This projects to the brainstem, reticulospinal and corticospinal tracts. It has a role in getting the body into to the right initial posture before movement. It also receives input from the cerebellum and basal ganglia via the thalamus, and from the primary sensory and visual cortices.

Basal ganglia

The basal ganglia comprise five structures on either side of the brain: caudate nucleus, putamen, globus pallidus, subthalamic nucleus and substantia nigra. The ganglia have the following functions:

- Regulation of complex/skilled patterns of motor activity in association with the corticospinal system, e.g. writing numbers sequentially, suturing a wound during surgery.
- Cognitive control of motor pattern sequences, e.g. seeing and recognizing that something is imminently dangerous and running away from it.
- Control of the timing and magnitude of movements: for example writing the word

'physiology' slowly or quickly; or even writing it in large and small letters. For both examples, the proportion of the letters will be the same.
- Saccadic eye movements via connections to the frontal eye fields.

Circuits of the basal ganglia (Fig. 4.14)

Collectively, the caudate and putamen are known as the striatum. They form the main input complex of the basal ganglia, containing identical cell types and receiving input from the motor, sensory, thalamus and limbic areas. They project to the globus pallidus and substantia nigra. The caudate regulates eye movements and cognition, whereas the putamen is concerned with preceding or anticipating body movements.

The globus pallidus is separated into two regions: internal (GP_i) and external (GP_e). This helps regulate muscle tone for specific bodily actions.

The subthalamic nucleus receives inputs from and projects to the GP_e. It also has an excitatory output to the GP_i.

Neurons from the substantia nigra release dopamine, which helps regulate subconscious activity in the muscles. An absence of these substantia nigra cells leads to Parkinson's disease.

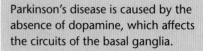
Parkinson's disease is caused by the absence of dopamine, which affects the circuits of the basal ganglia.

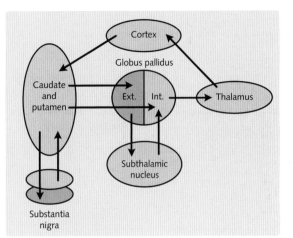

Fig. 4.14 Circuits of the basal ganglia.

Fig. 4.15 Anatomy of the cerebellum.

Cerebellum

The cerebellum (Fig. 4.15) is connected to the brainstem on either side by the superior, middle and inferior peduncles. Functionally, it has three parts:

1. Flocculonodular lobe/vestibulocerebellum: involved in equilibrium, posture and eye movements.
2. Spinocerebellum (vermis + paravermis): projects to the brainstem and regulates axial and proximal limb muscles.
3. Lateral part of the cerebellum/neocerebellum: interacts with the motor cortex to plan and programme movements.

Organization can be separated into input axons, processing interneurons and output neurons.

The cerebellum is involved in fine motor control. Many things can damage the cerebellum, alcohol being an example.

Input axons

These comprise the mossy and climbing fibres. Excitatory neurons send collaterals to the deep nuclei and pass to the cortex:

- Mossy fibres: arise from the spinocerebellar tract, dorsal column nuclei and pontocerebellar tract and terminate on granule cells.
- Climbing fibres: originate in the inferior olivary nuclei and project to the Purkinje cells. The inferior olivary nuclei receives input from all over the body.

Interneurons

- Granule cells: receive input from the mossy fibres and innervate the Purkinje cells. Their long axons are known as parallel fibres.
- Golgi cells: inhibitory interneurons located in the granular layer. Receive input from the parallel and mossy fibres and Purkinje cells. Axons output to the dendritic cells.
- Basket cells: inhibitory interneurons: receive signals from the parallel fibres and project to many Purkinje cells.
- Stellate cells: similar to basket cells but are located more superficially.

Output neurons

Output from the cerebellum is via Purkinje cells, which make GABAergic (inhibitory) projections to the deep cerebellar nuclei and then project to other parts of the CNS.

Peripheral motor control

Proprioceptors allow us to know the position of our head and limbs in space even with our eyes closed. These are present in tendons and muscles and give us an indication to the extent of the muscle contraction. There are two main types of proprioceptor found in the musculoskeletal system: muscle spindles and Golgi tendon organs (GTO).

Muscle spindles

These spindle-shaped organs lie parallel to the skeletal muscle fibres. They consist of intrafusal (within a spindle) muscle fibres surrounded by sensory nerve endings. This structure is then enclosed by a connective tissue capsule and attaches the spindle via the endomysium and perimysium. They are more concentrated in muscles that produce fine movements, such as the ocular muscles.

Muscle spindles measure the extent of muscle stretch (Fig. 4.16). They stimulate the attached sensory endings, which propagate to the cerebral cortex sensory areas and cerebellum. In the former region the incoming information tells the body where in space the moved part is. Information sent to the cerebellum is used to coordinate the contraction of muscle.

At the end of the intrafusal fibres are gamma (γ) motor neurons. The brain regulates the sensitivity of the muscle spindles through these nerves.

Extrafusal muscle fibres surround the muscle spindle. These are innervated by alpha (α) motor neurons, which are large-diameter A fibres. It is via these that the activation of the muscle spindles causes the muscle to contract.

Golgi tendon organs

These are present at the muscle–tendon junction (Fig. 4.17). They comprise bundles of collagen fibres encapsulated by a connective tissue layer. Sensory nerve endings enter the capsule and wrap around the collagen fibres. The GTOs protect the tendons and muscles from damaging excessive tension by tendon reflexes. These reflexes cause the muscle to relax and so decrease the tension on the muscle.

- Muscle spindles measure the extent of muscle stretch.
- Golgi tendon organs protect the tendons and muscles from excessive stretch.

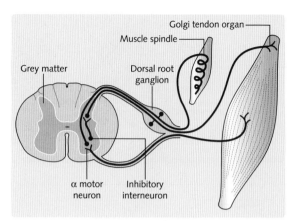

Fig. 4.17 Golgi tendon organ and muscle spindle.

Posture

This describes the relative positions of the trunk, head and limbs in space. The skeleton cannot stand straight without the support of coordinated muscle activity. These muscles are controlled by higher centres and reflex pathways, such as the stretch and crossed extensor reflex.

Balance is an important component of posture. Humans are relatively tall and balance on a small base represented by the feet. The centre of gravity is just above the pelvis. To maintain stability, the centre of gravity must be within the base of support. If the centre of gravity moves out of the base of support then one will fall over. Postural reflexes operate to counteract this.

Crossed extensor reflex

When one leg flexes and lifts off the ground, the opposite leg extends more strongly to support the shifted weight of the body. The centre of gravity is shifted so it remains over the base of support.

Stretch reflex

The muscle spindle stretches when the muscle does (Fig. 4.18), monitors the change in length and signals to the CNS. This instigates the stretch reflex, which attempts to resist the change in muscle length by causing the stretched muscle to contract.

The eyes, vestibular apparatus and somatic receptors provide afferent information for the postural reflexes. They send signals to the brainstem and spinal cord, which initiate muscular contraction via the α motor neurons.

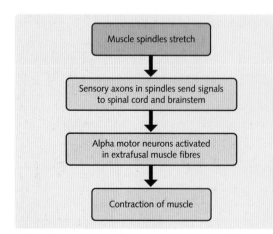

Fig. 4.16 Stretch reflex – order of events.

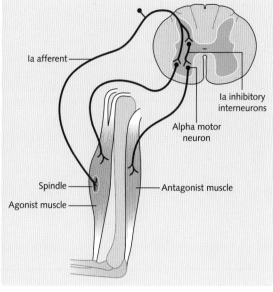

la afferent

la inhibitory interneurons

Alpha motor neuron

Spindle

Antagonist muscle

Agonist muscle

Fig. 4.18 Stretch reflex pathway.

The three afferent pathways of the postural reflexes send signals to the CNS to initiate muscular contraction.

Vestibular system

Changes in head position and acceleration are detected by the vestibular system. Afferent nerves from the neck muscles and cervical vertebrae direct impulses towards the vestibular nuclei, which can influence antigravity and axial musculature via the spinal cord. Eye movements are also initiated to help maintain balance.

Receptor system/vestibular apparatus

This consists of the three semicircular canals and the two sac-like swellings called the utricle and saccule, all of which lie within the temporal bone.

Semicircular canals

The semicircular canals (Fig. 4.19) contain the semicircular ducts, which detect angular acceleration during rotation of the head along three perpendicular axes. Receptor cells in the ducts contain hair-like stereocilia, which are enclosed by a cupula comprising gelatinous-like material in the dilated portion of the duct (the ampulla). During head movement, the semicircular ducts and hair cells move along with it. The endolymph within the ampulla

lags behind and causes the stereocilia to bend. This generates a receptor potential, which propagates along the vestibulocochlear cranial nerve (CN VIII).

Utricle and saccule

These structures provide information about linear acceleration and changes in head position relative to gravity. They contain stereocilia-covered receptors covered by a gelatinous material in which small stones (otoliths) are embedded. These otoliths move during movement and disrupt the stereocilia, causing a receptor potential signalling to CN VII.

Vestibular nuclei

The vestibular nuclei on the floor of the fourth ventricle and medulla receive input from the CN VIII. They then project to limb muscles via the medial and lateral vestibulospinal tracts. They also connect to the thalamus, cerebellum, contralateral vestibular nuclei and cranial nerve nuclei 3, 4 and 6. These help maintain eye position in the presence of head movement.

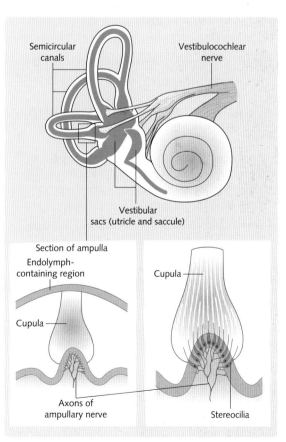

Semicircular canals

Vestibulocochlear nerve

Vestibular sacs (utricle and saccule)

Section of ampulla

Endolymph-containing region

Cupula

Cupula

Axons of ampullary nerve

Stereocilia

Fig. 4.19 The positional and functional appearance of the semicircular canals.

BONE

Bone types and functions

The skeletal system is usually divided into axial skeleton and appendicular skeleton. The axial skeleton includes the bones that form the axis of the body, such as skull, vertebrae, sternum and ribs. The appendicular skeleton includes the bones that form appendages of the body, such as femur, fibula, tibia, humerus, radius, ulna and phalanges.

Classification of bone

Bones are classified by their shape:

- Long bones: are longer than they are wide. Consist of a shaft (diaphysis) and variable number of ends (epiphyses). The ends are composed of spongy bone surrounded by a thin layer of compact bone. They are usually curved so as to absorb the stress of the body. Examples: femur, radius, ulna, tibia, fibula.
- Short bones: length and width are similar. Mainly cancellous bone surrounded by a layer of compact bone covered by periosteum. Hyaline cartilage covers the articulating surfaces. Examples: carpal bones (except pisiform), tarsal (except calcaneus).
- Irregular bones: cancellous bone covered with thin compact bone. Examples: vertebrae and sphenoid.
- Sesamoid bones: shaped like a sesame seed. These bones develop in close proximity to tendons where there is a lot of stress and friction. They protect the tendons from excessive wear and tear and also change the direction of pull of a tendon. Examples: quadriceps femoris and patella.
- Sutural bones: not classified by shape but by location. Located within joints called sutures. Examples: cranial, sagittal, coronal, occipital sutures.

Structure of long bones (Fig. 4.20)

- Diaphysis: bone shaft.
- Epiphysis: distal ends of the shaft.
- Metaphysis: mature bone – joins the diaphysis and epiphysis.

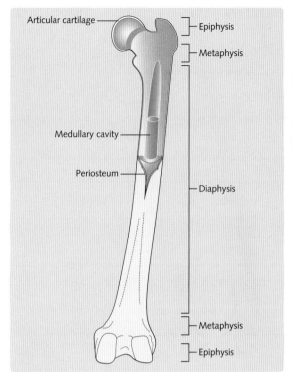

Fig. 4.20 Structure of a typical long bone (e.g. femur).

- Articular cartilage: thin layer of hyaline cartilage covering the epiphysis where the bones in the joint articulate. Reduces friction and acts as a shock absorber.
- Periosteum: surrounds that part of the bone that is not covered by non-articular cartilage. Contains bone-forming cells that allow the bone to thicken. Protects and nourishes bone. Provides attachment site for ligaments and helps in fracture repair.
- Medullary cavity: contains bone marrow in adults. Lined by a thin, single layer of bone cells and connective tissue (endosteum).

Histology of bone

Bones consists of a matrix surrounded by cells. The matrix is constituted of:

- 25% water.
- 25% collagen fibres, which are responsible for the flexibility and tensile strength of the bone.
- 50% crystallized mineral salts.

The salts are mainly hydroxyapatite (calcium phosphate and calcium carbonate). Also present are magnesium hydroxide, fluoride and sulphate. These salts are deposited in the matrix, crystallize and harden to form bone (calcification).

Bone cells

Osteogenic cells

- Mesenchymally derived unspecialized stem cells.
- Undergo mitosis to osteoblasts.
- Present along the inner portion of the periosteum and endosteum.

Osteoblasts

- Bone-forming cells that synthesize the matrix and initiate calcification.
- As they surround themselves in matrix they become entrapped and become osteocytes.
- Found on the inner periosteum and the endosteum.

Osteocytes

- Found in lacunae, which are small cavities in the bone.

Osteoclasts

- These are giant, multinucleated cells formed from the fusion of many monocytes.
- The plasma membrane releases lysosomal enzymes and acids that break down the bone matrix (resorption).

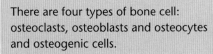
There are four types of bone cell: osteoclasts, osteoblasts and osteocytes and osteogenic cells.

Bone formation, growth and remodelling

All bones are derived from mesenchyme. They then ossify via one of two methods: intramembranous and endochondral ossification (see below). Following ossification, the immature bone is continually remodelled by osteoclasts and osteoblasts until it matures. Hormones regulate these mechanisms.

Ossification

This is the conversion of fibrous tissue or cartilage into bone.

Intramembranous ossification

- Mesenchymal cells migrate to the site of eventual bone formation.
- The cells condense, align and secrete an organic framework of extracellular matrix (ECM), i.e. the osteoid (or ground substance).
- Osteoblasts (differentiated mesenchymal cells) line the osteoid and begin to deposit calcium salts (bone matrix).
- The bone matrix is now a mixture of organic ECM and the inorganic salt component of the developing bone; the mineralized organic strands are termed trabeculae.
- Both the matrix and the trabeculae have the properties of strength and flexibility.
- As the trabeculae increase in thickness, consecutive growth rings called lamellae are seen.
- Cycles of osteoid secretion and mineralization by the mesenchymal cells and osteoblasts add to the lamellae.
- A lattice structure forms when multiple trabeculae within the developing bone contact one another.
- Bones that do not completely fill-in and contain lattice structures are called primary cancellous bones; bones that fill-in are called compact bones. Most – but not all – bones are a mixture of the two, containing a compact outer surface and a cancellous interior.

Endochondral ossification

- Mesenchymal cells migrate to the site of eventual bone formation and transform into chondrocytes.
- These produce cartilage, which takes the shape of the ensuing bone.
- Chondrocytes secrete a loose ECM comprising collagen and mucopolysaccharides.
- The chondrocytes are prised apart and become encapsulated.
- The cartilage is also surrounded by a layer of connective tissue cells (the perichondrium) that is also derived from mesenchyme.
- Within the body of cartilage, the encapsulated cells die and the matrix erodes.
- At this point, the cartilage begins to be replaced with bone.
- Blood vessels invade the cartilage.
- The outer layer of mesenchyme cells is now called the periosteum.
- The periosteum is identical to the perichondrium except for its location.

- As the cartilage is degraded, strands of remaining cartilage act as templates for osteoblasts, which secrete more ECM that undergoes calcification.
- Those areas that are completely calcified are compact bone, whereas those that are not are cancellous.

Functions of bone

Calcium regulation

Calcium is important for blood coagulation, muscle contraction and the stability of the nervous system. Ninety-nine percent of the body's calcium is in bone incorporated in hydroxyapatite. In plasma 45% is in the ionised form (Ca^{2+}) with the remainder being bound to plasma proteins and other solutes. It is imperative that there is tight control of body calcium and so there is a continuous interchange between plasma and bone. Hormones are inherent in this regulation.

Parathyroid hormone

Parathyroid hormone (PTH) is a peptide secreted from the parathyroid glands in response to low blood calcium concentration. It increases blood calcium by:

- Increasing osteoclast activity: so more calcium is released into blood.
- Increasing renal reabsorption of Ca^{2+}: so less is lost in urine.
- Increasing renal formation of vitamin D: increases Ca^{2+} absorption from the gut.

Calcitonin

This peptide is secreted by the parafollicular cells of the thyroid gland and decreases plasma calcium concentration by:

- Decreasing osteoclast activity.
- Decreasing Ca^{2+} reabsorption in the kidney.

Vitamin D

Vitamin D increases the absorption of calcium from the small intestine (mainly) and the kidney, and also increases osteoclast activity. Vitamin D can be ingested via foods but the majority is produced in the epidermis via a number of reactions mediated by ultraviolet light.

A lack of vitamin D can lead to demineralization and poor calcification of bone. This is called osteomalacia in adults and rickets in children.

Glucocorticoids

These inhibit osteoclast activity, decrease gut absorption and increase renal excretion of Ca^{2+} so plasma concentration is decreased.

Growth hormone

Even though this increases renal calcium excretion it makes up for this by increasing intestinal absorption and so overall gives a positive calcium balance.

Haemopoiesis

Haemopoiesis is the formation of red blood cells (RBCs) in the bone marrow. There are two types of bone marrow: red and yellow. Red marrow actively forms RBCs whereas yellow marrow is full of fat and thus inactive. When increased RBC production is required, yellow marrow can be converted to red.

Physiology of the cardiovascular system

Objectives

In this chapter, you will learn to:
- Discuss the organization of the cardiovascular system
- Describe the different layers and types of valves of the heart
- Describe the conduction system and how a cardiac action potential is generated
- Explain the regulation of cardiac output and the factors that influence it
- Discuss the structure of blood vessels and the factors by which their size is regulated
- Describe the lymph system
- Explain how the cardiovascular system is regulated via intrinsic and extrinsic mechanisms
- Describe how the cardiovascular system regulates blood flow in specialized tissues

OVERVIEW OF THE CARDIOVASCULAR SYSTEM

The cardiovascular system (CVS) consists of the heart and tissues (blood vessels). It is involved in circulating blood and lymph through the body.

Organization of the organs and structures

The heart

The heart is a hollow, muscular, pumping organ. It beats, on average, 72 times per minute in an adult and more quickly in newborn infants and young children. The heart has two sides: left and right. Each side is further separated into a ventricle and an atrium (Fig. 5.1). The right atrium receives deoxygenated blood from the venous system and delivers it to the right ventricle. The ventricle then contracts and sends the blood to the lungs, where it is re-oxygenated and carbon dioxide is released. The left atrium receives the oxygenated blood from the lungs and delivers it to the left ventricle, which subsequently contracts and pumps it to the organs of the body, where the oxygen is removed and carbon dioxide enters. The cycle then repeats. Valves are present in the heart to prevent the blood flowing the wrong way.

Blood vessels

Arteries primarily transport oxygenated blood away from the heart. They are highly elastic, tube-like structures that stretch and recoil to maintain blood flow and pressure (Fig. 5.2). Arteries branch and become smaller (arterioles) until they reach a precapillary sphincter that regulates the flow of blood into the capillaries (Fig. 5.3). Capillaries are thin-walled vessels that facilitate the exchange of nutrients and waste products, as well as oxygen and carbon dioxide, with the body tissues. Capillaries merge into very small veins (venules), which connect to form veins that transport blood back to the heart.

Pulmonary circulation

This differs from the systemic system described above in that the pulmonary artery carries deoxygenated blood to the lungs, and the pulmonary vein returns it to the heart.

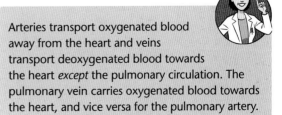

Arteries transport oxygenated blood away from the heart and veins transport deoxygenated blood towards the heart *except* the pulmonary circulation. The pulmonary vein carries oxygenated blood towards the heart, and vice versa for the pulmonary artery.

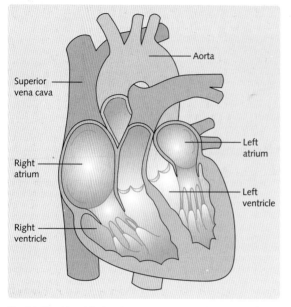

Fig. 5.1 Structure of the heart.

FUNCTIONS OF THE CARDIOVASCULAR SYSTEM

The functions of the cardiovascular system are outlined in Figure 5.4.

- Transports nutrients (oxygen, amino acids, glucose, water, etc.) and removes waste products (carbon dioxide, urea, creatinine).
- Protects the body against infection by circulating immunological cells and mediators to sites of disease.
- Circulation of clotting factors and cells to stop bleeding after injury.
- Transports hormones [insulin, antidiuretic hormone (ADH), etc.] to target cells and organs; the cardiovascular system also secretes its own hormone [atrial natriuretic peptide (ANP)].
- Contributes to regulation of body temperature.

THE HEART

Position of the heart

The heart is located in the mediastinum, which is a mass of tissue that extends from the sternum to the vertebral column and is flanked by the two lungs. The heart is placed in the middle of the mediastinum and has the following relations:

- Superior to the heart: great vessels and bronchi.
- Inferior to the heart: diaphragm.

Fig. 5.2 Basic structure of a vein and artery (see also Fig. 5.13).

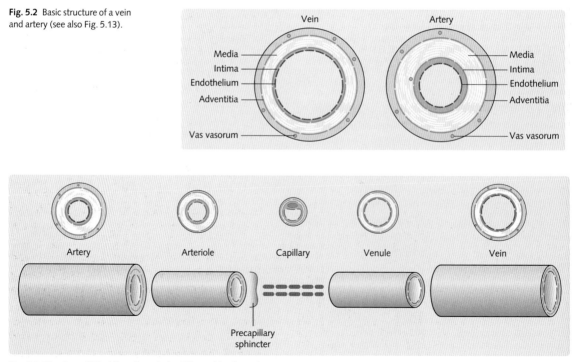

Fig. 5.3 Direction of blood flow through the different blood vessels.

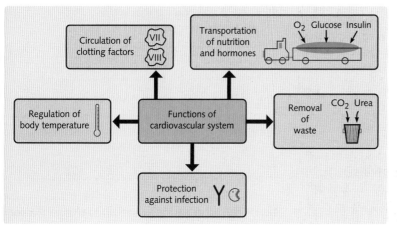

Fig. 5.4 Functions of the cardiovascular system.

- Lateral to the heart: pleurae and lungs.
- Anterior to the heart: thymus.
- Posterior to the heart: oesophagus.

Layers of the heart

The heart wall has three layers: pericardium, myocardium and endocardium (Fig. 5.5).

Pericardium

This comprises an outer fibrous sac, covering the whole heart, and an inner double layer of serous pericardium. The fibrous layer is a tough, inelastic connective tissue that protects the heart, preventing it from overstretching and stabilizing it within the mediastinum.

The serous pericardium has two layers. The parietal pericardium is attached to the fibrous layer. The visceral layer (the epicardium) adheres to the outer surface of the heart. Between these two layers, the serous pericardium produces a thin film of fluid that reduces friction as the heart moves while contracting.

Myocardium

This makes up the majority of the heart and consists of cardiac muscle cells (myocytes) that are responsible for its contractility. The thickest region of myocardium is in the left atrium.

Endocardium

This consists of three layers and is continuous with the endothelial lining of the large blood vessels attached to the heart:

- Outermost connective tissue layer containing nerves, veins and Purkinje fibres.
- Middle layer of connective tissue.
- Inner layer of endothelial cells.

Heart chambers

The heart has four chambers (see Fig. 5.1); two atria superiorly and two ventricles inferiorly. A series of sulci – grooves – on the surface of the heart mark the boundaries between the chambers. Within these grooves are the coronary vessels that carry the blood supply to the heart muscle. The thickness of the walls of the myocardium is related to function. The atria are thin walled, as they have only to deliver the blood to the adjacent ventricles. The walls of the ventricles are a lot thicker because they pump blood further; the walls of the left ventricle are the thickest because it transmits blood all around the body.

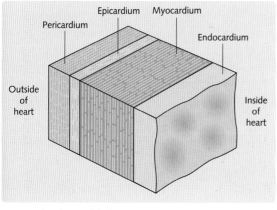

Fig. 5.5 Three layers of the heart.

Heart valves

The heart valves (Fig. 5.6) respond to pressure changes in the heart and ensure that blood travels in one direction. There are two main groups:

1. Atrioventricular valves

The atrioventricular valves are positioned between the atria and ventricles:

- Triscuspid valve: has three cusps and is situated between the right atrium and right ventricle.
- Biscuspid valve: has two cusps and is located between the left atrium and left ventricle.

2. Semilunar valves

These are located between the left ventricle and the aorta and between the right ventricle and the pulmonary artery. They prevent blood pumped to the aorta and lungs flowing back to the heart.

> Remember that the left ventricular wall is thicker than the right because it pumps blood all around the body.

Cellular physiology of the heart

The myocardium comprises cardiac myocytes (muscle cells) (Fig. 5.7), which have the following features:

- Striated and branched network.
- 50–100 micrometres in length and 10–20 micrometres in diameter.
- A single nucleus is (usually) present.
- Dense mitochondrial network.

The myocytes are connected by intercalated discs that help the cells adhere at desmosomes (proteoglycan bridges that glue cells together). Between these are gap junctions made of connexin proteins. These pores allow electrical conductivity through the myocardium.

Like skeletal muscle, the myocytes contain actin and myosin filaments – the contractile components of the heart. The actin and myosin form a similar network to that in skeletal muscle (see Chapter 4), with M and Z lines and A, H and I bands. At the Z line, the sarcolemma forms the transverse tubular structure, which enables rapid electrical conduction by activating the whole contractile apparatus. The sarcoplasmic reticulum stores the calcium required for muscle contraction.

Conduction system of the heart

The continual beating of the heart is due to autorhythmic fibres, which repeatedly generate action potentials. These fibres act as a pacemaker to

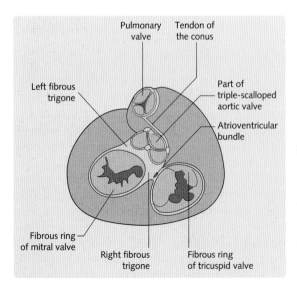

Fig. 5.6 Position, structure and leaflets of the heart valves.

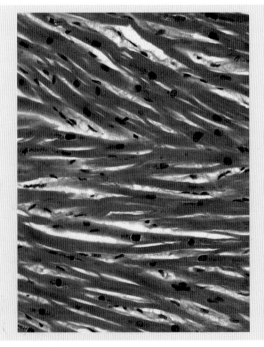

Fig. 5.7 Histology of heart muscle.

regulate the rhythm of contraction and ensure that the heart beats in a coordinated fashion.

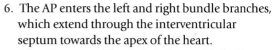

A thorough understanding of the conduction system of the heart allows the interpretation of clinical cardiac arrhythmias.

One beat of the heart is conducted as follows (Fig. 5.8):

1. Spontaneous depolarization in the sinoatrial node (SA node) results in a pacemaker potential. The SA node is located in the right atrium near the entrance of the superior vena cava.
2. When SA depolarization reaches threshold it generates an action potential (AP) in the myocytes.
3. The AP propagates to both atria via the gap junctions. This causes the atria to contract in synchrony and pump blood into the ventricles.
4. The AP travels along the cardiac muscle to the atrioventricular (AV) node, which is located in the septum, sandwiched by the two atria.
5. The AP then enters the bundle of His, which carries it from the atria to the ventricles.

6. The AP enters the left and right bundle branches, which extend through the interventricular septum towards the apex of the heart.
7. The AP eventually reaches the Purkinje fibres – a network of fine fibres through the ventricular walls. The AP causes synchronous contraction of the ventricles, pushing blood from the right ventricle into the pulmonary artery and from left ventricle into the aorta via the semilunar valves.

Action potentials

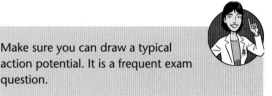

Make sure you can draw a typical action potential. It is a frequent exam question.

Myocyte action potential

The three key stages to the formation of an AP in the contractile fibres (Fig. 5.9) are as follows:

1. Rapid depolarization: the resting membrane potential of a cardiac cell is $-90\,mV$ because the cells are more permeable to K^+ than to Na^+ (see Chapter 3). When threshold is reached and an AP occurs, voltage-gated Na^+ channels open for a few milliseconds, allowing the entry of Na^+. This produces a

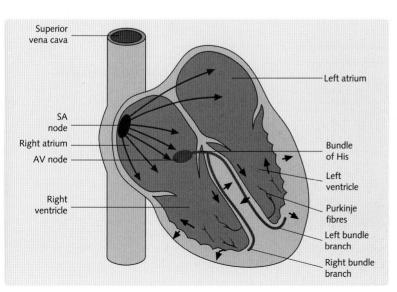

Fig. 5.8 Conducting system of the heart.

Superior vena cava

Left atrium

SA node

Right atrium

AV node

Bundle of His

Left ventricle

Purkinje fibres

Left bundle branch

Right ventricle

Right bundle branch

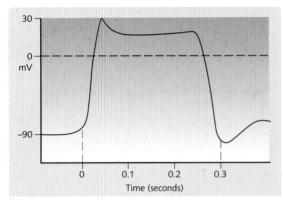

Fig. 5.9 Action potential in a ventricular myocyte.

rapid depolarization and the fast upstroke. The channels then close.

2. Plateau: voltage-gated slow Ca^{2+} channels then open in the membrane of the sarcoplasm reticulum, allowing the influx of Ca^{2+} to the cell. This phase influences the strength of contraction. Concurrently, K^+ channels close, causing a decrease in K^+ permeability and a brief depolarization. The Ca^{2+} influx is also responsible for the refractory period of the cell, making it less likely to start another contraction. This is important, as contraction needs to be co-ordinated or the heart will not have enough time to refill with blood.

3. Repolarization: voltage-gated K^+ channels open, which increases the efflux of K^+. This restores the resting membrane potential of $-90\,mV$ and causes the muscle to relax.

Sinoatrial node action potential

The SA node AP is different from the myocyte AP. It has an unstable resting membrane potential, so when it reaches the threshold value it triggers an AP. The upstroke is due to the entry of Ca^{2+}. The SA node is regulated by the autonomic nervous system and the hormones epinephrine (adrenaline) and thyroxine: these alter the rate at which the threshold potential is achieved.

THE CARDIAC CYCLE

The cardiac cycle is the sequence of pressure and volume changes that occur with one heartbeat.

Events of the cardiac cycle
(Fig. 5.10)

1. Atrial and ventricular filling (diastole): all the chambers of the heart are relaxed and there is passive filling of both the atria and ventricles.
2. Atrial systole (lasts for 0.1 s): the atria contract and add about 25 mL to the relaxed ventricles, producing a final volume of ~130 ml. The ventricles are still relaxed at this stage.
3. Ventricular systole (lasts for 0.3 s): both ventricles contract. The pressure inside the ventricles increases and closes the atrioventricular (AV) valves. For about 0.06 s both the semilunar (SL) and AV valves are shut. This is when the cardiac muscle fibres are contracting but not yet shortening (isovolumetric contraction).
4. When the ventricular pressure exceeds aortic and pulmonary pressure, both SL valves open (ventricular ejection) and blood is expelled into the aorta and pulmonary trunk. The amount of blood ejected per ventricle (around 70 mL) is called the stroke volume. The volume remaining in the ventricles after ventricular systole is called the end-systolic volume.

Remember: when listening to the heart sounds, the first is the *closure* of the atrioventricular valves; the second is the *closure* of the semilunar valves.

Regulation of cardiac output
Definitions

- Cardiac output (CO) = the volume of blood ejected by one ventricle into its respective artery each minute. Hence pulmonary blood flow = systemic blood flow in the aorta.
- Stroke volume (SV) = the volume of blood ejected in one ventricular contraction.
- Heart rate (HR) = number of ventricular contractions in one minute.
- End-diastolic volume (EDV) = volume of blood in ventricle immediately prior to contraction.

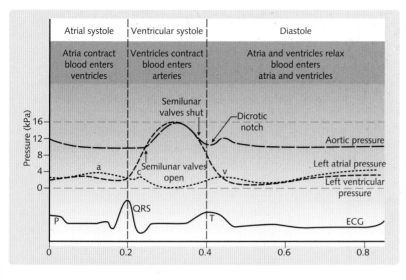

Fig. 5.10 Events in the cardiac cycle with the corresponding ECG and atrial and ventricular pressures. a, atrial systole; c, closure of the mitral valve and its bulging into the left atrium; v, atrial filling against closed mitral valve.

- End systolic volume (ESV) = volume of blood in ventricle after contraction.
- Central venous pressure (CVP) = pressure of blood in the great veins as it enters the right atrium.
- Total peripheral resistance (TPR) = resistance of blood flow in the circulatory system.
- Mean arterial blood pressure (MABP) = CO × TPR.
- CO (mL/min) = SV (mL/beat) × HR (beats/min).

Three things directly affect CO:

1. EDV of right heart.
2. Resistance to outflow.
3. Functional state of the heart–lung unit.

In a typical adult male, SV averages 70 mL/min and HR 75 mL/min, so CO = 5.25 L/min. This value is close to the total blood volume. So, effectively, it takes 1 minute for the whole circulation to flow through the pulmonary and systemic systems (Fig. 5.11).

SV is regulated by:

- Preload.
- Contractility.
- Afterload.

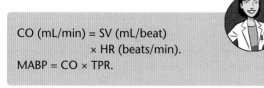

CO (mL/min) = SV (mL/beat)
            × HR (beats/min).
MABP = CO × TPR.

Preload

This is how much the heart stretches prior to contraction. Within reason, the more the heart stretches, the stronger will be the force of contraction. This is known as Starling's law of the heart. The more blood that enters the heart during diastole, the more the heart will stretch and hence the greater the contraction. Preload is determined by the duration of diastole and rate of venous return. The longer the diastole, the more the ventricles fill. Furthermore, the more blood is returned to the heart by the venous system, the more it will fill and strengthen contraction.

Contractility

Contractility can be affected by many factors, such as length of fibre and inotropic (referring to the strength of contraction of the heart) agents. Positive inotropes (increase contractility) usually work by increasing Ca^{2+} inflow during the action potential. These include:

- Norepinephrine (noradrenaline) binds to myocyte β_1-adrenergic receptors and increases Ca^{2+} via G protein activation.
- Cardiac glycosides, e.g. digoxin and ouabain.

Negative inotropes decrease contractility. Examples include:

- Ca^{2+} channel blockers (e.g. verapamil), β-blockers (e.g. atenolol), anaesthetic agents (e.g. halothane).

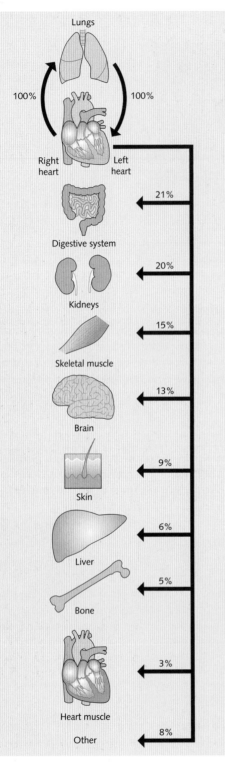

Fig. 5.11 Distribution of cardiac output.

Afterload

This is the pressure that must be overcome before the semilunar valve opens. Anything that increases afterload will cause stroke volume to decrease as less blood will be ejected and more will remain in the ventricles. Hypertension and atherosclerosis are conditions that increase the afterload. Any increase in afterload causes the ventricles to work harder to overcome the increased resistance and hence hypertrophy of ventricular muscle occurs. This can eventually lead to muscle weakness and failure. Right ventricular failure due to obstructed pulmonary circulation (*cor pulmonare*) is a common complication of emphysema and chronic bronchitis.

STRUCTURE AND FUNCTION OF THE BLOOD VESSELS

The main types of vessel in the circulatory system are the elastic arteries, muscular arteries, arterioles, capillaries, postcapillary venules, muscular venules and veins. Their properties are shown in Figure 5.12.

Arteries
Structure of the arterial wall

There are two types of artery:

- Muscular arteries: contribute a very small proportion to the overall resistance and so do not really regulate blood flow.
- Elastic arteries: the largest arteries are of this type.

The arterial wall has three layers:

- Tunica interna: endothelium lining – makes contact with blood.
- Tunica media: elastic fibres for compliance, smooth muscle fibres.
- Tunica externa: elastic tissue and collagen fibres.

The tunica media has a high proportion of elastic fibres and relatively thin walls (Fig. 5.13). Examples are the aorta, the brachiocephalic artery, and the common carotid artery.

Function of the arterial wall

When the ventricles are in diastole and the aortic valve is shut, blood in the circulation needs to be kept under pressure to move it around the system. This is achieved by the elastic arteries, which were stretched wide during systole.

Fig. 5.12 Features of blood vessels.

Fig. 5.12 Features of blood vessels

Feature	Structure	Function
Conductance	Large arteries with very low resistance	Deliver blood to more distal vessels
Resistance	Arterioles with smooth muscle walls	Regulate local blood flow by vasodilation (increases flow) or vasoconstriction (decrease flow)
Exchange	Capillaries with very thin walls	Transfer of materials between blood and tissue
Capacitance	Thin-walled, low-resistance veins and venules	Reservoir of blood and can increase blood to the heart when required e.g. during exercise.

Arterioles

These are smaller than the arteries, from which they branch. The walls of the arterioles contain variable amounts of smooth muscle and elastic tissue. They function to regulate – via vasodilation and vasoconstriction – how much blood goes to the capillaries. They are known as resistance vessels and act as precapillary sphincters.

Capillaries

These join the arterioles and venules and lie close to almost every organ in the body. The concentration of capillaries varies with the metabolic demand of the tissue. Those with high demands, e.g. heart muscle, have a far higher number than those with lower metabolic needs, e.g. tendons, ligaments.

Capillaries function to exchange fluids containing nutrients and waste products from the blood and adjacent cells.

Structure of the capillary wall

The capillary wall consists of a basement membrane and a single layer of endothelial cells. Thus the nutrients and waste products have to travel only a short distance to be exchanged. The endothelial cells have other functions:

- Minimizing friction so that blood travels smoothly.
- Clotting.
- Inflammatory responses.
- Regulation of vessel tone.
- Secretion of a variety of substances e.g. nitrous oxide.

Venules

These collect the blood from the capillaries and transmit it to the veins. The venules closest to the capillaries tend to be porous, allowing phagocytic white blood cells to exit the bloodstream to the site of infection.

Veins

These have the same structure as the arteries, i.e. three layers of tunica. However, the intima and media are a lot thinner and there is no external elastic lamina (see Fig. 5.13). The tunica externa, which consists of collagen, elastin and smooth muscle fibres, forms the majority of the vein wall.

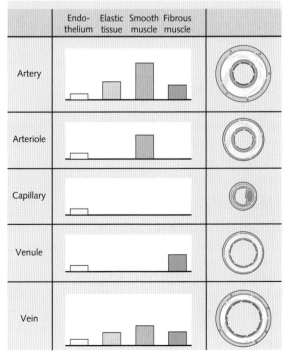

Fig. 5.13 Blood vessels and their relative constituents.

The lumen of a vein is far larger than the arterial lumen.

The lumen of veins in the lower limbs contains valves made from tunica externa. These function to prevent the backflow of blood and to encourage flow back to the heart.

Function of veins

- Distensible to allow large volumes of blood to be stored.
- Send blood back to the heart.

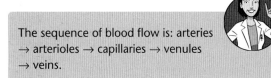

The sequence of blood flow is: arteries → arterioles → capillaries → venules → veins.

Vascular smooth muscle (Fig 5.14)

Smooth muscle is activated involuntarily and is found in the vascular system and many of the visceral organs, e.g. stomach, intestine, bladder.

Anatomy of vascular smooth muscle

- 30–200 micrometres long.
- Thicker in the middle and thinner towards the ends.
- Each fibre contains a single nucleus.
- No transverse tubules and rudimentary sarcoplasmic reticulum.
- Contains thick, intermediate and thin filaments.

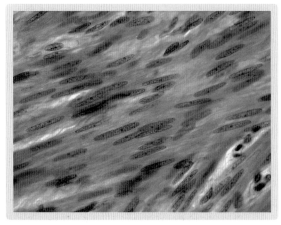

Fig. 5.14 Histology of smooth muscle.

Smooth muscle contraction

- Increase in Ca^{2+} from interstitium and sarcoplasm reticulum.
- The regulatory protein calmodulin binds to incoming cytosol Ca^{2+}.
- The calmodulin–Ca^{2+} complex activates the light chain of the enzyme myosin kinase.
- This phosphorylates the myosin head using ATP.
- The myosin head then binds to the actin and contraction is achieved.

Smooth muscle relaxation

For relaxation to occur, the intracellular concentration of Ca^{2+} must decrease. This can be achieved by the following mechanisms:

- Hyperpolarization of the resting membrane causes a decrease in the number of Ca^{2+} channels and a decrease in intracellular Ca^{2+} concentration. Causes of hyperpolarization are hypoxia, acidosis, calcitonin-related gene peptide and certain therapeutic agents (diazoxide, cromakalim).
- cAMP- and cGMP-mediated vasodilation.

Ca^{2+} is the most important ion in smooth muscle contraction.

Smooth muscle tone

Smooth muscle can contract for long periods of time. This is important in the gastrointestinal system, where prolonged contraction is needed for peristalsis. This is achieved by the Ca^{2+} entering and leaving the smooth muscle cells very slowly.

Haemodynamics
Regulation of blood flow

The volume of blood travelling through a tissue per unit time is deemed to be its blood flow. Ohm's law states:

Flow = (pressure difference)/resistance $[I=V/R]$

So where the volume ends up depends on two factors:

1. Pressure changes.
2. Resistance to blood flow.

Knowing that blood passes down the path of least resistance is a good foundation for understanding the haemodynamics of blood flow.

Pressure changes
Blood travels from an area of high pressure to one of low pressure. The pressure drops as blood moves from the aorta and arteries into the arterioles. Most of the pressure drop occurs at the arterioles. The pressure in the large arteries (e.g. the brachial artery, which is used when assessing blood pressure with a syphygmomanometer) is a good measure of pressure in the aorta.

Resistance to blood flow
As with all liquids, blood flows along the path of least resistance. This depends on three factors:

1. Lumen diameter: Poiseuille's law states that resistance is proportional to $1/r^4$ where r = radius. So the smaller the diameter of the lumen, the greater is its resistance. If the lumen radius is 5 mm, the resistance will be $1/5^4 = 0.0016$. If the vessel diameter was 2 mm then the resistance would be a lot more: 0.0625.
2. Blood viscosity: Newton defined viscosity as a 'lack of slipperiness'. Poiseuille's law states that resistance is proportional to viscosity. Viscosity is influenced by the ratio of red blood cells (RBC) and proteins (globulins and albumins) and the fluid component (plasma) in blood. The higher the RBC or protein concentration relative to the plasma, the greater the viscosity: an example is polycythaemia, in which there is an increase in numbers of RBCs. Dehydration can also increase viscosity as there is a decrease in the fluid component of plasma, thereby increasing the cell/protein concentration.
3. Vessel length: resistance to blood flow is directionally proportional to the length of the blood vessels. The longer the vessel, the greater the resistance.

Haemodynamics of the venous system
Venules and veins are thin-walled, distensible capacitance vessels that act as a reservoir of blood for cardiac filling. The volume in the venous system depends on the venous pressure and the active wall tension.

Venous pressure
Venous pressure causes blood to flow back to the heart. This pressure is generated by the contraction of the left ventricle. A pressure difference of 15–18 mmHg exists between the venules and the right ventricle, and this helps drive the blood to the right atrium.

Active wall tension
Sympathetic nervous stimulation makes the smooth muscle vasoconstrict and also regulates the volume of the vessels. Dilation of blood vessels, e.g. during a hot bath, can cause dizziness when standing as blood pools in the legs and does not return to the heart.

Posture and gravity
When a person moves from supine to a standing position, gravity increases the blood pressure in the vessels below the heart. The body has a number of mechanisms to push blood back up to the heart:

- Venous valves: these permit the blood to flow in one direction towards the heart. The blood pressure in the veins rises because blood is continually flowing into the veins from the capillaries. The increase in pressure will cause pooling of blood in the legs, which will decrease central venous pressure and cardiac output. Baroreceptor-mediated reflex mechanisms that cause vasoconstriction to maintain blood pressure are brought into play.
- Skeletal muscle pump: this is based on an alternating compression and relaxation of veins. Contraction of the leg muscles compresses the veins, which pushes the blood towards the heart and maintains central venous pressure. Drainage of blood back to the feet is prevented by valves in the veins.
- Respiratory pump: this is based on similar principles to the leg muscle pump. During inspiration the diaphragm flattens and descends. This causes negative intrathoracic pressure and an increase in intra-abdominal pressure. Consequently, the veins in the abdomen are compressed and cause blood to flow into the thoracic cavity, thereby maintaining cardiac output.

Blood in veins below the heart is returned to the heart by venous valves, the respiratory pump and muscles in the legs.

Capillary dynamics and transport of solutes

The majority of capillaries comprise a single layer of endothelial cells resting on a basement membrane: no smooth muscle or elastic tissue is present in the walls. The aim of the cardiovascular system is to maintain blood flow through capillaries and allow capillary exchange.

Starling's forces and capillary exchange

The plasma and the interstitial fluid (ISF) exchange readily through the capillary system. The primary forces that govern this exchange are hydrostatic pressure (the blood pressure within the capillaries) and the colloid osmotic pressure (COP) of plasma (see Fig. 5.15 and Fig. 2.5).

Hydrostatic pressure

The capillary wall acts as a filtration barrier. Most of the fluid within the capillaries is retained, although some filters through pores between the cells, driven by the pressure difference between the capillary plasma and the ISF. Water and small solutes pass readily through these pores. The net effect of the hydrostatic pressure alone is movement of water and solute from plasma to the ISF. However, the capillary walls are impermeable to the plasma proteins and lipids. Under normal conditions, these stay within the plasma.

At the arteriolar end of the capillary, the hydrostatic pressure is higher than the oncotic pressure, so there is fluid movement from plasma to interstitium. The hydrostatic pressure decreases along the length of the capillary. At the arteriolar end, the pressure is usually about 35 mmHg; at the venule end of the capillary, the pressure is around 15 mmHg.

Make sure you can draw the forces acting on the capillary and indicate the flow of fluid. This is a common exam question.

Osmotic forces in the capillaries

The capillary wall is impermeable to plasma proteins so they generate an osmotic pressure, drawing fluid into the capillary. Furthermore, as these proteins are negatively charged, they tend to hold additional cations in the plasma, further enhancing an osmotic gradient between the plasma and the interstitial fluid. The combined effect results in a pressure that draws water out of the interstitium and into the plasma. This pressure (COP) is also known as the oncotic pressure (Fig. 5.15).

So hydrostatic pressure tends to cause fluid to leave the plasma, and oncotic pressure pulls it back; the two forces tend to balance each other. They can be altered by certain conditions:

- An increase in pulmonary capillary pressure from cardiac failure will prevent fluid from leaving lung tissue, causing pulmonary oedema (left-sided failure).
- In liver disease or extreme starvation, plasma protein concentration will decline, resulting in a fall in COP. More fluid will move out into the interstitium, causing oedema.
- Capillary damage will result in an inflammatory response, with mast cells releasing histamine. This causes an increase in permeability, allowing protein to leave and so decreasing the COP. Again, this results in oedema of the site involved.

Capillary transport

The exchange of substances between the capillaries and the ISF occurs via three mechanisms:

1. Simple diffusion through the endothelial cell membrane

This applies to simple substances, such as CO_2, O_2, glucose, amino acids and hormones.

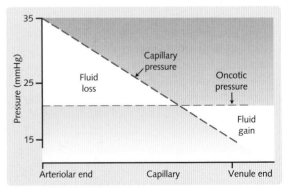

Fig. 5.15 Forces acting across a capillary wall.

2. Diffusion through pores and fenestrations

Pores and fenestrations are present in the cell membrane and permit the diffusion of water-soluble substances such as glucose, amino acids and ions. Lipid-soluble substances, such as steroid hormones, can diffuse directly through the lipid bilayer.

The blood–brain barrier is unusual in that the endothelial cell junctions are arranged tightly. This allows greater regulation of substances obtaining access to the brain.

Capillary pores in the liver, where many substances are synthesized, are large (fenestrated capillaries) to allow easy exchange of materials and secretion.

3. Active transport by transcytosis

Substances in the blood plasma are transported across the endothelial cells by vesicles. They exit at the other end by exocytosis. This is imperative for large, lipid-insoluble molecules, e.g. the hormone insulin.

LYMPH AND THE LYMPHATIC SYSTEM

Approximately eight litres of fluid comprising solutes and plasma proteins are filtered from the microcirculation daily. This is returned to the blood via the lymphatic system.

Structure

The lymphatic system comprises lymph fluid, lymphatic vessels, lymphatic tissue and red bone marrow (Fig. 5.16). The lymph capillaries are blind-ending, bulbous tubes with a monolayer of endothelial cells between the cells. The endothelial cells contain a door-like structure that permits lymph to flow in the lymph capillaries, but in only one direction.

The lymph capillaries merge to form a network of collecting lymphatics vessels. These contain smooth muscle cells, which help to move the lymph along. During its passage through the lymph vessels, lymph flows through lymph nodes containing B and T cells, where the lymphatic fluid is presented to the immune system. Efferent vessels from the nodes unite to form lymph trunks (lumbar, intestinal, mediastinal, subclavian, jugular, bronchomediastinal). The lymph then passes either through the left (thoracic) or right lymphatic ducts into venous blood.

The sequence of lymphatic fluid flow is: blood capillaries → interstitial spaces (IF) → lymphatic capillaries (lymph) → lymphatic vessels (lymph) → lymphatic ducts (lymph) → subclavian veins (blood).

The skeletal muscle and respiratory pump enhance the lymph flow.

Fig. 5.16 Anatomy of lymphatic capillaries with respect to blood capillaries.

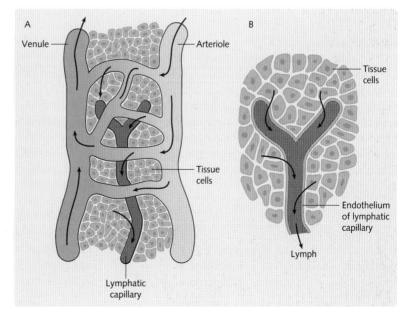

Functions of the lymphatic system

- Drainage of surplus interstitial fluid and returning this to the blood.
- Transporting dietary lipids: lymphatic vessels are responsible for transporting the fat-soluble vitamins (A, D, E and K) from the gut into the blood.
- Immunity: the lymph presents foreign antigens to the immune system, which reacts appropriately.

CONTROL OF THE CARDIOVASCULAR SYSTEM

Blood flow is regulated by intrinsic and extrinsic mechanisms.

Control of blood vessels

The vascular system delivers nutrients to tissues in the body. The amount of flow required, and the consumption, vary depending on the region involved. Mechanisms that regulate blood flow through the arterioles must therefore be present.

Intrinsic control

Intrinsic control mechanisms are independent of nerves and hormones and are the means by which tissues and organs alter their own arteriolar resistances.

Local temperature

This mainly regulates flow through skin. High temperature causes vasodilatation in skin arterioles and veins. At temperatures of $12-15\,^{\circ}C$ α_2-adrenoceptors are stimulated and vasoconstriction occurs. Below $0\,^{\circ}C$, neurotransmitter release is paralysed and the vessels dilate slowly, resulting in paradoxical cold vasodilatation; eventually the vessels warm up owing to the flow of blood, and vasoconstriction occurs.

Transmural pressure

This is the pressure across the wall of the vessel. It can be affected by external and internal pressures:

- External: high external pressure, such as during muscle contraction, can impair blood flow.
- Internal: initially, an increase in blood pressure causes the vessel to distend. The smooth

muscle stretches to produce a contracted response. The vessel constricts, reducing flow within the vessel. This is known as the myogenic response and is a mechanism of autoregulation.

Local metabolites

Changes in the concentration of various metabolites can cause vasodilatation:

- Hypoxia: decreased arterial blood PO_2 (PaO_2).
- Acidosis: due to CO_2 and lactate.
- ATP breakdown products.
- K^+: from contracting muscle and active brain neurons.
- Increase in osmolality.

Different tissues react differently to these factors. For example, the coronary vessels mainly react to hypoxia and adenosine, whereas cerebral vessels are influenced by K^+, H^+ and PCO_2.

Cytokines

The old name for cytokines is autacoids. They are chemical substances that are produced and are effective locally:

- Histamine: this inflammatory mediator causes vasodilatation in arterioles (H_1 receptor mediated) and vasoconstriction and increased permeability in veins (H_2 receptor mediated).
- Bradykinin: release of this causes vasodilatation via nitric oxide (NO) and an increase in vessel permeability via Ca^{2+}.
- 5-hydroxytryptamine (5-HT or serotonin): this causes vasoconstriction and is present in platelets, intestinal wall and the CNS.
- Prostaglandins (PGs): macrophages, leukocytes, fibroblasts and endothelium produce these inflammatory mediators. They are synthesized from arachidonic acid by cyclo-oxygenase and are inhibited by non-steroidal anti-inflammatory drugs (NSAIDs) such as aspirin. PGF causes vasoconstriction whereas PGE and PGI_2 are vasodilators.
- Thromboxane A2: this activates platelets and is important in haemostasis by causing vasoconstriction.
- Leukotrienes: leukocytes produce these inflammatory mediators from arachidonic acid catalysed by lipoxygenase. They cause vasoconstriction and increased vascular permeability.

- Platelet-activating factor (PAF): this inflammatory mediator causes vasodilatation, increased vascular permeability and vasospasm in hypoxic coronary vessels.

Endothelium-dependent relaxation and contraction

The endothelium produces an array of products that influence the patency of blood vessels:

- Endothelium-derived relaxing factor (EDRF): this endothelium-produced factor was discovered to be NO, the production of which is stimulated by thrombin, bradykinin, substance P, adenosine diphosphate, acetylcholine and histamine. NO is cleaved from L-arginine by the enzyme NO synthase. It then diffuses into smooth muscle cells and activates the cGMP messenger system, causing relaxation and vasodilatation.
- Endothelin: a peptide that has a potent, long-lasting vasoconstrictor action that is released in response to stretch, thrombin and epinephrine. It acts locally but appears to have a role in regulation of systemic blood pressure.
- Endothelium-derived constricting factor (EDCF): a putative factor produced in larger arteries to cause vasoconstriction in response to stretch and/or hypoxia.

> Autoregulation is an important concept that aims to maintain tissue perfusion despite changes in blood pressure.

Autoregulation

This is where tissue perfusion remains relatively constant despite blood pressure changes. Consider the following:

Flow $\propto$ pressure difference/resistance

So, to maintain flow, any pressure change must be opposed by a resistance change. For example, an increase in pressure causes arteriolar vasoconstriction, thereby increasing resistance. It takes 30–60 s for the effect to take place, so there is usually an increase in flow with a pressure increase before steady state is achieved.

Autoregulation occurs only over a limited pressure range. It is an intrinsic feature of the vessels and is independent of nervous control. Autoregulation can be reset to work at a new level by, say, an increase in sympathetic drive. The mechanisms for autoregulation are:

- Myogenic response: increased pressure produces constriction of the vessel, opposing the rise in pressure and stabilizing blood flow.
- Vasodilator washout: this is the effect that blood flow has on the concentration of local vasodilator metabolites. If blood flow increases, these metabolites are washed away faster, causing the vessel to constrict, thereby increasing resistance and slowing flow.

In the heart, any increase in coronary arterial pressure causes a rise in tissue PO_2. This leads to vasoconstriction, which autoregulates coronary flow.

Autoregulation is also important in the renal artery to maintain the glomerular filtration rate (GFR). If the GFR is too fast, the nephron will not be able to reabsorb sufficient ions and water.

Hyperaemia

Hyperaemia means an increase in blood flow; there are two types:

- Metabolic/functional/active hyperaemia: is due to metabolites released by exercising muscle (e.g. CO_2, H^+, K^+, adenosine) causing vasodilatation and a decrease in vascular resistance locally.
- Reactive/post-ischaemic hyperaemia: when the blood supply to an organ or tissue is completely occluded, a profound transient increase in blood flow occurs as soon as the vessel becomes patent again. During the period of absent flow, the arterioles in the affected organ dilate due to the local metabolites, e.g. prostaglandins. Reactive hyperaemia enables resupply to ischaemic tissue. It is a temporary process and decays exponentially. It can also be painful after a prolonged period of ischaemia.

Ischaemic reperfusion injury

Reactive hyperaemia is impaired after a prolonged interruption of blood flow. Reperfusion of the ischaemic tissue causes formation of superoxide (O_2^-) and hydroxide radical (OH^-), K^+ increase from damaged tissue and an increase in Ca^{2+} ions in the cell. All these damage the tissue and vessel wall, resulting in further occlusion.

It is thought that reperfusion injury exacerbates damage to the myocardium, intestine and brain following ischaemia.

Extrinsic control

The nervous system is mainly responsible for the short-term control of the vasculature (Fig. 5.17).

The sympathetic nervous system

Vasoconstrictor nerves These innervate the vascular smooth muscles of the resistance and capacitance vessels. A basal level of activity of these nerves is responsible for vessel tone at rest. The neurotransmitter involved is noradrenaline (norepinephrine), which acts on α_1-receptors on vascular smooth muscle to cause contraction. During an increase in sympathetic drive:

- Vasoconstriction decreases local blood flow.
- Venoconstriction decreases local blood volume.
- Arteriolar constriction decreases capillary pressure, leading to greater resorption of fluid from the interstitium back into the blood.

Vasodilator nerves These innervate some skeletal muscle and sweat glands. In skeletal muscle, the neurotransmitter acetylcholine (ACh) acts on muscarinic receptors and causes vasodilatation.

In sweat glands, stimulation of these nerves via the neurotransmitter vasoactive intestinal peptide (VIP) produces sweating and cutaneous vasodilatation.

The parasympathetic nervous system

Parasympathetic vasodilator nerves innervate the blood vessels of:

- Head.
- Salivary glands.
- Pancreas.
- Gastrointestinal mucosa.
- Genitalia.
- Bladder.

The effect of these nerves on the total peripheral resistance is small owing to their limited innervation. Their postganglionic neurons release acetylcholine, which relaxes vascular smooth muscle.

Nitrergic vasodilator nerves These nitric oxide (NO)-releasing neurons act on smooth muscle to cause relaxation. Nitrergic innervation is found in the gut and in both male and female genitalia.

Fig. 5.17 Overview of nervous control of the vasculature. The sympathetic tracts are shown in black and the parasympathetic tracts in red.

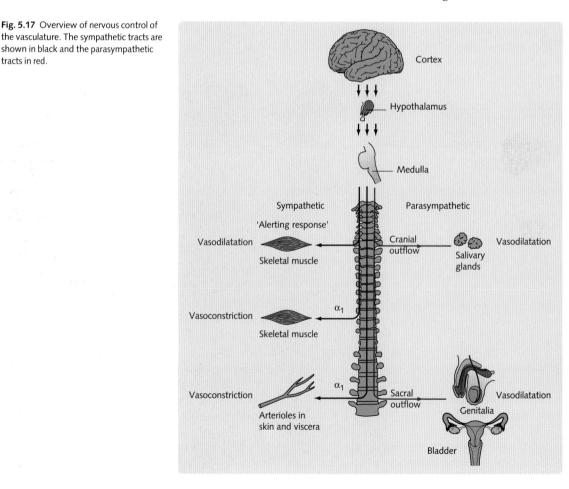

Hormonal control

The following is a brief account of the hormones involved in regulating vascular tone. Further details of these hormones can be found in *Crash Course: Endocrine and Reproductive Systems*.

Adrenaline (epinephrine) and noradrenaline (norepinephrine)

The old names for epinephrine and norepinephrine are adrenaline and noradrenaline, respectively. These hormones are secreted by the adrenal medulla in response to flight, fight or frolic situations, with epinephrine secretion being three times as much as that of norepinephrine. Both are β-adrenoceptor agonists, and so increase heart rate and myocardium contractility. They also stimulate alpha-receptors, causing vasoconstriction.

Epinephrine is a potent vasoconstrictor in most organs and a vasodilator in skeletal muscle, myocardium and liver. This is because there are more β-receptors in these latter tissues, and epinephrine has a higher affinity for these receptors. Thus it can be used to stimulate the heart without causing an overall increase in peripheral resistance and consequently blood pressure.

Norepinephrine has higher affinity for α-receptors and so causes vasoconstriction. The effects on the heart include increased contractility, stroke volume and heart rate. Blood pressure rises as a result because the vasodilatory effects of epinephrine do not fully counteract the vasoconstrictor effects combined with the increased cardiac output.

Antidiuretic hormone

ADH is a peptide produced by the hypothalamus in response to an increase in plasma osmolality; it is released into the bloodstream by the posterior pituitary gland. ADH causes the kidney to retain water and results in vasoconstriction in most tissues. However, vasodilatation occurs in the heart and brain, which ensures these vital tissues remain perfused.

Renin–angiotensin–aldosterone system (RAA)
(Fig. 5.18)

The enzyme renin, produced by the juxtaglomerular cells of the kidney, converts liver α-globulin,

Fig. 5.18 Renin–angiotensin–aldosterone system.

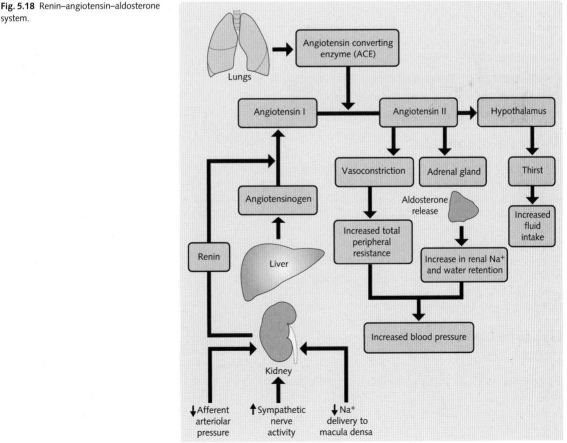

angiotensinogen, to angiotensin I. In the lungs, angiotensin-converting enzyme (ACE) converts angiotensin I to angiotensin II. This:

- Stimulates the adrenal cortex to secrete aldosterone, which increases salt and water retention in the kidneys.
- Acts on the vascular smooth muscle and causes vasoconstriction.
- Increases norepinephrine release.
- Increases cardiac contractility.

Renin production is increased by:

- A decrease in afferent arteriolar pressure to the glomeruli.
- Increased sympathetic activity.
- Decreased Na^+ in the macula densa of the adjacent tubule.

Atrial natriuretic peptide

The stretching of the atria that occurs when there is an increase in blood volume stimulates the atria to secrete atrial natriuretic peptide (ANP). ANP acts to relax the glomerular mesangial cells, increase GFR and hence salt and water excretion. Decreased salt and water reabsorption through its actions on proximal convoluted tubules and, more importantly, medullary collecting ducts have been reported. ANP also has a small vasodilating effect.

CARDIOVASCULAR RECEPTORS AND CENTRAL CONTROL

Medulla oblongata

The cardiovascular centre (CVC) in the medulla oblongata regulates heart rate, stroke volume and regional blood flow via neural, hormonal and local negative-feedback systems. Impulses from the CVC travel via sympathetic and parasympathetic nerves (nucleus ambiguus) to stimulate or inhibit the heart, respectively. Similarly, the vasomotor centre of the CVC can vasoconstrict or dilate vessels via sympathetic nerves. Information from the CVC can be relayed to the hypothalamus.

Hypothalamus

This has four areas of interest:

1. Depressor area: this can produce the baroreceptor reflex.
2. Defence area: this is responsible for the fight/flight response and governs sympathetic outflow.

3. Temperature-regulating area: this regulates skin vascular tone and sweating.
4. Vasopressin-secreting area: vasopressin travels through nerve axons to the pituitary.

Cerebellum

The main role of the cerebellum is muscle coordination, especially during exercise.

Baroreceptor reflexes

Baroreceptors are sensitive to pressure and play a fundamental role in short-term blood pressure control; they also respond to vessel wall stretch. They are found in the aorta, internal carotid arteries and other large arteries in the neck and chest. The two main baroreceptor reflexes are the carotid sinus and aortic reflexes, which produce continuous impulses at normal vessel-wall tone. The impulses from the receptor travel via the glossopharyngeal nerve (carotid sinus) and vagus nerve (aortic arch) to the CVC in the medulla oblongata. An increase in firing causes the medulla to decrease the heart rate and total peripheral resistance by increasing parasympathetic and/or decreasing sympathetic drive.

The above mechanisms serve to reduce the blood pressure. Carotid sinus massage can be used on patients to apply pressure to the carotid bodies and so slow the heart rate if the patient has paroxysmal supraventricular tachycardia.

When the baroreceptors are not stretched, for example because of rapid blood loss, the CVC is not as stimulated and the end result is an increase in sympathetic drive, which causes:

- Heart rate and contractility increase.
- Peripheral vasoconstriction and venoconstriction.
- Increased secretion of catecholamine and renin.

Ultimately, these increase cardiac output, total peripheral resistance and the circulating volume, all of which serve to return blood pressure to normal.

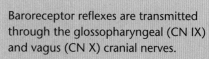
Baroreceptor reflexes are transmitted through the glossopharyngeal (CN IX) and vagus (CN X) cranial nerves.

The sensitivity of the arterial baroreceptors is decreased by:

- Age: the compliance of the arterial vessels falls so that it is more difficult to stretch.
- Chronic hypertension: distensibility of the arterial wall is decreased.

The level of blood pressure that the baroreceptor takes as the set point can be adjusted by central or peripheral mechanisms:

- Central mechanisms: during exercise, the corresponding increase in blood pressure does not cause a bradycardia as a new blood pressure set point has been initiated. The neurons that drive inspiration inhibit cardiac vagal nerves, so blocking baroreceptor impulses and causing a decreased vagal drive.
- Peripheral: persistent hyper- or hypotension results in the resetting of the blood pressure level.

Cardiopulmonary receptors

Numerous cardiopulmonary receptors are connected to afferent fibres innervating the heart, great veins and pulmonary arteries. When these are stimulated they cause bradycardia, vasodilatation and hypotension. There are three main types:

1. Venoatrial stretch receptors

Located where the great veins join the atria, venoatrial stretch receptors are connected to myelinated vagal fibres. Stimulation produces a reflex tachycardia by increasing sympathetic drive to the pacemaker and increasing salt and water excretion.

2. Unmyelinated mechanoreceptor fibres

These are located in the atria and left ventricle and travel in the vagal and sympathetic nerves. Large distension stimulates these receptors, causing an inhibitory effect. Reflex bradycardia and peripheral vasodilatation occurs.

3. Chemosensitive fibres

Some unmyelinated vagal and sympathetic afferents are chemosensitive. They can be stimulated in response to substances such as bradykinin. Respiration also increases, causing bradycardia and peripheral vasodilatation.

4. Chemoreceptors

Chemoreceptors are located in close proximity to the baroreceptors of the carotid sinus and aortic arches in small structures known as the carotid and aortic bodies. They are nerve terminals, which are sensitive to changes in the level of O_2 (hypoxia), CO_2 (hypercapnia) and H^+ (acidosis). They send impulses to the CVC, which stimulates arterioles and veins to vasoconstrict and so increases the blood pressure. Impulses are also sent to the respiratory centre (RC) to modify the rate of breathing.

REGULATION OF CIRCULATION IN INDIVIDUAL TISSUES

Coronary circulation

The coronary arterial system comprises the right and left coronary arteries, which supply the bulk of the myocardium (Fig. 5.19).

Oxygen demand in the mycocardium is very high, at 9 mL/min/100 g, and this rate can increase fivefold during exercise. The coronary circulation adapts to this via:

- High capillary density: the large number of capillaries means that there is a higher degree of flow and a shorter distance for exchange of O_2.
- Myoglobin: this red-coloured protein is found in muscle cells, where it binds diffusing oxygen molecules. The myocardium has a high proportion of myoglobin, so oxygen extraction from the capillaries is high.
- Anastomoses between the coronary arteries: these ensure adequate blood supply if some arteries become occluded.

Skeletal muscle

The delivery of nutrients and removal of waste products needs to increase during exercise. This is aided by:

- Capillary density: this varies depending on the muscle type. White muscle fibres involved in phasically active muscles (e.g. biceps) have a lower capillary density than postural muscles (e.g. soleus).
- Myoglobin: levels are higher in tonic muscles than in phasic muscles.
- Hormones: epinephrine (adrenaline) is a potent vasodilator, which acts on β_2-adrenoceptors in smooth muscle cells.
- The skeletal muscle pump: this encourages venous return to the heart by pushing blood through one-way valves in the veins.

Fig. 5.19 Sternocostal external view of the heart.

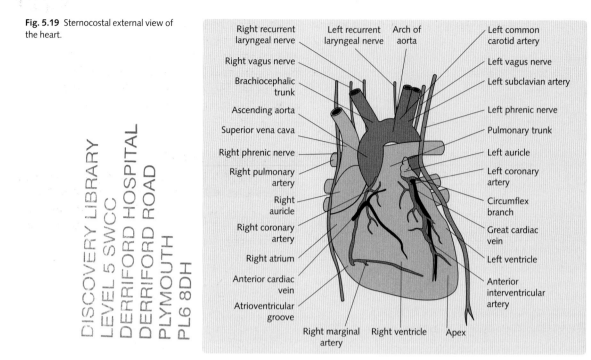

Right recurrent laryngeal nerve
Right vagus nerve
Brachiocephalic trunk
Ascending aorta
Superior vena cava
Right phrenic nerve
Right pulmonary artery
Right auricle
Right coronary artery
Right atrium
Anterior cardiac vein
Atrioventricular groove
Right marginal artery
Left recurrent laryngeal nerve
Arch of aorta
Right ventricle
Apex
Left common carotid artery
Left vagus nerve
Left subclavian artery
Left phrenic nerve
Pulmonary trunk
Left auricle
Left coronary artery
Circumflex branch
Great cardiac vein
Left ventricle
Anterior interventricular artery

Cutaneous circulation

The skin can be a source of blood reserve. Initially, during exercise, epinephrine and norepinephrine cause blood vessels in the skin to constrict and divert more blood to skeletal muscle and the heart.

Temperature is regulated by the rate of blood flow through the skin. Exposed areas of skin – the fingers, toes, palms, soles, lips, nose and ears – contain numerous, well-innervated arteriovenous anastomoses (AVA). The hypothalamus regulates these via the sympathetic nerves. If the core temperature is too high, the AVAs open, blood flow in the skin increases and heat loss occurs via conduction and convection. This occurs in the later stages of exercise as the body core warms up. The converse happens if the core temperature drops.

In exposed areas, such as the extremities, prolonged exposure to low temperatures causes a paradoxical vasodilation. This is probably due to the low temperature impairing sympathetic vasoconstriction; it ensures the skin is not damaged. However, cold vasodilation can also cause a rapid heat loss.

Hypovolaemia results in the release of angiotensin II, ADH and epinephrine, which causes blood vessels in the skin to vasoconstrict and produces the pale skin colour seen in shock.

Reactive hyperaemia occurs when there is a period of obstruction to blood flow to the skin. Once the obstruction is overcome and blood flow is re-established, the rate of flow increases to supply the starved region. This, along with the skin's high tolerance to hypoxia, prevents ischaemic damage.

Cerebral circulation

As an organ, the brain has the third highest rate of oxygen consumption (3.3 mL/100 g/min) in the human body. Grey matter has little tolerance to hypoxia and loss of consciousness follows within as little as 9 s of ischaemia. As a result, the cerebral circulation is strictly regulated at the expense of other tissues to ensure an adequate flow. The factors affecting cerebral blood flow are:

- Intracranial pressure (ICP): a rise in ICP puts pressure on the arteries and restricts blood flow. The venous pressure also rises, which effectively compresses the cerebral vessels.
- Cerebral arteriole resistance: this is determined by metabolites, hormones, circulating peptides and vasomotor nerves.

Autoregulation is widespread in the brain and aims to keep flow at a relatively constant level despite variations in perfusion pressure. However, autoregulation ceases when blood pressure falls below 60 mmHg.

The anatomy of the cerebral circulation is an important component in the maintenance of blood flow (Fig. 5.20). The circle of Willis is the site of anastomosis of the large blood vessels supplying the brain. Its circular structure ensures that if blood flow through one vessel is obstructed blood can access the deficient area. Furthermore, capillary density in the brain is also very high.

> Oxygen consumption is greatest in the heart, then the kidneys and then the brain.

Pulmonary circulation

The pulmonary circulation has the following features:

- High capillary density.
- Pulmonary arterioles are shorter, thinner walled and more distensible than systemic vessels so the pulmonary vascular resistance and pressure are very low.
- Thin-walled vessels facilitate gaseous exchange.
- There is no autoregulation of blood flow, although the pulmonary vessels still react to systemic mediators, e.g. epinephrine (adrenaline).

Renal circulation

The kidneys have the second highest oxygen consumption (6 mL/100 g/min) after the heart. Renal blood flow is autoregulated within certain limits to ensure adequate glomerular filtration rate and is dependent on:

- Hormones: norepinephrine (noradrenaline) constricts renal blood flow.
- Local factors: renally manufactured dopamine causes dilatation and natriuresis:
  - prostaglandins increase flow to the renal cortex but decrease it to the medulla
  - neurotransmitters: acetylcholine produces renal vasodilation
- Diet: a high-protein diet increases renal blood flow.

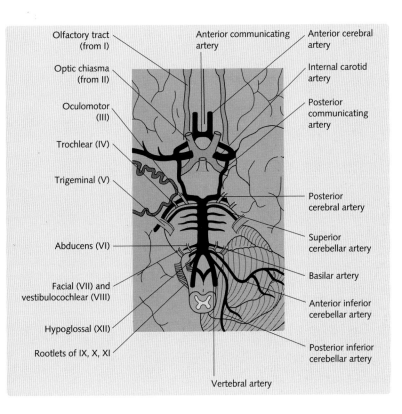

Fig. 5.20 Arteries of the brain with the circle of Willis centrally.

Olfactory tract (from I)
Optic chiasma (from II)
Oculomotor (III)
Trochlear (IV)
Trigeminal (V)
Abducens (VI)
Facial (VII) and vestibulocochlear (VIII)
Hypoglossal (XII)
Rootlets of IX, X, XI
Anterior communicating artery
Anterior cerebral artery
Internal carotid artery
Posterior communicating artery
Posterior cerebral artery
Superior cerebellar artery
Basilar artery
Anterior inferior cerebellar artery
Posterior inferior cerebellar artery
Vertebral artery

Mesenteric circulation

The intestines are supplied by the superior and inferior mesenteric arteries, of which there are extensive anastomoses. Blood flow to the mucosa responds to changes in metabolic activity. The arrival of food causes a hyperaemia and is due to local hormones (gastrin, cholecystokinin), products of digestion (amino acids, glucose) and an increase in vagal activity.

This rise in blood flow produces a tachycardia and an increase in cardiac output. Conversely, blood flow to skeletal muscle decreases.

Physiology of the respiratory system

OVERVIEW OF THE RESPIRATORY SYSTEM

Structure of the respiratory tract

Figure 6.1 shows the components of the respiratory tract, their functions and the divisions within the respiratory tract.

Functions of the respiratory system

Apart from its primary concern with gaseous exchange, the respiratory system has a number of tasks, which are outlined below. Particularly important is the regulation of body fluid pH.

$PaCO_2/PaO_2$	partial pressure of CO_2/O_2 in arterial blood
$PvCO_2/PvO_2$	partial pressure of CO_2/O_2 in mixed venous blood
P_ACO_2/P_AO_2	partial pressure of CO_2/O_2 in alveolar air
P_ICO_2/P_IO_2	partial pressure of CO_2/O_2 in inspired air
P_ECO_2/P_EO_2	partial pressure of CO_2/O_2 in expired air
PCO_2/PO_2	partial pressure of CO_2/O_2 in general

Acid–base regulation

The control of breathing is regulated by the concentration of carbon dioxide ($[CO_2]$) and hydrogen ions ($[H^+]$) via chemoreceptors in both the central and peripheral nervous systems. By controlling $PaCO_2$, the respiratory system contributes to pH regulation by the following reaction:

$$CO_2 + H_2O \leftrightarrow H_2CO_3 \leftrightarrow H^+ + HCO_3^-$$

Diseases of the lung can produce an acidosis, or sometimes an alkalosis, depending on CO_2 accumulation in that particular condition. In clinical settings, measurement of $PaCO_2$ and arterial blood pH and HCO_3^- concentration is used to investigate acid–base disorders.

Metabolism

The respiratory system is involved in the maintenance of metabolism through the:

- Uptake of vascular substances from mixed venous blood (common to the general circulation): compounds taken up include some prostaglandins (PGs), 30% of norepinephrine (noradrenaline) and leukotrienes.
- Production of functional substances:
  - substances released into the respiratory system: many different substances, e.g. mucus and other tracheobronchial secretions, enzymes and immunologically active material,

93

Fig. 6.1 The respiratory tract.
(A) Components and function.
(B) Divisions of the respiratory tree, represented by their generation number (0-23).

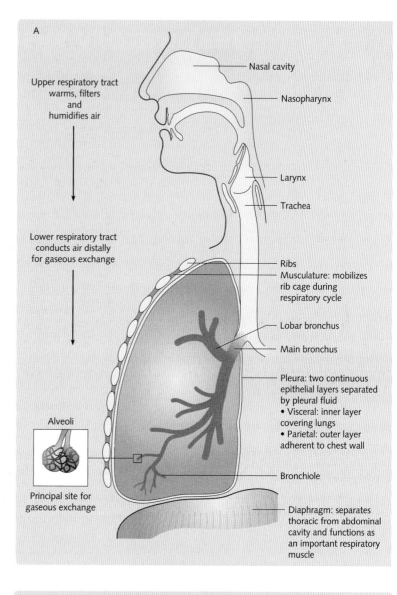

A

Upper respiratory tract warms, filters and humidifies air

Lower respiratory tract conducts air distally for gaseous exchange

Alveoli

Principal site for gaseous exchange

Nasal cavity

Nasopharynx

Larynx

Trachea

Ribs
Musculature: mobilizes rib cage during respiratory cycle

Lobar bronchus

Main bronchus

Pleura: two continuous epithelial layers separated by pleural fluid
• Visceral: inner layer covering lungs
• Parietal: outer layer adherent to chest wall

Bronchiole

Diaphragm: separates thoracic from abdominal cavity and functions as an important respiratory muscle

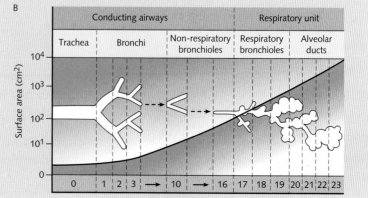

B

Conducting airways

Respiratory unit

Trachea | Bronchi | Non-respiratory bronchioles | Respiratory bronchioles | Alveolar ducts

Surface area (cm²)

10^4

10^3

10^2

10^1

0

0   1 2 3 → 10 → 16 17 18 19 20 21 22 23

are released into the airways and alveoli. An important example is surfactant, produced by type II pneumocytes (alveolar epithelial cells), which decreases surface tension and is important in stabilizing the alveoli.

- substances released into the blood: including histamine, serotonin and bradykinin.

Hormones

- Conversion of angiotensin I to angiotensin II by angiotensin-converting enzyme (ACE) occurs in the lungs.
- PGI_2, which prevents platelet aggregation, is released by damaged lung tissue.

Filter

Effectively, the pulmonary circulation acts as a sieve for mixed venous blood entering the systemic circulation. Potential blockages caused by blood clots, emboli, cancer cells or clumps of platelets embed in the walls of the pulmonary capillaries. They are removed by a combination of factors, involving macrophage ingestion, lymphatic drainage and lysis by vascular endothelial cells.

Reservoir for left ventricle

The volume of the pulmonary circulation is 500 mL in an adult male. This acts as a blood reserve to minimize differences in left and right ventricular output.

Body temperature

Insensible heat loss occurs with exhalation, which, if altered, might affect body temperature. However, the effect of the respiratory system on body temperature is marginal except in extreme conditions.

Excretion

In addition to CO_2, the lungs excrete some drugs that are administered through them, such as general anaesthetics.

Phonation

CNS controllers acting on respiratory muscles cause speech, singing and other sounds by the movement of air through the vocal cords.

VENTILATION AND GASEOUS EXCHANGE

Ventilation

Ventilation is the rate of airflow in and out of the lungs (volume per unit time). It is brought about by intrapleural pressure changes and employs the respiratory muscles – in particular the diaphragm – to ventilate the alveoli for gaseous exchange.

Lung disease can affect factors contributing to the work of breathing. This can impair ventilation in terms of air delivery (O_2) and/or removal of excreted substances (namely CO_2). Hence, gaseous exchange will be affected.

Anatomical dead space

The anatomical dead space (V_D) is that volume of the airways that does not constitute alveoli and therefore does not contribute to gaseous exchange. It comprises the conducting airways down to and including the terminal bronchioles.

The usual volume of V_D is 150 mL or 2 mL/kg (in a person with a normal body mass index); it is estimated according to sex, age, height and weight. Inspiration causes widening and lengthening of the conducting airways, which will therefore increase the V_D.

Dead space causes inspired air to be re-used: on expiration, the V_D exits first, with the remainder of the alveolar space moved to the conducting airways, which thus always contain air. When new air is inspired, air in the V_D (old alveolar air) is pushed down into the alveoli and is mixed with some of the fresh air (because alveolar capacity exceeds that of the V_D).

V_D can be measured using Fowler's method. This involves the subject taking a normal-sized breath of pure oxygen, which fills the entire dead space and mixes with some of the alveolar air (not pure O_2). The person then exhales through a meter that reads N_2.

V_D containing pure O_2 would be exhaled with a zero N_2 reading; once alveolar air starts to be exhaled, increasing amounts of N_2 will be detected by the meter. The volume at the midpoint of the rapid rise in N_2 is taken as V_D (Fig. 6.2).

Physiological dead space

Physiological dead space refers to the sum of V_D (anatomical dead space) and the volume occupied by unperfused alveoli, which is known as the alveolar dead space (V_S).

Physiological dead space = $V_D + V_S$

Absent or poor blood flow through pulmonary capillaries result in the adjacent alveoli being unable to function as gaseous exchange membranes. This value is normally very small in healthy people (<5 mL) and is due to ventilation inequality in different parts of the lung. However, the ageing process and lung disease cause this figure to rise,

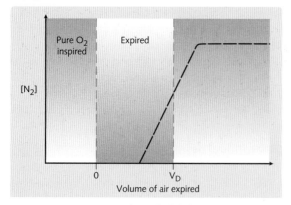

Fig. 6.2 N_2 concentration vs volume of exhaled air, depicting Fowler's method. V_D, anatomical dead space.

especially disease that alters the ventilation–perfusion relationships.

Measurement of physiological dead space

The Bohr approach makes use of the concept that as inspired air contains almost 0% $[CO_2]$, the only CO_2 in mixed expired air comes from ventilated and perfused

$$V_T = V_D + V_A$$

alveoli. If the fractional concentration (F) of CO_2 is applied to each volume in the equation:

where V_T = tidal volume,
$\quad V_D$ = dead space and
$\quad V_A$ = alveolar volume
$V_T \times F_E = (V_D \times F_D) + (V_A \times F_A)$
If we substitute $F_D = 0$
$V_T \times F_E = (V_A \times F_A)$
Since $V_T = V_D + V_A$ or $V_A = V_T - V_D$
$V_T \times F_E = (V_T - V_D) F_A$
$$\frac{V_D}{V_T} = \frac{(F_A - F_E)}{F_A}$$
Hence for CO_2
$$\frac{V_D}{V_T} = \frac{(P_A CO_2 - P_E CO_2)}{P_A CO_2}$$

This is the Bohr equation where $P_A CO_2$ = partial pressure of CO_2 in alveolar air, $P_E CO_2$ = partial pressure of CO_2 in mixed expired air.

The fractional concentration (F) of a gas in a mixture is directly related to its partial pressure (P). In healthy people, the value for PCO_2 is the same for alveolar gas and arterial blood, therefore $PaCO_2$ can be used in this equation. However, if $PaCO_2$ > mixed $PACO_2$, there is significant alveolar dead space. A component balance can be used

because atmospheric CO_2 is 0.03% (negligible), which works on the principle of mass conservation.

Minute ventilation

The minute ventilation ($\dot{V}_E$) is the total volume of air moved into the lungs each minute. It is the product of:
1. Respiratory rate (f): the number of breaths per minute; normal values are 12–20/min.
2. Tidal volume (V_T): the volume of air inspired with each breath. At rest, it is 500 mL.

So:

$$\dot{V}_E = V_T f$$

Therefore, for someone with a respiratory rate of 16/min at rest:

$$\dot{V}_E = 500 \times 16 = 8000\,mL\,/\,min$$

Alveolar ventilation

The alveolar ventilation ($\dot{V}_A$) is the volume of inspired air reaching the alveoli in 1 minute. Its value is normally less than the minute ventilation ($\dot{V}_E$) because the last portion of inspired air is occupied by the anatomical dead space/conducting pathway. Recalling that $V_A = V_T - V_D$, then:

$$\dot{V}_A = f(V_T - V_D)$$

$\dot{V}_A$ determines the content of O_2 and CO_2 in the alveolar gas, together with the body's O_2 consumption and CO_2 production.

Variation of ventilation in the lung

The lung is subject to regional inequalities of ventilation, the lower regions being better ventilated than the upper (when standing or sitting upright). This can be demonstrated using an inhaled mixture of O_2 and radioactive xenon-133 (^{133}Xe) to determine alveolar ventilation at different levels of the lung by appropriately placed scintillation counters, which detect the amount of radioactivity given off by the ^{133}Xe. It is assumed that PO_2 and $P^{133}Xe$ remain the same throughout the respiratory system so that regions receiving larger amounts of the air mixture will give off more radiation.

Lung volumes

There are four lung volumes whose total gives the lung's maximum volume of expansion (total lung capacity) (Fig. 6.3):

1. Tidal volume (V_T): volume of air inspired or expired per breath (500 mL at rest).

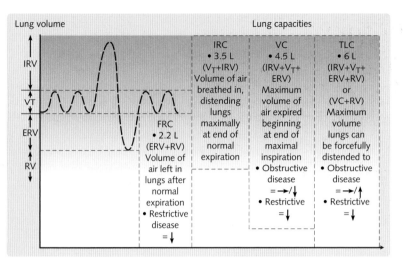

Fig. 6.3 Spirometry trace illustrating lung volumes and capacities. ERV, expiratory reserve volume; FRC, functional residual capacity; IRC, inspiratory reserve capacity; IRV, inspiratory reserve volume; RV, residual volume; TLC, total lung capacity; VC, vital capacity; V_T, tidal volume.

2. Inspiratory reserve volume (IRV): volume of air inhaled into the lungs by a maximum inspiration at the end of a normal expiration (2000–3200 mL at rest).
3. Expiratory reserve volume (ERV): maximum volume of air that can be forcefully expired at the end of a normal expiration (750–1000 mL at rest).
4. Residual volume (RV): volume of air left in the lungs after maximal expiration (1200 mL).

In addition, there are four standard lung capacities, combining two or more lung volumes.

Measurement of lung volumes

Lung volumes can be measured by the following methods:

Spirometry
- Breathing, which takes place through a closed space, moves a bell floating in a closed device. The top of the device is connected to a pen, which marks a moving chart.
- Breathing normally will record tidal volume; a deep breath in indicates the inspiratory reserve volume and a deep exhalation demonstrates expiratory reserve volume.

Nitrogen washout
- Measures initial lung volume by calculating the total volume of nitrogen (N_2) in the lung at the beginning of inspiration.
- Collects expired gas from a person breathing 100% O_2 through a one-way valve. The concentration of expired N_2 is monitored until it reaches zero (all the N_2 is washed out).

- Total volume of expired air × the percentage of N_2 in mixed air = original volume of N_2 in the lungs. As N_2 makes up 80% of the atmospheric air, the original volume of N_2 in lungs × 100/80 = original lung volume.

Helium dilution
- Measures functional residual capacity (FRC): but only ventilated lung volume.
- The person breathes in and out of a spirometer of a known volume that is filled with O_2 and helium (He; this is virtually insoluble in blood) until the concentration of expired He = the concentration of inspired He. The test is stopped at the end of a normal tidal expiration.
- Initial amount of He = Amount of He at the end of test. So:

initial [He] × initial volume of spirometer = Final lower [He] × (lung FRC + spirometer volume)

Body plethysmograph
- Measures FRC as total volume of air in lungs inclusive of trapped air.
- The person sits in an airtight box and makes an inspiratory effort against a closed airway. This causes a small increase in lung volume and box pressure with a decrease in airway pressure.
- Box pressures and volume, and mouth pressures–both pre- and postinspiratory effort – are used to calculate FRC by applying Boyle's law (Pressure × Volume = Constant at a given temperature). The pressure change in the box is proportional and opposite to that of the mouth pressure.

Effects of disease on lung volume

Many diseases affect lung volumes, in particular residual volume (RV) and functional residual capacity (FRC). These parameters are particularly important in obstructive (e.g. chronic obstructive pulmonary disease; COPD) and restrictive (e.g. fibrosis) disease.

Residual volume and functional residual capacity

Residual volume A certain amount of air always occupies the lungs, even after maximal forced expiration, permitting easier reinflation in terms of energy and effort. This is RV.

On contraction, the muscles of expiration generate a positive intrapleural pressure, causing dynamic airway compression during forced expiration. Some of the smaller airways collapse, trapping some air in the alveoli.

Functional residual capacity FRC is the volume of gas remaining in the lungs at the resting expiratory level. This level is reached when there is no net pressure causing air to move into or out of the lungs, i.e. alveolar pressure = atmospheric pressure. It is at this point that the forces trying to collapse the lung equal the forces trying to expand the chest wall.

Patterns of lung function

The lung volumes are affected in different ways, depending on the type of lung disease (see Fig. 6.3):

- Obstructive disease: refers to airway obstruction with air trapping and $\uparrow$RV (e.g. COPD).
- Restrictive disease: refers to reduced lung expansion and $\downarrow$FRC (e.g. fibrosis).
- Mixed disease: refers to disease displaying both obstructive and restrictive features.

Obstructive lung disease Normal airflow is obstructed by airway narrowing, which can be due to blockage of the lumen, constriction of the walls or external compression. Air is therefore trapped behind closed airways: $\uparrow$RV and $\uparrow$RV:total lung capacity (TLC). Examples include:

- Bronchiectasis: a mucous plug blocks the airway.
- Cystic fibrosis: viscous secretions block the airway.
- Asthma: bronchospasm and inflammation of the smaller airways cause obstruction to airflow.
- COPD: persistent inflammation and smooth muscle constriction of larger airways result in airflow obstruction.
- Tumour: this can be intraluminal, with variable occlusion of the airway, or – if outside the airway – can cause external compression and hence narrowing.

If severe, vital capacity (VC) is reduced because of extensive air trapping.

Restrictive lung disease Stiffness of the lungs with restricted expansion occurs. This results in reduction in all the standard lung volumes and capacities (because they are sums of lung volumes). If VC is reduced disproportionately to RV, then RV:TLC will actually be increased, even though both volumes are decreased. Patients with an inability to expand the lungs due to spinal deformity would be classified with a restrictive lung disease. Examples include:

- Pulmonary fibrosis.
- Silicosis.
- Asbestosis.
- Sarcoidosis.

The mechanics of breathing

Flow of air into the lungs

For air to move between the lungs and the atmosphere, there has to be a difference between atmospheric pressure (P_{ATM}) and pressure inside the lungs/alveolar pressure (P_A). As P_{ATM} is generally fixed, this is achieved by altering P_A:

- At functional residual capacity (FRC) there is no airflow: $P_A = P_{ATM}$.
- On inspiration, airflow is from the atmosphere $\rightarrow$ lungs: $P_A < P_{ATM}$. As relative P_{ATM} is taken as $0\,cmH_2O$, P_A is negative.
- On expiration airflow is from lungs $\rightarrow$ atmosphere: $P_A > P_{ATM}$.

The changes in alveolar pressure are achieved by altering the alveolar volume. As the alveoli cannot expand themselves, this is accomplished by the respiratory muscles enlarging the thoracic cavity. The inverse relationship between volume and pressure are explained by Boyle's law. Therefore:

Contraction of respiratory muscles $\rightarrow$ chest expansion $\rightarrow \downarrow$intrapleural pressure lung $\rightarrow$ expansion and $\uparrow$ alveolar volume $\rightarrow \downarrow P_A \rightarrow$ air drawn into the lungs

Relaxation of chest wall muscles $\rightarrow$ facilitated elastic recoil of lung $\rightarrow$ reduced lung volume $\rightarrow$ expulsion of gas

Intrapleural pressure

Fluid in the intrapleural space effectively glues the lungs to the chest wall, the respective forces of which pull in opposite directions. This creates a subatmospheric pressure in the intrapleural space, with values of −3 to −5 cmH$_2$O. This intrapleural pressure (P$_{PL}$ is negative compared to the alveoli (which communicate with the atmosphere):

Alveolar pressure = P$_{PL}$ + Alveolar elastic pressure

The P$_{PL}$ fluctuates with breathing and is also called negative intrathoracic pressure. It sets up a pressure gradient across the lung walls – the transmural pressure – thus ensuring that there is partial lung expansion in the thorax.

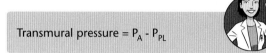

Transmural pressure = P$_A$ - P$_{PL}$

- Just before inspiration: respiratory muscles are relaxed. This situation corresponds to FRC because the elastic recoils of the lungs and chest, respectively, are balanced. P$_{PL}$ is −5 cmH$_2$O and there is no airflow.
- During inspiration: inspiratory muscles contract, increasing intrathoracic pressure and lowering intrapleural pressure (making it more negative). Hence, transmural pressure increases with alveolar distension. This causes alveolar pressure to be lower than atmospheric pressure, so that air flows into the lungs (Fig. 6.4).
- During expiration: chest wall muscles relax, causing intrathoracic volume to decrease while increasing P$_{PL}$ (i.e. making it less negative). Hence, transmural pressure decreases with alveolar pressure greater than atmospheric pressure and exhalation occurs.

Intrapleural pressure is only positive in the case of forced expiration.

Differences in intrapleural pressure between apex and base

The lungs can be considered to be suspended from the trachea, with each layer of lung resting on the layer below and hanging from the layer above. Hence, the lung at the apex stretches or distends, whereas at the base it is fairly compressed. Therefore, the alveoli at the apex have a greater volume than those at the base of the lung.

Because of gravity, the lungs behave rather like a water tank: pressure increases the further below the surface you go. In the lungs, P$_{PL}$ differs according to the level of the lung, i.e. it has a gradient, being less negative in the lower regions and increasing by 0.2–0.5 cmH$_2$O relative to the apex.

The magnitudes of P$_{PL}$ changes that occur during breathing are the same regardless of the level of the lung. These changes in P$_{PL}$ lead to differences in ventilation:

Recall
- Pressure α 1/Volume (Boyle's law).
- Transmural pressure = P$_A$ − P$_{PL}$

Apex
- Larger alveoli so lower P$_A$
- More negative P$_{PL}$.
- Relatively higher transmural pressure.

Base
- Smaller alveoli so higher P$_A$
- Less negative P$_{PL}$.
- Relatively lower transmural pressure.

Muscles of respiration

The muscles involved in respiration bring about the necessary change in volume via movement of the chest wall. These movements enable P$_{PL}$ to be adjusted for ventilation purposes.

Inspiratory muscles
Diaphragm

This is the most important inspiratory muscle. It comprises a sheet of muscle, the nervous supply to which comes from the:

- Phrenic nerves: C3–C5.
- Internal thoracic artery branches: pericardiophrenic and musculophrenic.

The diaphragm has three components:

1. Costal region: made up of muscular fibres, which insert into the sternum and six lower ribs.
2. Crural region: comprising two crura, which pass either side of the oesophagus and attach to the ligaments along the vertebrae:
   i. left crus attaches to ligaments of L1 and L2
   ii. right crus attaches to ligaments of L1–L3.
3. Central tendon: into which the converged costal and crural fibres insert. It lies level with the T8 vertebral segment with its upper surface attached to pericardium.

The dome of the diaphragm extends high up into the thoracic cavity. The right side is higher than the left because of the position of the liver.

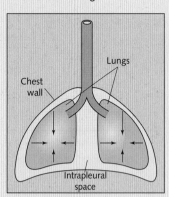
A Elastic recoil of lungs

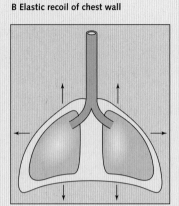
B Elastic recoil of chest wall

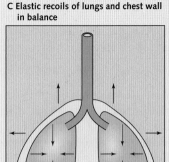
C Elastic recoils of lungs and chest wall in balance

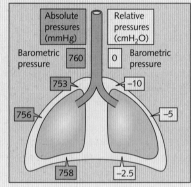
D Intrapleural pressures

Fig. 6.4 Absolute and relative atmospheric and intrapleural pressures.

The diaphragm separates the thorax from the abdomen. Contraction of the diaphragm results in downward movement of its dome into the abdominal cavity. Paralysis of the diaphragm results in paradoxical movement: it moves up instead of down with inspiration.

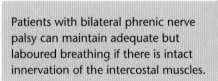
Patients with bilateral phrenic nerve palsy can maintain adequate but laboured breathing if there is intact innervation of the intercostal muscles.

External intercostal muscles
These muscles span each rib space, sloping downwards and forwards (Fig. 6.5A). They are innervated by nerves from the T1–T11 anterior rami. Contraction both elevates the lower ribs and pulls them forward: known as a pump-handle action (Fig. 6.5C). As the ribs rotate on their 'hinge' at the back, the movement produced resembles the action of a bucket handle (Fig. 6.5B).

Contraction also serves to reinforce intercostal spaces, preventing the recession of the thoracic wall through these gaps, which would otherwise decrease thoracic volume. Intercostal paralysis does not exert a large effect on breathing provided the diaphragm is unaffected.

Accessory muscles
These muscles have a primary role that is non-respiratory and generally only aid with laboured breathing. They include the:

- Scalene muscles: which raise the first two ribs.
- Sternomastoids: which raise the sternum.

Expiratory muscles
Active expiration makes use of the internal intercostal and abdominal muscles.

Internal intercostal muscles
These muscles run perpendicular to the external intercostals, sloping downwards and backwards (see

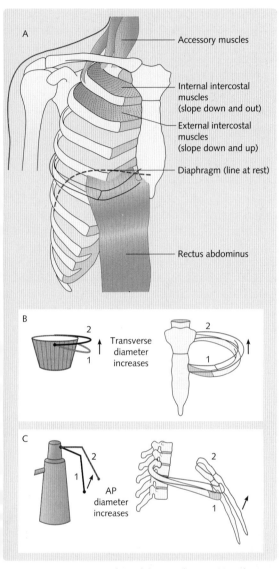

Fig. 6.5 (A) Respiratory muscles and their attachments. (B) Bucket-handle action of the inspiratory muscles on ribs to increase transverse diameter. (C) Pump-handle action of inspiratory muscles on lower ribs and sternum to increase AP (anterior-posterior) diameter.

Fig. 6.5A). The internal layer blocks deformation of the ribs during respiration and prevents the protrusion of thoracic contents through the rib spaces during expiration. It shares the same nerve supply as the external intercostal layer. Contraction of the internal intercostal muscles produces downward and inward movement of the ribs, i.e. the opposite movement to the external intercostals.

Abdominal muscles

Contraction of the following abdominal wall muscles raises intra-abdominal pressure and forces the diaphragm upwards:

- Rectus abdominis. ↑ AP diameter (Fig. 6.5C).

- Internal and external oblique muscles ↑ transverse diameter (Fig. 6.5B).
- Transversus abdominis.

In addition, the quadratus lumborum fixes the rib cage downwards, further aiding with expiration.

Function of the respiratory muscles

Both inspiration and expiration can be considered as either quiet (normal) or forced:

Quiet respiration

The diaphragm is the primary muscle, aided by the external intercostals:

- Diaphragm: contraction of 1–2 cm results in ↑ length of thorax and accounts for 70% of intrathoracic volume change.
- External intercostals:
  - push the lower end of the sternum forward, resulting in ↑ anterior–posterior diameter
  - lift and pull out the ribs, resulting in ↑ transverse diameter.

These actions increase intrathoracic volume, thereby lowering intrapleural pressure and drawing air into the lungs.

- The diaphragm and external intercostals muscles contract simultaneously.
- If only the diaphragm contracts: retraction occurs (sucking-in of rib space muscles).
- If only the external intercostals contract: the diaphragm is pulled into the thorax.

Expiration during normal breathing is passive, with the elastic recoil of distended alveoli driving air out. This occurs against a background of controlled relaxation of the inspiratory muscles, ensuring the gentle shift from inspiration to expiration.

Forced inspiration

This usually occurs with exercise or hyperventilation, or in some pathological conditions, such as asthma. The following occur:

- Recruitment of the accessory muscles: including the scalenes and sternomastoids.
- More active movement of the rib cage by the external intercostals.

101

- Increased descent of the diaphragm (up to 10 cm with deep inspiration).

The erector spinae might also be employed to increase intrathoracic volume by arching the back.

When breathing becomes increasingly laboured, the following additional measures to those above are adopted:

- Scapulae are fixed: by the levator scapulae, trapezius and rhomboid muscles.
- Ribs are raised: by the pectoralis minor and serratus anterior muscles.

In addition, if the person leans over a table, the pectoralis major might be used because the arms are now fixed.

Forced expiration

This usually occurs in the context of exercise, speech, coughing or sneezing (expiratory phase) or disease such as chronic bronchitis. The main muscles involved are the internal intercostals and abdominal muscles. In addition, the quadratus lumborum, which is attached to the twelfth rib, further adds to rib depression by aiding with decreasing intrathoracic volume.

Elastic properties of the lung

The elastic recoil of the lungs and chest that is overcome by the forces generated by the respiratory muscles is due to the collagen and elastin fibres embedded in the lung tissue and surface tension forces within the lungs.

> Elastic recoil is the ability of an object to return to its original dimensions on removal of a distorting force.

Compliance

Compliance describes the ease with which an object can be stretched. It is defined as change in volume per unit change in pressure.

Compliance (C) is also the inverse of elasticity (E), which refers to an object's resistance to stretch when a pressure is applied. Hence:

$$C = 1/E$$

If volume is plotted against pressure on a graph, a substance with a high compliance will have a large volume change when a small increase in pressure is applied (Fig. 6.6). Compliance can be measured under:

- Static conditions (no breathing): values are independent of time.
- Dynamic conditions (with breathing): values will change with time.

Lung compliance

This refers to the change in lung volume per unit change in transmural or transpulmonary pressure. Recall that transmural pressure is the pressure gradient across the lungs: $P_A - P_{PL}$ (i.e. alveolar – intrapleural pressure) and is responsible for partial expansion of the lungs at FRC, therefore acting to generally inflate the lungs.

In static conditions, pressure changes, including those in deflation, can be measured over different lung volume ranges by two means:

1. In-vitro method: involves expansion of an excised animal lung from complete collapse to total lung capacity. Recall that, in vivo, there is residual volume within the lungs.
2. In-vivo method: involves inflation between FRC (or RV) and TLC. Transpulmonary pressure is calculated as $P_A - P_{PL}$, with measurements of P_{PL} taken by a swallowed oesophageal balloon.

In both cases, the line is not straight, i.e. compliance varies with the lung volume; there is a lower compliance at volumes below RV. At high volumes, the distensible alveolar compartments have already been stretched, so compliance is again reduced. Hence, lung compliance can more accurately be referred to as specific lung compliance (sp.C_L):

$$sp.CL_L = \frac{C_L}{V_L}$$

The alveolar volume, and hence compliance, varies with the level of the lung:

- At the apex alveoli are larger, corresponding to the less steep right-sided portion of the pressure–volume graph. Compliance here is lower than at the base, where the smaller volume fits the steeper part of the slope.
- The same magnitude of change in P_{PL} occurs during breathing regardless of height within the

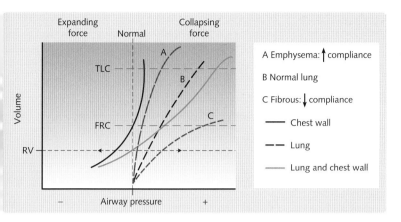

Fig. 6.6 Pressure–volume curves in health and disease for the lungs and the respiratory system. Note decrease in compliance at extremes of lung volume.

lung. So, for the same pressure change, the more compliant base will expand more relative to its initial volume compared to the apex.

The second feature of note from these compliance graphs is that they are different for inflation and deflation: a phenomenon known as hysteresis (see p. 105). Finally, the lungs lie in parallel with each other. Hence, their compliance can be added to give the total compliance of the lungs.

Chest wall compliance

As previously mentioned, the elastic forces of the chest wall act in an outward direction (expansile) and are equal and opposite to those of the lung at FRC: chest wall compliance (C_{CW}). When transmural pressure = 0, the 'resting position' of the chest wall is about two-thirds of TLC:

- Below this point, atmospheric pressure and intrapleural pressure push/pull in the chest wall, so inflation pressure is negative, although the pressure–volume gradient (compliance) is still positive.
- Above this point, in order to stretch the chest wall beyond its resting point, a positive pressure is needed.

The lung and chest wall lie in series with each other, so their compliances are added as reciprocals to give an inverse value of total compliance:

$$\frac{1}{C_{TOT}} = \frac{1}{C_L} + \frac{1}{C_{CW}}$$

For normal individuals, chest wall compliance (C_{CW}) is about 0.1 L/cmH$_2$O at FRC.

Effect of disease on compliance

The lungs are more subject to compliance change than the chest wall, which can be altered by structural back deformity, such as scoliosis and kyphosis, or by obesity. Lines A and C in Figure 6.6 represent disease states affecting lung compliance.

In line A, compliance is increased. This typically occurs with emphysema owing to destruction of interalveolar septa, including elastic lung tissue constituents that are important for elastic recoil. In addition, there is air trapping, so RV is increased.

In line C, compliance is reduced, usually secondary to pulmonary congestion or fibrosis, which make the lung stiffer. In the latter case, elastic recoil is additionally increased, thereby providing further resistance to lung expansion. The presence of air or blood in the intrapleural space will also interfere with the stretching of the lung and cause a similar picture to that in line C.

SURFACE TENSION AND SURFACTANT

Surface tension

Surface tension is a physical property of liquid that minimizes the surface area of the liquid at the gas–liquid interface. It is generated by attractive intermolecular forces in the liquid. These forces are balanced by equal repulsive forces of other molecules within the liquid. However, at the surface, they remain unopposed by the much weaker intermolecular forces of the gas.

In the alveolus, which is spherical in shape, surface tension acts to decrease the radius (Fig. 6.7). The alveolus would collapse if there were not an equal and opposite expansible force, i.e. the positive pressure from within. This pressure can be calculated from Laplace's law for a sphere with a single fluid-lined surface:

$$P = \frac{2T}{r}$$

Where P = pressure within sphere, T = surface tension and r = radius. A smaller radius = higher pressure.

From this equation, it follows that a small alveolus in communication with a larger one would collapse, as a result of airflow from a higher-pressure (small) to a lower-pressure region (larger alveolus). In addition, by increasing elastic recoil, surface tension would be responsible for huge forces involved in breathing. Hence, it would seem impractical to have fluid in the alveoli.

It is now known that surface tension in the alveoli is lowered by a phospholipid known as surfactant, which is secreted by specialized cells – alveolar type II epithelial cells – thereby reducing the magnitude of the forces involved in breathing. The presence of surfactant is also the reason why smaller alveoli do not collapse in the lung.

Surfactant

Surfactant is composed mostly of phospholipids – notably dipalmitoyl phosphatidodylcholine (DPPD) – and some protein, including four specific surfactant proteins. Unlike water, the surface tension of surfactant alters with the surface area of the liquid–air interface, as can be demonstrated with the Wilhelmi balance (Fig. 6.8).

The special properties of surfactant that are concerned with lowering surface tension also contribute towards:

- Alveolar stability.
- Prevention of pulmonary oedema.
- Hysteresis.

Lowering surface tension

When added to saline, surfactant greatly reduces surface tension, which is extremely low when the area is small. This is enabled by the structure of the phospholipids, in particular DPPD, which has both a hydrophobic and hydrophilic end. As the molecules align themselves in the surface, their intermolecular forces, being repulsive, interfere with surface tension, which requires attractive or cohesive forces.

When the surface area of the liquid–air interface is decreased, the proximity of the surfactant molecules is increased, further repulsing each other and reducing surface tension.

Alveolar stability

It follows from the above that surface tension will differ with alveolar size. Large alveoli will have a higher surface tension, and therefore higher inner pressure, because the surfactant molecules are

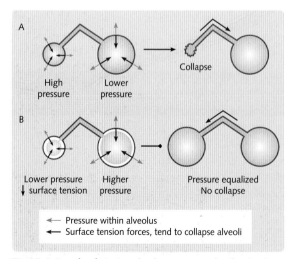

Fig. 6.7 Action of surfactant on alveolar pressures and surface tension. (A) Without surfactant. (B) With surfactant.

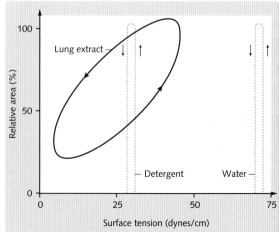

Fig. 6.8 The Wilhelmi balance.

more spread out. This compensates for differences in pressure due to size, so that the pressure within all the alveoli is about the same; this explains why smaller alveoli do not empty into larger alveoli or collapse.

Alveolar stability is also maintained by mechanical interdependence. A collapsing alveolus pulls on the walls of its neighbours, which hold it open. Similarly, when one alveolus of a collapsed lobule is reinflated, it pulls on the walls of its neighbours, helping to open them.

Prevention of pulmonary oedema

Surfactant decreases surface tension, which would otherwise be accompanied by an unopposed 20 mmHg force favouring fluid transudation from the capillary.

Hysteresis

The straight structure of surfactant molecules results in closer packing during expiration and unpacking in inspiration. Hence, surface tension differs between these two phases of breathing, with a difference seen in the compliance. This partly explains the hysteresis in the pressure–volume curves (see Fig. 6.6). Additionally, under dynamic conditions, this difference is widened, with overall values less than with the static compliance.

Respiratory distress syndrome

Respiratory distress syndrome (RDS) is the result of a lack of pulmonary surfactant in premature babies (<32 weeks). Normal surfactant is produced in large amounts from about 34 weeks' gestation. In RDS, surface tension is high and compliance is low, resulting in rapid and laboured breathing, often with an expiratory grunt. In addition, proteinaceous exudate forms a membrane – referred to as hyaline membrane (because it is nearly transparent). Treatment is with synthetic surfactant and adequate oxygen therapy to reverse the hypoxaemia.

Dynamics

When there is air movement in the respiratory system there is dynamic resistance to inflation (P_{DYN}), comprising:

- 80% airways resistance (P_{AR}): resistance of airways to airflow.
- 20% viscous tissue resistance (P_{VTR}): due to friction between lung tissues with expansion.

So:

$$P_{DYN} = P_{AR} + P_{VTR}$$

The sum of pressure required to overcome dynamic resistance and lung compliance, is equal to the value required to inflate the lungs:

$$P_{TOT} = P_{COM} + P_{DYN}$$

Airway resistance

This is defined as the resistance to flow of air within the airways. It can be pathologically increased in the cases of asthma and COPD.

Pattern of flow

The rate and pattern of airflow are responsible for setting up a pressure difference between the ends of the tube through which the air will travel. Airflow can occur either as laminar or turbulent patterns, depending on the velocity and physical properties of the gas.

Laminar flow This occurs at low flow rates, and describes streamlined airflow moving in parallel with the vessel walls (Fig. 6.9A). The velocity of air varies across the cross-section of the vessel; it is slowest at the walls (due to friction) and fastest at the centre, with a velocity twice the average value. Poiseuille's law applies for laminar flow:

$$V = \frac{P\pi r^4}{8\eta l}$$

where r = radius, V = volume flow rate, P = driving pressure, η = viscosity (which is not the same as density of a gas!) and l = length. As the pressure difference (P) = flow (V) × resistance (R):

$$R = \frac{8\eta l}{\pi r^4}$$

The radius of the vessel is critical to both volume flow rate and resistance to airflow. Hence doubling the radius of a tube will increase the volume flow rate 16-fold, and similarly decrease resistance to airflow (recall 2^4 = 16). Altering length has a much smaller effect (one-fold).

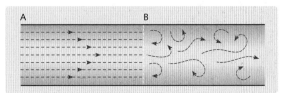

Fig. 6.9 Laminar (A) and turbulent (B) flow.

Turbulent flow This describes a breakdown in the streamlines, some of which move at right-angles to the wall, with eddy currents (small streams moving in random directions). In contrast to laminar flow, it may be erosive and noisy. Turbulent flow occurs in the following settings:

- High flow rates.
- High gas density.
- Wide-bore vessel (large radii).

The relationship between pressure difference, flow and resistance is different than for laminar flow:

$$P \propto V^2 R$$

Flow pattern can be largely determined from Reynold's number:

$$\text{Reynold's number} = \frac{2rv\rho}{\eta}$$

Where v = average velocity, r = radius (and 2r = diameter), ρ = density and η = viscosity. When Reynold's number > 2000 then turbulent flow is more likely.

Flow at branch points or at points distal to partial obstruction usually consists of a mixture of turbulent and laminar flow; this is known as transitional flow.

Sites of airway resistance

Total airway resistance can be split into:

- ≤40% in the upper airways: mouth breathing can lower this value.
- 60% in the lower respiratory tract: 80% of this figure is from the trachea and bronchi and 20% from the smaller airways.

There is turbulent or transitional airflow in the upper airways. It would be assumed that the very small radii in the smaller airways, where laminar flow occurs, would offer great resistance to airflow according to Poiseuille's law. However, the parallel arrangement of the small airways means that total resistance is very low under normal conditions, because the resistances of the small airways are added together as reciprocals:

$$\frac{1}{R1} + \frac{1}{R2} + \frac{1}{R3} = \frac{1}{R_{Total}}$$

During normal quiet breathing, the medium-sized bronchi offer the greatest resistance to airflow.

Factors determining airway resistance

Factors affecting airway resistance include:

- Lung volume.
- Bronchial smooth muscle contraction.

Lung volume Increasing lung volume decreases airway resistance (Fig. 6.10). This is because the bronchi and smaller airways form attachments to lung parenchyma and alveolar walls, respectively. Hence they are supported by radial traction of surrounding lung; airways open by transmitted elastic recoil through attachment from the tissue upon deep inspiration.

Bronchial smooth muscle The autonomic nervous system controls tone, with the motor innervation by the vagus nerve. Bronchoconstriction (increasing airway resistance) is caused by:

- Stimulation of cholinergic parasympathetic postganglionic fibres. This also produces additional mucus secretion.
- Inflammatory mediators, such as leukotrienes and some prostaglandins.
- Stimulation of local irritant receptors by smoke or dust.
- Decreased $[CO_2]$ in the conducting airways.

Bronchodilation (decreasing airway resistance) is caused by:

- Stimulation of adrenergic sympathetic fibres acting on β_2 receptors.
- Nitric oxide.
- Increased $[CO_2]$ in the airways and in alveolar gas.

The physical properties of a gas can also affect airway resistance; gas density has greater importance than

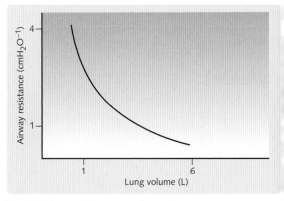

Fig. 6.10 Airway resistance versus lung volume.

viscosity. Hence, during deep-sea diving, there is more airway resistance as gas density is raised by the increased gas pressures.

Effect of transmural pressure on airway resistance

Importantly, the transmural pressure holds the small airways in radial traction. As these airways have no cartilaginous support, this pressure can effectively distend and compress them. Remember that:

Transmural pressure = $P_A - P_{PL}$ (normally negative)

Inspiration During normal inspiration, transmural pressure is positive and so distends the airways. During deep inspiration, P_{PL} becomes more negative, thereby increasing the transmural pressure, which distends the airways further.

Expiration During normal passive expiration, the alveoli and the small airways are patent because P_{PL} is increased, but to less than the alveolar pressure. However, transmural pressure is dictated by expiratory flow rate and P_{PL}.

Forced expiration The goal of forced expiration is to achieve a very low lung volume, during which a positive P_{PL} is generated (Fig. 6.11B). This results in a negative transmural pressure, which is transmitted to the airways where it acts to constrict them. Hence, there is increased resistance to airflow, which causes a pressure difference between alveoli (high) and airways (low):

↑ Expiratory flow rate = ↑ Resistance = ↑ Pressure difference between alveoli and airway

If the pressure in the lumen (keeping it patent) is so low that the negative transmural pressure overpowers it by radial traction, the airways cannot remain open and collapse.

Dynamic compression Dynamic compression of the airways refers to the increased airway resistance during forced expiration. This phenomenon is responsible for limiting maximum expiratory flow rate, as there will be more airway collapse with increasing effort to exhale beyond a certain level. Below this rate, effort determines airflow rate, which is therefore described as effort dependent.

The effect of dynamic compression is greatest at low lung volumes because radial traction forces keeping the airways patent will be greatly reduced.

Dynamic compression in disease Exaggeration of dynamic compression so that maximum airflow occurs at lower levels can occur in conditions in which there is:

- An increase in peripheral airway resistance: multiplying the pressure difference causing decreased intrabronchial pressure.
- Reduced driving pressure/radial traction forces (keeping the airways patent): can occur with:
  - low lung volume
  - decreased recoil pressure (destruction of architecture, including collagen and elastic fibres).

Airway collapse and air trapping can occur even in tidal breathing. Such patients display pursed-lip breathing in an attempt to increase expiratory pressure and remove some of the trapped air.

Measuring airway resistance

Everyday estimates of airway resistance can be made using:

- Peak expiratory flow rate: subjects take a deep breath then place their mouth around the seal of a meter into which they blow out as quickly and as hard as possible. This measures maximum airflow during a rapid forced expiration.
- Spirometry: the subject measures the rate and volume exhaled. Of particular use in measuring airway resistance is the forced expiratory volume in 1 second (FEV_1), or forced expiratory volume (FEV), especially when the first value is expressed as a percentage of the second (Fig. 6.12).
- Plethysmography: this is a more accurate method but not as practical as the two above.
- Interrupter method: the subject breathes through a rapidly closing shutter in a tube. Pressure at the

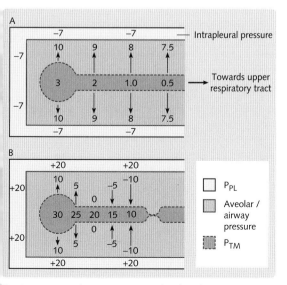

Fig. 6.11 Pressures during (A) passive and (B) forced expiration.

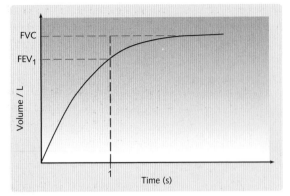

Fig. 6.12 Spirometer traces.

lips is measured just before airflow is transiently blocked. Clinical assessment of airways resistance is calculated by examining PEF or FEV_1.

The work of breathing

This is the work (W) done to move the lung and chest wall, which is proportional to the pressure change × the volume change. Thus work done for normal inspiration is:

$$W = P \times V_T$$

where $P = P_{TOT}$ (the pressure needed to overcome dynamic resistance and lung compliance) and $V_T =$ tidal volume. The work of breathing is increased if:

- Volume is increased: e.g. in COPD or chronic severe asthma. Note that in these conditions increased airways resistance requires a higher pressure gradient for airflow.
- Pressure is increased: due to reduced compliance (lung fibrosis) or increased dynamic resistance as a result of:
  - increased airway resistance (asthma, COPD)
  - turbulent airflow (exercise).

The work of breathing can be reduced by the use of bronchodilators, which reduce dynamic resistance. The volume–pressure curve for the lung (Fig. 6.13) illustrates the work of breathing. The efficiency of breathing can be worked out from the formula below, with a normal value of 5–10%:

$$\text{Efficiency} = \frac{\text{Useful work}}{\text{Total energy expended (or } O_2 \text{cost)}} \times 100\%$$

In health, quiet breathing requires 5% of total body energy expenditure. This figure rises with COPD and can reach 30% during vigorous exercise.

GASEOUS EXCHANGE IN THE LUNGS

Once the alveoli have been ventilated, gas transfer takes place between them and the bloodstream. This exchange occurs by diffusion and is dependent on the partial pressure or fractional concentration of the gas involved in both media. These concepts need to be addressed, along with their relevant laws.

Diffusion

All gases diffuse passively between alveolar air and the pulmonary capillaries across the alveolar capillary wall. The term 'diffusion' refers to the net movement of substances from an area of high concentration to an area of lower concentration until their concentration is equalized, i.e. equilibrium is reached. Random movement of molecules will still occur, although there is no net exchange. This is known as dynamic equilibrium.

In the lungs, net movement is dictated by the difference in partial pressures between alveolar air (P_A) and arterial blood (Pa). This diffusion is described by Fick's law of diffusion (Fig. 6.14). The rate of diffusion (J) for a gas through a membrane is given by the following equation:

$$J = \frac{A \times D \times (P_1 - P_2)}{T}$$

where A = surface area of the membrane, T = membrane thickness, ($P_1 - P_2$) = partial pressure difference of gas across a membrane, D = diffusion coefficient of the gas (dependent on size and solubility: D = solubility/$\sqrt{\text{molecular weight}}$ – Graham's law).

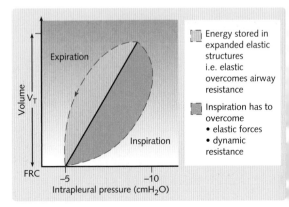

Fig. 6.13 Dynamic volume-pressure graph during respiration.

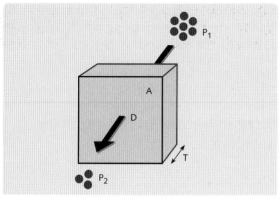

Fig. 6.14 Fick's law of diffusion.

The rate of diffusion of a gas is increased with:

- An increase in:
  - surface area of membrane
  - partial pressure difference/gradient
  - solubility of the gas.
- A decrease in:
  - thickness of the membrane
  - molecular weight.

The alveolar capillary membrane is particularly suited to its gaseous exchange function. The terminal respiratory branch division gives it a huge surface area (50–100m^2) that is generally 0.3 µm thick.

The much higher solubility of CO_2, even with its higher molecular weight, means that it should diffuse 20 times more readily in the opposite direction to O_2 when all other variables are the same. However, CO_2 diffusion equilibrium is very similar to the rate for O_2 because:

- There is slow release of CO_2 from the blood.
- The partial pressure gradient is 10 times less for CO_2 than it is for O_2 in the opposite direction.

Partial pressures of respiratory gases

In a mixture of gases, each gas behaves as if it were alone, i.e. the individual gases do not affect one another's exerted pressure. This is exemplified by Dalton's law, which states that the partial pressure of a gas (P_G) is equal to its fractional concentration multiplied by the sum of all the partial pressures of the separate gases in the mixture (the total partial pressure; P_T):

$$P_G = \text{Fractional concentration} \times P_T$$

At sea level, where P_T is 760 mmHg with a fractional concentration of 21%:

$$PO_2 = 21/100 \times 760 = 159 \text{ mmHg}$$

Hence, the partial pressure is proportional to fractional concentration and to P_T. Therefore, at a high altitude, where P_T (atmospheric pressure) is less, the PO_2 will also be less than normal.

Water vapour pressure

When air is warmed and humidified on inspiration, the added water vapour dilutes the gases. In a closed container of gases, the pressure exerted by the extra water vapour would be added to the total pressure of the gas mixture. However, in the body, according to Boyle's law, the gas will expand:

$$\text{Pressure of a mass of gas at a constant temperature} \propto \frac{1}{\text{Volume}}$$

This water vapour pressure is subtracted from the atmospheric pressure:

$$PO_2 = 21\% \times (760 - \text{water vapour pressure})$$

The water vapour pressure depends on:
- Temperature: ↑water vapour pressure with ↑temperature.
- Saturation of the gas (water vapour is 100% water).

The partial pressure of water is 47 mmHg with 100% saturation at 37°C so:

$$PO_2 = 21\% \times (760 - 47) = 150 \text{ mmHg}$$

Diffusion limitation and perfusion

Exchange of a gas across the alveolar capillary membrane into venous blood in the pulmonary capillaries, and vice versa, can be limited by either diffusion or perfusion of the blood, depending on the component affected:

- Crossing the alveolar capillary membrane.
- Transit time of blood in the capillaries.

Perfusion limitation:

- Diffusion in blood plasma.
- Chemical combination with haemoglobin.

Diffusion limitation

Diffusion is the rate-determining step for uptake of these molecules, which readily perfuse the blood (plasma soluble and combine readily with haemoglobin). Fick's law of diffusion clearly defines the limitation of a gas's diffusion according to:

- Diffusion coefficient depending on solubility and molecular weight.
- Partial pressure gradient: this also depends on partial pressure in mixed venous blood.

The membrane surface area and thickness can be altered by changes in cardiac output, capillary volume or with pulmonary artery pressure.

Normally, the transit time for an erythrocyte with its plasma through the alveolar capillary is 0.7–1.2 s on average. If this transit time is very short (<0.2 s), equilibrium between capillary and alveolar partial pressures might not be reached. Hence, any gas transfer becomes diffusion limited, especially those in which transfer is dependent on the rate of diffusion of the gases.

Carbon monoxide (CO) is such a gas: its diffusion into blood plasma occurs rapidly but with strong chemical combination with haemoglobin, so that its arterial partial pressure rises slowly (as most is bound to haemoglobin).

Perfusion limitation

This refers to the transfer of gases that readily diffuse across the alveolar capillary membrane but the transfer of which is limited by plasma insolubility and weak or no binding with haemoglobin.

If a gas does not bind well with haemoglobin, even with low plasma solubility, transfer into the liquid component of blood will be limited. A rapid rise in the Pa occurs, causing the partial pressure gradient to diminish and with it the driving force for diffusion.

Nitrous oxide (N_2O) is a good example of a perfusion-limited gas. It traverses the alveolar capillary barrier with ease, it has low plasma solubility and it does not combine chemically with haemoglobin or any other blood component.

The time course for O_2 transfer lies between that of CO and N_2O, although, overall, it is normally perfusion limited because, although O_2 combines with haemoglobin, the rise in partial pressure induced is far greater than that for CO. Furthermore, at rest, equilibrium between alveolar and capillary PO_2 occurs when the erythrocyte is one-third of the way along the capillary, so no partial pressure gradient after this point = no driving force.

In some diseases where the diffusion properties of the lung are impaired – typically by thickening of the diffusion membrane or reduced surface area (e.g. emphysema) – equilibrium is not reached by the time blood has reached the end of the capillary ($PaO_2 < P_AO_2$) so there is some diffusion limitation.

Oxygen uptake in the capillary network

Pulmonary PaO_2 (40 mmHg) equilibrates with P_AO_2 (100 mmHg) in 0.25 s, i.e. one-third of the transit time of capillary blood (0.75 s) at normal resting cardiac output. Therefore, at rest, the transit time is surplus to the requirements for adequate O_2 transfer. During strenuous exercise, the transit time is decreased to 0.25 s because ↑ cardiac output results in ↑ flow rate through the pulmonary capillaries. In the absence of disease, there is very little change in end capillary PaO_2 – as can be deduced from Figure 6.15.

During exercise there is increased total O_2 transfer from:

- Improved ventilation–perfusion matching.
- Greater gaseous exchange surface area due to an increased pulmonary capillary network: there is capillary distension and use of previously unperfused capillaries at rest.

If there is a diffusion limitation of O_2 transfer owing to greater diffusion distance (fibrotic thickening or interstitial oedema), PaO_2 values cannot be maintained with exercise, even with normal P_AO_2.

Carbon dioxide transfer

As already mentioned (page 108), the diffusion coefficient (D) component of Fick's law of diffusion implicates solubility and molecular weight (solubility/molecular weight) in the rate of diffusion. Graham's law is the basis for this coefficient with respect to molecular weight.

Pulmonary $PaCO_2$ (45 mmHg) equilibrates with alveolar P_ACO_2 (40 mmHg) in 0.25 s – the same time as for O_2 (see Fig. 6.15). This is because the lower partial pressure gradient for CO_2 and its slow release from blood negate its much higher diffusion coefficient than O_2.

CO_2 transfer is normally perfusion limited, although an abnormal gaseous exchange membrane as is seen with some diseases may cause it to be diffusion limited.

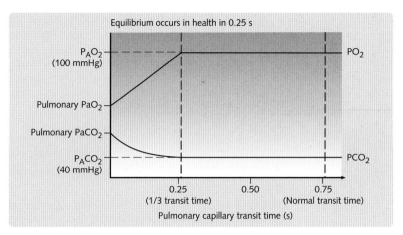

Measuring diffusion

Recall Fick's law of diffusion:

$$J = \frac{A \times D \times (P_1 - P_2)}{T}$$

The structure of the blood-gas exchange barrier is too complex to allow measurements of area and thickness during life. So the equation is rewritten substituting D_L, which is the diffusing capacity of lung, or the transfer factor, which incorporates A, D and T:

$$J = D_L (P_A - Pa)$$

D_L is the rate at which O_2 or CO is absorbed from alveolar gas into the pulmonary capillaries (mL/min) per unit of pressure gradient:

$$D_L = \frac{J}{P_A - Pa}$$

For CO, D_L is the rate of uptake of CO divided by the difference between alveolar and arterial PCO.

CO is most frequently used to study the diffusing capacity by the single-breath or steady-state techniques. CO is used because the average PaCO is nearly zero when safe $P_A CO$ are used. In addition, it is not perfusion limited, unlike O_2. So if PaCO ≈ 0 then:

$$D_L = \frac{J}{P_A}$$

The single-breath method involves a single inhalation of dilute CO–air mix, after which the breath is held for 10 s and the rate of CO disappearance from alveolar gas is calculated. Helium is often used as part of the air mix, enabling lung volume measurement and calculation of the transfer coefficient (K_{CO}) or diffusion rate per unit of lung volume.

D_L is sensitive and is decreased with the following:

- ↑Surface tension (T):
  - ↑thickness of alveolar capillary barrier as seen in alveolar fibrosis (sarcoidosis, scleroderma, pneumoconiosis).
  - ↑diffusion distance (interstitial oedema).
  - ↑fluid in alveoli (alveolar pulmonary oedema).
- ↓Surface area of the membrane (A):
  - blood: low cardiac output/pulmonary circulatory volume
  - gas: emphysema.
- ↓Perfusion of blood:
  - low pulmonary circulatory volume.
- ↓P_A:
  - hypoventilation.
  - obstructive lung disease – air trapping.

PERFUSION AND GAS TRANSPORT

Presentation of blood for effective gaseous exchange is a complicated process. Most of the cardiac output is directed through the pulmonary circulation, where it flows in very close contact with air in the alveoli. As well as a thin membrane between blood and air, a low-pressure circulation is needed to avoid fluid transudating into the alveolar space. There must also be a ventilation–perfusion match, as gaseous exchange will be ineffective if non- or underventilated alveoli are perfused or, similarly, if there is very little perfusion of well-ventilated alveoli (i.e. if either of the two media – blood or air supply – is inadequate).

Pulmonary blood flow

Anatomy of the pulmonary circulation

The body can be thought to have two circulatory systems – pulmonary and systemic – each with its own ventricular contraction and flow. Pulmonary circulatory vessels differ from their systemic counterparts, as detailed below:

- Thinner walls.
- Less vascular smooth muscle.
- No particular muscular vessels.
- Greater distensibility and compliance.
- Terminal branches have a greater internal diameter.
- Rapid subdivision.

In reality, the lung receives blood from two circulatory sources:

1. Bronchial blood flow: 2% of the left ventricular output supplies oxygenated arterial blood to the tracheobronchial tree and other lung structures to the level of the terminal bronchioles.
2. Pulmonary blood flow: the entire right ventricular output supplying mixed venous blood, which drains into the pulmonary veins.

Only the pulmonary circulation undergoes gaseous exchange, and so in this context will be discussed further in this section.

- The pulmonary circulation begins at the pulmonary artery, containing partially deoxygenated blood from the right ventricle.
- This then splits into right and left main branches, which go to their respective lungs.
- These main arteries branch successively, closely following the airways.
- Beyond the terminal bronchioles, the vessels form the capillary bed, starting with smaller arterioles that function as capillaries, as gaseous exchange starts at this level.
- The capillary bed forms a dense network in the walls of the 300 million alveoli; this bed is made up of approximately 280 billion capillaries. Hence, a rich mesh structure, likened to a flowing blood sheet, produces a huge surface area for gaseous exchange; the area is estimated at 50–100 m^2.
- As the average pulmonary capillary diameter is 6 μm, erythrocytes with a diameter of 8 μm have to squeeze through, thereby reducing the diffusion distance and therefore facilitating efficient gaseous exchange.

- Oxygenated blood then flows into the pulmonary veins, which pass between lung lobules. These are also short, and unite to form four large veins, which empty into the left atrium.

Mechanics of the circulation

At rest, the pulmonary circulation constitutes 3–5 L blood/min/m^2 of body surface area. A large compliance enables the pulmonary arteries to accommodate two-thirds of the right ventricular stroke volume. In addition, the very small amount of smooth muscle in the pulmonary arterial walls means that the pulmonary circulation offers little resistance to the low-pressure blood leaving the thin-walled right ventricle (25 mmHg systolic). The pulmonary arteries are therefore very compliant and distensible.

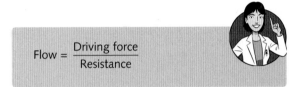

$$\text{Flow} = \frac{\text{Driving force}}{\text{Resistance}}$$

Pressure for this system (driving force) is low, so it follows that resistance must also be low, as flow is considered equal to the systemic value.

Pulmonary arterioles and capillaries are the main resistors. However, they permit further drops in pressure because their large number amount to a considerable total cross-sectional area for flow.

At rest, pulmonary vascular resistance (PVR) is split evenly among pulmonary arteries, capillaries and veins.

Two groups of pulmonary vessels may be considered to operate within this circuit: alveolar and non-alveolar vessels, which lie either within or outside of the alveolar walls. Pulmonary blood flow and PVR are subject to:

- Vascular effects: including hydrostatic pressure within the vessel and vascular smooth muscle.
- Extravascular effects: including lung volume and intrapleural pressure, the influences of which are based on the concept of a transmural pressure gradient.

Hydrostatic pressure

The hydrostatic pressure dictates pulmonary arterial or venous pressure and provides the driving force for flow. When this pressure rises, PVR decreases further by two mechanisms:

1. Recruitment: blood is conducted through previously unused capillaries, which were either closed or open without blood flow. When they open the cross-sectional area to flow is increased, thereby lowering resistance further. Recruitment occurs when pulmonary artery pressure increases from low levels.
2. Distension: individual capillary segments dilate, thereby increasing the radius and lowering the resistance four-fold (Poiseuille's law). Distension occurs at higher hydrostatic pressures than recruitment.

Recruitment and distension usually occur in unison, with each being the predominant mechanism for lowering PVR at the different pressures indicated above. These effects are best illustrated by the higher hydrostatic pressures in the gravity-dependent (lower) regions of the lung.

Vascular smooth muscle

Vascular smooth muscle determines the calibre of extra-alveolar vessels: contraction will decrease vessel radius and increase PVR. Smooth muscle is subject to both neural and humoral influences. Constriction, and hence ↑PVR, is caused by:

- Catecholamines: epinephrine (adrenaline) and norepinephrine (noradrenaline).
- Serotonin.
- Histamine.

Dilatation, and therefore ↓PVR, is caused by:

- Acetylcholine.
- Isoproterenol(a β-adrenergic agonist).
- Nitric oxide.

Effects of transmural pressure gradients

As mentioned earlier, the two different groups of pulmonary vessel – alveolar and extra-alveolar vessels – exhibit different behaviours related to PVR and are affected by different factors. Both are subject to a transmural pressure gradient so that $P_{\text{inside the vessel}} - P_{\text{outside the vessel}}$ dictates vessel diameter:

- ↑Transmural pressure = ↑Diameter = ↓PVR.
- ↓Transmural pressure = ↓Diameter = ↑PVR.
- Negative transmural pressure = compression or collapse of the vessel!

Alveolar vessels These include capillaries residing within alveolar walls. The transmural pressure gradient related to their calibre is determined by the pressure within the capillaries and alveolar pressure outside.

During a normal inspiration, increased alveolar volume and stretching causes elongation of the capillaries with ↑ in length and, importantly, ↓ in radius. Furthermore, the higher alveolar pressure compresses pulmonary capillaries. Hence, there is greater vascular resistance of the alveolar vessels. At low lung volumes, resistance to blood flow is low.

Extra-alveolar vessels

- All arteries and veins that run through lung parenchyma.
- Corner vessels/extra-alveolar capillaries located at the junction of alveolar septae.

The first group (the parenchymal vessels) are subject to intrapleural pressure, which is the other factor involved in determining the transmural pressure. As this entity is usually always negative – more so during inspiration – transmural pressure is elevated; these vessels will distend by radial traction.

The second group – corner vessels – also distends by radial traction forces with the increased wall tension from alveolar expansion (Fig. 6.16).

A common problem encountered during mechanical positive-pressure ventilation is increased alveolar and extra-alveolar resistance to blood flow during inflation. This is because both alveolar pressure and intrapleural pressure are positive, so both types of vessel are compressed with increasing lung volume.

These two groups of pulmonary vessels lie in series with each other so that total PVR is the sum of the two resistances.

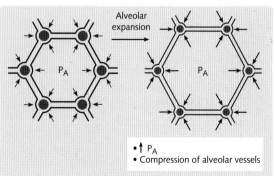

Fig. 6.16 Alveolar walls showing the effects of expansion on the alveolar vessels. P_A, alveolar pressure.

Lung volume

This affects PVR (Fig. 6.17). The following must be taken into account:

- The effect of lung volume on transmural pressure.
- The individual effects on alveolar and extra-alveolar vessels.

At low lung volumes The PVR of extra-alveolar vessels is large as a result of effective smooth muscles in their walls, which resist distension. However, alveolar pressure and distension are minimal, so alveolar vessels have low levels of resistance.

During deep inspiration, alveolar pressure rises and there is stretching of alveolar walls, both of which increase alveolar PVR. In addition, capillary hydrostatic pressure falls due to more negative intrapleural pressure surrounding the heart. Hence, the transmural pressure gradient is greatly decreased, possibly negative, with squashing of the alveolar vessels. Without the effects of transmural pressure, PVR would still be elevated by stretching and thinning of the alveolar walls. Meanwhile, extra-alveolar vessels are distended by increased radial traction.

As the resistance of the vessels adds up, total PVR will be highest at extremes of lung volume corresponding to maximal PVR of either group.

Distribution of blood within the lung

A gradient of regional perfusion exists, from the apex with the least blood flow to the base where blood flow is highest. This inequality is due to hydrostatic pressure differences in the upright lung, whose blood vessels are subject to gravity.

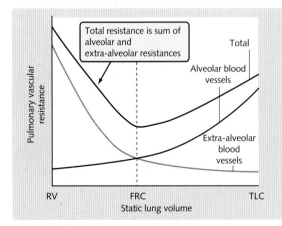

Fig. 6.17 Pulmonary vascular resistance (PVR) variation with lung volume of alveolar and extra-alveolar vessels.

The pulmonary system can be considered as a continuous column of blood whose hydrostatic pressure can be calculated by the following equation:

$$P = \rho g h$$

Where P = hydrostatic pressure, ρ = density of substance, g = gravitational acceleration, and h = height of column (distance from the top).

This hydrostatic pressure is greater at lower regions and responsible for lowering PVR in these areas by distension and recruitment of capillaries. In addition, the capillary transit time is less in lower regions (i.e. increased blood flow rate).

From the above equation, it would be expected that ventilation would be better at the base where alveolar compliance is higher. However, comparatively air has a very low density, so that the hydrostatic pressure differences are much less than for blood.

The alveolar pressure (P_A), which is the same throughout the lung, interacts with the hydrostatic pressures of blood vessels:

- P_A> downstream (i.e. venous pressure) = Alveolar pressure determines blood flow.
- P_A> arterial pressure = compression of that vessel with no blood flow.

Zones of the lung

Blood flow within the lung can be split vertically into four sections or zones from apex to base (Fig. 6.18), with the following pressures taken into account:

- P_A: alveolar pressure.
- Pa: arterial pressure.
- Pv: venous pressure.

Zone 1

- P_A> Pa: capillary compression.

This refers to ventilated but not perfused space, i.e. alveolar dead space. This does not exist normally, as although the lung apex is approximately 15 cm above the right ventricle, pulmonary arterial pressure here is just enough to allow blood flow. This zone can include the tops of the lungs if:

- Hydrostatic pressure drops (haemorrhage).
- Alveolar pressure increases (positive-pressure ventilation).

Zone 2

- Pa> P_A: there is blood flow in this region, which is lower down the lung than zone 1. Pa is increased

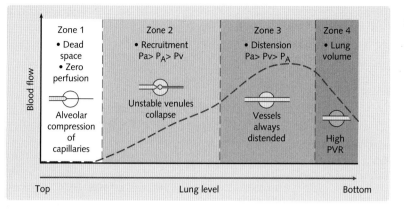

Fig. 6.18 Blood flow within the vessel zones and levels of the lung.

due to higher hydrostatic pressure, which causes recruitment.

- $Pv < P_A$: Pa and P_A still determine blood flow. Opening and closing of postcapillary venules occurs with systole and diastole, respectively.

Zone 3

- $Pa > Pv > P_A$

The arteriovenous pressure difference accounts for blood flow, which is continuous. Distension, which decreases PVR, is particularly important in this zone, with a resultant increase in mean vessel width.

Zone 4

A small section at the base of the lung is referred to as zone 4. The volume here is smaller than elsewhere because the lungs are compressed, which causes extra-alveolar vessel compression and therefore high PVR, so that blood flow is less than in zone 3.

Measurement of pulmonary blood flow

Various methods have been used to calculate:

- Total pulmonary blood flow: taken to be equivalent to cardiac output.
- Regional pulmonary blood flow.

Total pulmonary blood flow is measured using Fick's principle. This measures average blood flow and is based on the following equation:

$$VO_2 = Qt(CaO_2 - CvO_2)$$

where VO_2 = amount of O_2 taken up by blood in lungs per minute, Qt = volume of blood passing through the lungs (mL/min: taken to be cardiac output), Ca_{O2} = O_2 concentration of arterial blood, and Cv_{O2} = O_2 concentration of venous blood (pulmonary artery).

The difference in arteriovenous $[O_2]$ is equivalent to O_2 consumption by the body.

An alternative method of measurement is the indicator dilution technique. This measures average blood flow. The arterial concentration of a dye, which is confined to the vascular compartment, is measured after intravenous injection of a known volume. The calculation is based on the following equation:

$$CO = \frac{\text{Amount of dye (mg)}}{[\text{dye}] \times \text{time(s)}}$$

where CO = cardiac output.

Regional pulmonary blood flow can be determined using a combination of pulmonary angiography and radioactive-labelled macroaggregates of albumin, most of which are trapped in small perfusing pulmonary vessels. Hence, perfused areas of lungs can be demonstrated by scanning for radioactivity.

Control of pulmonary blood flow

As already mentioned, blood flow within the lung can be altered by passive or active means:

- Passive effects: due to change in hydrostatic pressure and include recruitment and distension of capillaries to alter PVR.
- Active effects: include modification of vascular smooth muscle tone and include humoral effects, such as contraction by catecholamines, increasing PVR and decreasing pulmonary blood flow.

Apart from these mechanisms, reduced alveolar PO_2 is also important in controlling pulmonary circulation. A decrease in P_AO_2 to <70% of normal values results in adjacent precapillary smooth muscle contraction. This is probably due to direct action by hypoxia. The function of this phenomenon is that blood is not wasted in perfusing poorly ventilated areas where there is air trapping and where no effective gaseous exchange would occur.

Pulmonary water balance

It is essential for gaseous exchange that the alveoli are kept free from fluid that may escape through the leaky endothelium of capillary blood vessels only 0.3 µm away. This is achieved by:

- Surfactant in the alveoli: this lowers surface tension and decreases liquid transudation into alveoli.
- Capillary oncotic pressure: this draws fluid back into the blood vessels.
- Interstitial fluid (ISF) hydrostatic pressure.
- Lymphatics: transudate drains through the interstitial space into the lymphatics.

Fluid movement across the capillary is governed by Starling forces (see Chapter 2; Fig. 6.19) and can be determined using the following equation:

$$\text{Net flow of fluid} = K(P_{CAP} - P_{ISF}) - \sigma(\pi_{CAP} - \pi_{ISF})$$

where K = filtration coefficient (refers to membrane permeability), P_{CAP} = capillary hydrostatic pressure, P_{ISF} = interstitial fluid hydrostatic pressure, σ = reflection coefficient (refers to membrane impermeability to solute particles), π_{CAP} = capillary oncotic pressure and π_{ISF} = ISF oncotic pressure.

In reality, it is likely that a small outflow of fluid from the capillaries occurs. This fluid moves through the ISF to perivascular and peribronchial spaces within the lung, where there is a rich lymphatic channel supply, transporting the fluid to hilar lymph nodes.

Pulmonary oedema refers to the extravascular accumulation of fluid in the lung, the earliest form of which is known as interstitial oedema. This refers to the accumulation of fluid in the peribronchial and perivascular spaces, which, if they become engorged, will result in increased diffusion distance and can result in fluid transudation into alveoli, further impairing gaseous exchange. This condition can occur in cases of lymph duct blockage due to tumour or scarring.

In addition, the following components from the equation above may result in pulmonary oedema:

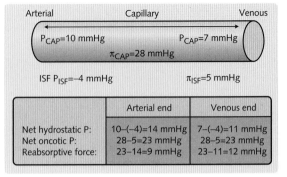

Fig. 6.19 Starling forces and pressures (P) of capillaries and interstitial fluid (ISF) within the lung.

- $\uparrow P_{CAP}$:
  - this will favour transudation
  - increased with excess intravenous fluid administration or outflow obstruction: increased left atrium and left ventricular pressure, with occlusion of pulmonary veins.
- $\downarrow P_{ISF}$:
  - more negative if there is rapid evacuation of chest fluid.
  - raises alveolar surface tension.
- $\downarrow \sigma$: decreased if capillary endothelium becomes more leaky, allowing more solutes to pass through, as with septic shock.
- $\downarrow \pi_{CAP}$: occurs in hypoproteinaemic states.

The ventilation–perfusion relationship

The ventilation–perfusion ratio is an important measure of gaseous exchange.

Basic concepts

Alveolar partial pressures are dictated by:

- Alveolar perfusion (Q).
- Ventilation rate (V).

However, neither of these two factors is uniformly distributed in the lungs. Instead, both increase towards the base, owing to hydrostatic pressure. As blood is denser than air, the increase will be more marked for blood (remember: hydrostatic pressure α density). Therefore, the V/Q ratio will vary throughout the lung.

In general, alveolar minute ventilation and pulmonary blood flow can be substituted as values for V and Q, respectively, to give a normal V/Q value of 0.84. V/Q values are bordered at either end by extreme numbers:

- V/Q = 0: there is no ventilation (V = 0). This can be caused by shunts (see below).
- V/Q = ∞: there is no perfusion (Q = 0). This corresponds to dead space.

Shunts

This refers to mixed venous-type blood bypassing ventilated regions of the lung. Shunts can be either physiological or pathological:

Physiological shunts

Blood from the bronchial circulation and thebesian vessels (coronary venous blood) drain directly into the pulmonary veins and left ventricle, respectively, thereby avoiding the pulmonary capillaries. The consequence of adding this quality of blood – known as venous admixture – is a small depression of PaO_2 and only slight elevation of $PaCO_2$, as the shunt accounts for 2–5% of cardiac output only. The venous admixture can be calculated from the following equation:

$$Qs = \frac{Cc'_{O_2} - Ca'_{O_2}}{Q_T Cc'_{O_2} - Cv'_{O_2}}$$

Where Q_S = venous admixture flow, Q_T = total blood flow, Cc'_{O2} = O_2 content of capillary blood, Ca'_{O2} = O_2 content of arterial blood, Cv'_{O2} = O_2 content of mixed venous blood.

The larger the Q_S/Q_T value, the greater the amount of blood shunted and hence the more blood that is not fully oxygenated.

Pathological shunts

This can occur with one of the following:

- Intrapulmonary shunt: whereby totally unventilated or collapsed alveoli are perfused (≤1% in normal individuals).
- Anatomical shunt: this is usually described as a right–left shunt. Classically, it involves structural abnormalities of the heart and great vessels. In the case of septal defects, mixed venous blood from the right side of the heart might mix directly with arterial-quality blood from the left side.

As with all shunts, the result of a larger than normal venous admixture is to reduce the concentration of O_2. This disproportionately reduces the PO_2 of the arterial blood, as demonstrated by the O_2 dissociation curve (see Fig. 2.12, p. 25).

Unexpectedly, $PaCO_2$ is not usually raised despite the relatively high PCO_2 of the venous admixture. Chemoreceptors sense the initial elevated $PaCO_2$, stimulating ventilation. This decreases the PCO_2 of unshunted blood below its normal value. Hence

$PaCO_2$ is not grossly elevated. However, PaO_2 is not corrected even if the patient is given 100% O_2.

Hypoxic shunt does not undergo gaseous exchange with higher P_AO_2.

Unshunted blood is already almost maximally saturated with O_2. Although it has a higher than normal level of dissolved O_2, which serves to increase PaO_2, this is somewhat below normal values.

Effect of V/Q mismatch on alveolar partial pressures and gaseous exchange

- Inspired air: PO_2 = 150 mmHg; PCO_2 = 0 mmHg.
- Mixed venous blood: PO_2 = 40 mmHg; PCO_2 = 45 mmHg.
- Alveolar gas (P_A): PO_2 = 100 mmHg (i.e. lower than inspired air) due to removal of some of the inspired O_2 by the blood.

The easiest way to imagine the effects of V/Q on P_A is to think of Q (blood flow) as a raging river, sweeping away the trees (O_2) from the river bank (alveolar gas) and dumping its debris and silt (CO_2) as it flows on the river bank (i.e. more CO_2 exchanged). Hence:

- A lower V/Q (Q>V) means more trees/O_2 swept away by the capillary, and more silt/CO_2 deposited so the effects on the riverbank or alveolar gas are: $\downarrow P_AO_2$ and $\uparrow P_ACO_2$.
- Higher values of V/Q (V>Q) mean fewer trees being taken away by the river (less O_2 exchanged with blood) and less silt /CO_2 deposited on the riverbank/alveolar gas, so $\uparrow P_AO_2$ and $\downarrow P_ACO_2$.

The consequences of extreme values of V/Q on P_A are discussed below and illustrated in Figure 6.21, below

V/Q = 0: no ventilation

No ventilation can occur when the airway leading to the alveolus is occluded (Fig. 6.20). Gas trapped in the alveolus will undergo exchange with blood until the Pa and P_A equilibrate for respective PO_2 and PCO_2 values. Over time, the alveolus will collapse and this unit (alveolus with its blood) will function as an intrapulmonary shunt.

V/Q = ∞: no perfusion

Non-perfusion of a ventilated alveolus can occur when the blood vessel is occluded, as with pulmonary embolus. No blood means no gaseous exchange; no O_2 is swept away and no CO_2 released into the alveolus. Physiologically, this occurs with dead space/zone 1 (see above). Therefore, there is no difference between alveolar and inspired gas.

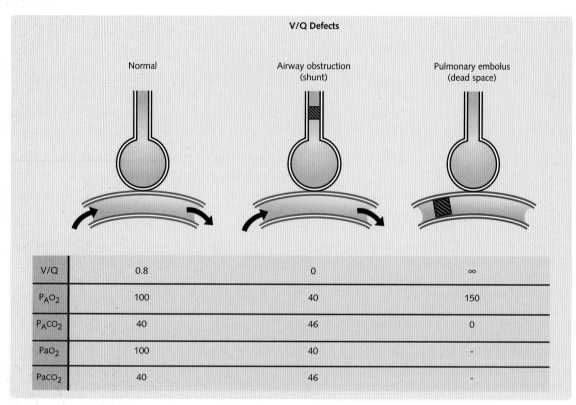

Fig. 6.20 Effect of ventilation/perfusion defects on gaseous exchange in the lungs.

Gaseous exchange

The CO_2 removal : O_2 uptake ratio at the blood–gas interface, known as the exchange ratio (R), varies with V/Q values (Fig. 6.21). This is explained by the dissociation curves of haemoglobin for these gases (see Fig. 2.12, p. 25 and Fig. 2.16, p.29). For this reason:

- R is high when V/Q is high: well-ventilated alveoli will remove more CO_2 from the blood in comparison to O_2 uptake by the blood, as haemoglobin is already almost fully saturated.
- R is low when V/Q is low: blood exchanging with less well-ventilated alveoli will take up more O_2 in comparison to CO_2 released into the alveoli because the gradient of the O_2 dissociation curve corresponding to lower PO_2 is steeper than that for the CO_2 dissociation curve.

Regional variation of V/Q

Although the lower regions of the lung receive more ventilation than the higher regions, greater hydrostatic effects of blood mean that (Fig. 6.22):

- Lower regions are relatively overperfused ($\downarrow$V/Q).
- Upper regions are relatively underperfused ($\uparrow$V/Q).

Lung apex : high V/Q

V/Q can be as high as 3 in the upper regions of the lung. This means R is also high, so more CO_2 is released than O_2 is taken up.

Carbon dioxide As can be seen in Figure 6.21, the smaller CO_2 load in the blood (consequent upon lower volume) diffuses out into the alveolus, from where it is promptly removed. Hence, P_ACO_2 and $PaCO_2$ are reduced compared with the base.

Oxygen Less perfusion removes less O_2 from the alveoli, so P_AO_2 is high. PaO_2 is also increased owing to diffusion.

Lung base : low V/Q

V/Q here can be as low as 0.6. R is also low, so that relatively less CO_2 is removed compared to O_2 uptake. The base corresponds to the left side of the graph in Figure 6.21: low PO_2 and relatively high PCO_2.

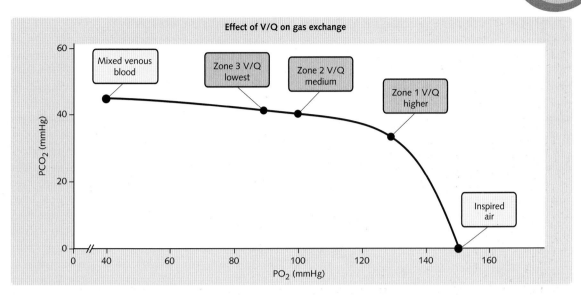

Fig. 6.21 Effect of V/Q on gaseous exchange.

Carbon dioxide Unmatched ventilation does not clear the relatively large load of CO_2 presented by the rich perfusing blood supply. Hence, P_ACO_2 tends to rise and is relatively high compared to the apical regions. Consequently, $PaCO_2$ will also be higher due to diffusion. P_ACO_2 is still less than $PaCO_2$ because V/Q is greater than zero (in which case the two values would be the same after equilibrating).

Oxygen Greater perfusion transports away relatively more O_2 from the alveolus than at the apex. As ventilation is less in lower regions, P_AO_2 is lower at the base than at the apex. Consequently, PaO_2 will be lower.

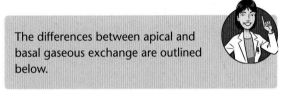

The differences between apical and basal gaseous exchange are outlined below.

Fig. 6.23 Differences between apical and basal gaseous exchange.

Gaseous exchange at apex	Gaseous exchange at base
V > Q	Q > V
Less blood flow	More blood flow
↑P_AO_2	↓P_AO_2
↓P_ACO_2	↑P_ACO_2
↑PaO_2	↓PaO_2
↓$PaCO_2$	↑$PaCO_2$

Remember, more gaseous exchange takes place in more basal regions.

Fig. 6.23 Differences between apical and basal gaseous exchange.

Effect of regional variation in V/Q on overall gaseous exchange

Importantly, there is more blood flow through the basal regions, which comprise a greater volume of the lung. This is of no consequence for CO_2 as the combined blood from different levels of the lungs has a 'normal' CO_2 content. However, for O_2 it is

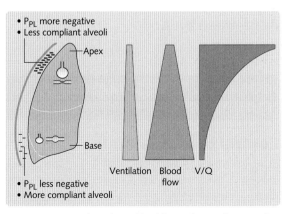

Fig. 6.22 Variation of ventilation, blood flow and V/Q values according to lung level.

• P_{PL} more negative
• Less compliant alveoli
Apex

Ventilation Blood flow V/Q

Base
• P_{PL} less negative
• More compliant alveoli

critical that V/Q values do not differ too greatly as a result of to unequal perfusion.

A relatively small quantity of blood with a high PaO_2 and low $PaCO_2$ leaves the apex. However, as haemoglobin is already saturated, the O_2 content of this blood is not significantly higher than normal values. This then mixes with a large volume of blood with a lower O_2 content from the bases.

The narrow range of V/Q ratios have to be maintained and ventilation best matched to perfusion to avoid depression of PaO_2 in the left atrium (majority of which drains from the lung bases).

> Inequalities in V/Q result in reduction of either uptake or elimination of gases by the lung.

Distribution of ventilation and blood flow follows a bell-shaped curve, with the greater proportion of all ventilation and perfusion directed to regions where V/Q is nearer 1. This can change in disease, when areas with low V/Q ratios are perfused well.

Measurement of ventilation and perfusion

V/Q scan

This consists of two separate scans of ventilation and perfusion, respectively, which are then compared for any mismatch or 'filling defects'. Typically, such mismatches occur in pulmonary embolus, where this investigation is particularly useful.

Ventilation scan   This involves whole-lung scintillation images after inhalation of radioactive gas mixture such as ^{133}Xe or technetium-labelled diethylene triamine penta-acetic acid (^{99m}Tc DTPA). Ventilated areas will take up the radioactive gas and will therefore show up on the scan.

Perfusion scan   As with the measurement of pulmonary blood flow, ^{99m}Tc-labelled macro-aggregates of albumin (MAA) are used to demonstrate any perfusion defects. A gamma camera is used to scan the positions of the MAA. As these have a larger diameter than the pulmonary capillaries, they are wedged within the capillaries for several hours.

Disordered perfusion or ventilation can be seen in respiratory disease such as asthma or COPD. However, in such cases, any defects in ventilation are geographically matched to defects on the perfusion scan.

Multiple inert gas procedure

Specific V/Q ratios are measured using a mixture of six different inert gases, each with a different partition coefficient (i.e. different solubility in gas and blood). The procedure is based on the concept that variations in V/Q ratios in different lung units, and the solubility of gas in the blood, affect the elimination of the gas by the lungs from mixed venous blood.

The gas mixture is continuously infused via a peripheral vein at a constant rate for about 20 min (until there is a steady rate of gaseous exchange). For measurement purposes, the lung can be thought of as containing 50 different sections, each of which will have the gases split between its blood and alveolar air. Gas chromatography is used to determine the concentration of each gas from samples of arterial blood and expired air to give the following measures:

- Retention:

$$\frac{\text{Arterial [gas]}}{\text{Venous [gas]}}$$

- Excretion:

$$\frac{\text{Expired [gas]}}{\text{Venous [gas]}}$$

When retention or excretion is plotted against the gases' solubility, perfusion or the ventilation versus V/Q ratio is shown (Fig. 6.24). In healthy subjects, V and Q distributions are narrowly spread about a mean of approximately 1. With old age or disease, this spread becomes wider, deviating further from units with ideal V/Q ratios.

Gas transport in the blood

See Chapter 2.

Acid–base balance

See Chapter 8.

CONTROL OF RESPIRATORY FUNCTION

Control of ventilation

Control of ventilation is vital for maintaining arterial blood gases while meeting the hugely variable requirements of the body. Generally, control of ventilation is involuntary or automatic, resulting from impulses originating in the brainstem

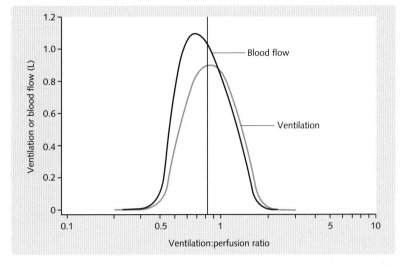

Fig. 6.24 Normal human V/Q curves.

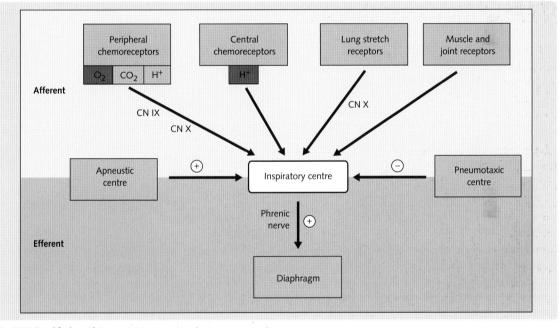

Fig. 6.25 Simplified ventilatory system.

(controller), which generates the normal quiet breathing cycle of rhythmic inspiration and expiration. The automatic nature of breathing is achieved through feedback mechanisms of a control system.

The control system adjusts its output through effectors in response to alterations in ventilation or chemical markers such as [H⁺] or $PaCO_2$ at values beyond the normal limits as detected by sensors (Fig. 6.25).

The respiratory control system consists of three basic components:

1. Central control: produces a modifiable breathing pattern and sends command discharges to the effectors.
2. Sensors: detect changes and feedback information to the central control.
3. Effectors: perform ventilation and modifications to breathing as dictated by the central control.

Tidal volume is determined by the frequency of neural signals and number, and how many motor units are activated.

As can be seen from Figure 6.25, chemical and neural controls of ventilation exist:

- Neural: concerned with adjusting individual breaths.
- Chemical: involves the use of chemoreceptors:
  - peripheral chemoreceptors: project to neural central control; adjust individual breaths
  - central chemoreceptor: adjust minute ventilation.

Central control of breathing

Brainstem collections of neurons also known as respiratory centres generate the breathing cycle. Several functional groups concerned with breathing control are recognized in the medulla and pons:

- Medullary groups: located in the reticular formation beneath the floor of the fourth ventricle bilaterally:
  - dorsal respiratory group (DRG): mainly involved with inspiration
  - ventral respiratory group (VRG): largely concerned with expiration.
- Pontine groups:
  - apneustic centre: modifies inspiration
  - pneumotaxic centre (also known as pontine respiratory group): modifies inspiration.

Medullary respiratory centre

This is believed to be essential to generate breathing. Both DRG and VRG are not completely anatomically discrete, with intermingling of their neurons.

Dorsal respiratory group   The DRG is located mostly in the nucleus of tractus solitarius (NTS), which is where the sensory parts of the vagus and glossopharyngeal nerve (cranial nerves X and IX) terminate. The information the DRG receives from receptors/sensors is thought to be integrated, and breathing control is influenced by this area as well as by the VRG, which receives collaterals from the DRG (Fig. 6.26).

The DRG is believed to be primarily responsible for establishing rhythmic breathing by the periodic, repeated production of action potentials (APs) that excite lower motor neurons. Early evidence suggested that these APs were generated in the absence of any known afferent signal. However, more recent evidence is that cells of the pre-Botzinger complex are responsible.

APs from the DRG neurons, which project to the contralateral portions of the spinal cord, stimulate the inspiratory muscles – diaphragm and external intercostals– in a ramp-like style. Thus lung volume steadily increases with increasing signal intensity.

Normal quiet breathing consists of several stages (Fig. 6.27):

- A: start of impulses = start of inspiration.
- B: increase in impulses in crescendo style = increasing muscle activity and lung volume.
- C: abrupt cessation of impulses = diaphragm contraction ceases with limit reached in lung volume.
- D: no impulses = expiration with diaphragm relaxation permitting elastic recoil of chest wall.

Ventral respiratory group   The VRG is loosely located anterolateral to the DRG, caudally to rostrally in the:

- Nucleus retroambiguus.
- Nucleus ambiguus.
- Retrofacial nucleus.

There are both inspiratory and expiratory neurons in each of the above.

NUCLEUS RETROAMBIGUUS   Contains two types of inspiratory neuron, which:

1. Project within the medulla to other inspiratory cells.
2. Stimulate contralateral external intercostals and the phrenic nerve supplying the diaphragm.

Expiratory neurons in the caudal portion of the nucleus excite internal intercostals and abdominal muscles.

NUCLEUS AMBIGUUS   Contains expiratory and inspiratory vagal motor neurons. Parasympathetic innervation of upper respiratory structures concerned with airway patency (laryngeal, pharyngeal and tongue muscles).

RETROFACIAL NUCLEUS   Consists mainly of expiratory neuron clusters – known as the Botzinger complex – the functions of which include:

- Inhibition of DRG inspiratory cells.
- Inhibition of some phrenic motor neurons.

Pontine groups   The main function of the pontine respiratory groups is to fine-tune or modify the breathing rhythm.

Apneustic centre   This is located in the lower pons and receives information via the vagus nerves. It stimulates the medullary inspiratory regions to control the degree of inspiration and prolongs the inspiratory ramp of APs so that the abrupt halt (point C in Fig. 6.27), which is associated with the start of expiration, does not occur on time; instead, apneustic breathing occurs (i.e. prolonged inspiration with lung hyperinflation separated by short gasps of expiration). This type of breathing also occurs when vagal nerves are not intact.

Pneumotaxic centre   This is located in the upper pons, in the nucleus parabrachialis medialis and in the Kolliker-Fuse nucleus. It receives information from pulmonary afferents and inhibits inspiratory neurons in the brainstem, including those in the apneustic centre, as a function of fine-tuning the respiratory rhythm. Normal eupnoeic breathing occurs in the absence of this centre. In addition, it might also modify the response of the respiratory control centre to chemical stimuli.

The pneumotaxic centre appears to limit inspiration by altering the cut-off point of the inspiratory ramp. A strong signal from the pneumotaxic centre results in dramatic shortening of inspiration, with a secondary effect of increasing the respiratory rate. A weak signal from this region will cause lung hyperinflation and a very slow respiratory rate.

Effectors and spinal pathways

Impulses from the respiratory centres travel in the spinal cord; the inspiratory and expiratory fibres are separate. Before reaching the respiratory muscles (effectors), respiratory motor neurons integrate descending information, and are themselves subject to spinal reflexes. Hence, once this information reaches the effectors, coordination between the muscle groups is assured.

- During inspiration: expiratory motor neurons are inhibited.
- During expiration: inspiratory motor neurons are inhibited.

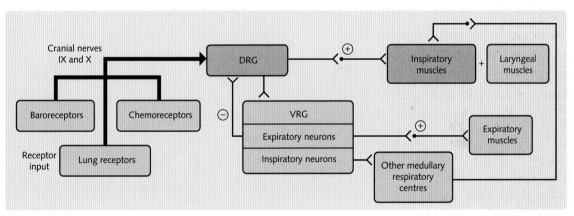

Fig. 6.26 Dorsal respiratory group (DRG) inspiratory signals during a breathing cycle.

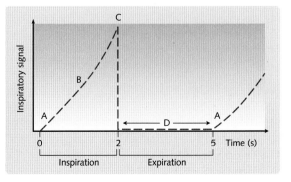

Fig. 6.27 Schematic representation of dorsal (DRG) and ventral (VRG) respiratory group circuits.

Both of the above occur as a direct result of the respiratory control centres. Effectors consist of:

- Inspiratory muscles: diaphragm, external intercostals and accessory muscles (e.g. sternomastoid).
- Expiratory muscles: abdominal muscles and internal intercostals.

Sensors

Sensors, or receptors, gather information related to ventilation and send it to the central control. They include:

- Chemoreceptors:
  - both central and peripheral
  - report chemical changes associated with matching between ventilation to the body's needs.
- Lung receptors:
  - three main types of receptor: pulmonary stretch, irritant and J receptors
  - travel via the vagus nerve.
- Others:
  - chest wall receptors, which use muscle spindle reflex
  - include baroreceptors and pain.

Chemoreceptors

These report variations in blood chemistry, stimulating the respiratory control system to ensure homeostasis of PaO_2, $PaCO_2$ and plasma $[H^+]$ as markers of ventilation. Chemoreceptors are either:

- Central: lie within the CNS; their effects bypass respiratory centres.
- Peripheral: lie outside the CNS; they send afferents to the respiratory centres.

In general, the effects of the chemoreceptors are to increase ventilation in response to the following:

- Hypercapnia ($\uparrow PaCO_2$): detected mainly by central chemoreceptors.
- Hypoxia ($\downarrow PaO_2$): mostly mediated by peripheral chemoreceptors.
- Acidosis ($\uparrow$ plasma H^+): central and peripheral chemoreceptors

Central chemoreceptors Chemosensitive areas on the ventral medullary surface, separate from the VRG and DRG, provide 80% of the ventilatory drive and mediate hyperventilation in response to surrounding extracellular fluid (ECF) increase in $[H^+]$ because of:

- $\uparrow PaCO2$.
- $\uparrow [H^+]$.

Low $[H^+]$, by reversing the conditions above, inhibits ventilation.

Factors affecting the ECF content are cerebrospinal fluid (CSF; most important), local blood flow and local tissue metabolism. These are not in direct contact with arterial blood and do not respond to hypoxia.

RESPONSE TO $PaCO_2$ CO_2 has an indirect effect on the central chemoreceptor by changing $[H^+]$. CO_2 diffuses freely across the blood–brain barrier into the interstitial fluid (ISF) and CSF. Hence, $[CO_2]$ in ISF and CSF mirror those in the blood. In these areas, increased CO_2 liberates H^+ by reacting with H_2O, displacing the following reaction to the right:

$$CO_2 + H_2O \leftrightarrow H_2CO_3 \leftrightarrow H^+ + HCO_3^-$$

Although both acid and base are generated, HCO_3^- buffer cannot be used since this would necessitate displacing the reaction to the left, in the direction of raised CO_2 (see p. 193).

Importantly, CSF contains very little protein so does not buffer pH changes as well as blood (plasma proteins and RBC haemoglobin are buffers). Hence, for a fixed change in PCO_2, the effect on pH is greater in CSF than blood.

H^+ is released into the chemosensitive area, increasing ventilation.

RESPONSE TO PLASMA H^+ The effect of increased plasma $[H^+]$ on central chemoreceptor stimulus is less than that of PCO_2 because of the relative impermeability of the blood–brain barrier to H^+ and HCO_3^-, which penetrate very slowly.

CSF has a normal pH of 7.32, which varies more easily than blood and returns more quickly to normal levels (compared to renal compensatory effects on blood pH). However, if the pH of the CSF is altered for a long time, as is the case in persistent hypercapnia, it returns to nearly normal values. This is because slow HCO_3^- diffusion through the blood–brain barrier buffers the protons; it explains why some patients with COPD have inappropriately low ventilation.

Peripheral chemoreceptors These are also known as arterial chemoreceptors, and are located in the carotid and aortic bodies. They are responsible for all increases in ventilation due to hypoxaemia. They are also sensitive – to a lesser extent – to the following:

- Hypercapnia.
- pH change.
- Blood flow.

Information from the peripheral chemoreceptors is sent to the respiratory centres and ventilation is altered. The hypercapnic stimulus is less important than it is with the central chemoreceptor, contributing 10–20% of the steady-state response to elevated PCO_2.

Carotid bodies   In humans, the carotid bodies are the principal source of chemoreceptors; these have the most significant effect on respiration. The carotid bodies are located at the bifurcation of the common carotid arteries. They measure arterial values only. A high blood-flow rate (estimated at 200 mL/100 g tissue per minute) means the arteriovenous difference is small despite its own very high metabolic rate.

The response of the carotid bodies is sensitive and rapid: information relating to variation in arterial blood gases between each breathing signal is relayed via the glossopharyngeal to the respiratory centres.

Each carotid body contains a rich blood supply and islands consisting of two different types of cell:

- Type 1 (glomus) cells.
  - monitor PaO_2
  - lie in close proximity to the carotid sinus nerve(afferent) – a branch of the glossopharyngeal nerve
  - hypoxia → release of its catecholamine granules and excitation
  - principal signal transmitter appears to be dopamine, which excites dopamine receptors.
- Type 2 cells:
  - supportive function believed to be comparable to that of CNS glia
  - surround type 1 cells.

RESPONSE TO PaO_2 The carotid bodies are sensitive to PaO_2 which ranges from 0 to 500 mmHg (Fig. 6.28).

- Response is increased below $PaO_2 > 100$ mmHg.
- Most sensitive when 30 mmHg $< PaO_2 < 60$ mmHg, with the greatest change in ventilation.

As carotid bodies derive their O_2 supply from dissolved O_2, they respond to vascular stasis (decreased rate of O_2 delivery) and cyanide (an inhibitor of cellular aerobic respiration). They do not respond to factors affecting O_2 carriage by haemoglobin, such as anaemia.

RESPONSE TO OTHER STIMULI   The carotid, but not aortic, bodies respond to a fall in arterial pH. Also,

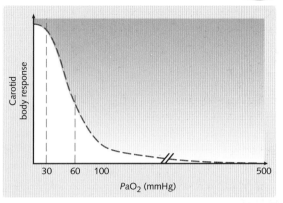

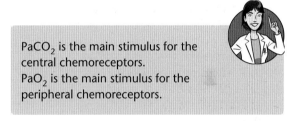

Fig. 6.28 Effect of PaO_2 on carotid body impulse with constant $PaCO_2$ and pH values.

> $PaCO_2$ is the main stimulus for the central chemoreceptors.
> PaO_2 is the main stimulus for the peripheral chemoreceptors.

the peripheral chemoreceptor stimuli interact so that hypoxic ventilatory drive is increased when there is concurrent hypercapnia and acidosis.

The carotid bodies have both parasympathetic and sympathetic innervations, which can modulate the hypoxic response. In addition, sufficient nicotine will activate chemoreceptors and possibly increase $[K^+]$, as shown by experimental studies and supported by the positive effects of exercise, which increase plasma $[K^+]$.

Lung receptors
Three main types of receptor in the lung relay information, via the vagus nerve, to the respiratory control centre:

1. Pulmonary stretch receptors.
2. Irritant receptors.
3. J receptors: including bronchial C fibre receptors.

Pulmonary stretch receptors   Located in the bronchi and bronchiolar smooth muscle, these receptors respond to change in transmural pressure as a function of stretch or lung inflation/deflation. Stretch receptors project to the DRG and pontine groups, providing them with information on lung volume; increasing lung volume increases the firing of APs.

Because they respond only to lung inflation, displaying very little adaptability, these receptors are known as slowly adapting pulmonary stretch receptors. Their role has been described as generally to limit the tidal volume by terminating inspiration, and extending the expiratory phase. Two reflexes are associated, providing negative feedback on respiratory activity:

1. Hering Breuer inflation reflex

Lung inflation inhibits the activity of the inspiratory muscles. This appears to be more important in minimizing the work of breathing in infants and babies, as the reflex occurs within their normal tidal volume. In adults, this reflex is triggered by tidal volume > 1 L – more than occurs with normal quiet breaths, but probable in exercise.

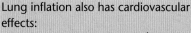

Lung inflation also has cardiovascular effects:
- Moderate lung volume → ↑Heart rate (slight vasoconstriction).
- Very large lung volumes → (↓Heart rate) ↑Blood pressure (BP) due to ↑systemic vascular resistance.

2. Hering Breuer deflation reflex

Deflation stimulates inspiration. Hyperpnoea results from decreased receptor activity, with the possible involvement of other lung receptors. This reflex is thought to be important in maintaining the FRC in infants who would have a greater tendency to collapse the lungs due to (inward) lung recoil >> (outward) chest wall recoil.

Irritant receptors

These are thought to be located between epithelial cells throughout the respiratory tract (possibly including the alveoli). They respond to inhalation of noxious gases and vapours, such as cigarette smoke, dust and ammonia, sending impulses via the vagus nerve (except the nasal receptors, which send afferents via the trigeminal and olfactory nerves).

Rapidly adapting pulmonary stretch receptors

Those stretch receptors located at branch points in the tracheobronchial tree have mechanoreceptor functions. These are known as rapidly adapting pulmonary stretch receptors, and they respond to the rate of change in lung volume. Hence, there is transient firing of impulses with lung volume change and also rapid adaptability, as the impulse frequency quickly drops to background levels if the volume is maintained.

Stimulation of these rapidly adapting receptors has two effects:

1. Shortening of expiration, with rapid shallow breathing in between.
2. Deep slow breaths: which have a refractory period (compulsory length of time during which the breath cannot be repeated).

In humans, normal eupnoeic breathing is interrupted approximately every 10 minutes by a deep slow breath, which serves to prevent the slow lung collapse that would otherwise occur. These mechanoreceptors can mediate symptoms of respiratory disease in which changes in architecture stimulate the receptors to produce reflex bronchoconstriction.

Mechanical or chemical stimuli can also elicit the following reflexes:

- Sneeze: occurs in response to stimulation of nasal receptors, which send afferents via the trigeminal and olfactory nerves. Such stimulation brings about sneeze, bronchoconstriction and ↑BP.
- Cough: occurs as a result of stimulation of receptors in the upper and lower airways, which send afferents via the vagus nerve. Responses include bronchoconstriction and cough.

J receptors and bronchial C-fibres

Both these receptor types comprise non-myelinated C-fibre endings that project to the respiratory centres via the vagus nerve.

J (juxtacapillary) receptors

- Located in the pulmonary capillary walls or interstitium. They are stimulated by:
  - pulmonary vascular congestion
  - ↑ISF (oedema)
  - factors released by lung damage: histamine, prostaglandins and bradykinin.
- Reflex action: rapid, shallow breathing (tachypnoea) and cardiovascular effects: ↓BP and bradycardia. Intense stimulation results in apnoea.
- Probably responsible for dyspnoea (sensation of difficulty in breathing) in pulmonary oedema and pulmonary vascular congestion secondary to heart failure.

Bronchial C-fibres

- Supplied by the bronchial circulation.
- Stimulated by injection of chemicals into the bronchial circulation, such as capsaicin.
- Reflex action: tachypnoea, bronchoconstriction and mucus secretion.

Other receptors

Chest wall The receptors of the chest wall are muscle spindles, which function via a feedback mechanism. Information on muscle elongation is fed back and used to adjust the strength of muscular contraction. This can be important in altering ventilatory muscle output to meet increased workloads, such as decreased compliance.

Joint and muscle receptors Limb proprioceptive and tendon organ receptors can send information to the respiratory centres. In addition, proprioceptive receptors stimulated by moving limbs are believed to increase ventilation, especially during the early stages of exercise.

Arterial baroreceptors These function as stretch receptors in the carotid sinuses and aortic arches. They are stimulated by increased arterial BP:

- Increase in BP results in:
  - hypoventilation or a brief period of apnoea
  - bronchodilation
  - bradycardia.
- Decrease in BP results in hyperventilation.

Pain The effects of pain on respiration depend on the type of pain:

- Visceral pain (e.g. internal organ distension) causes decreased ventilation or apnoea.
- Somatic pain (e.g. biceps tear) causes rapid shallow breathing.

Voluntary control of breathing

Automatic control of breathing is subject to voluntary control from higher centres (within limits). Information from the cortex bypasses the brainstem structures and is relayed directly to the spinal nerves driving respiratory muscles via the corticospinal tract. Higher brain centres mediate control in the case of emotion, speech or singing.

Coordinated responses of the respiratory system

Response to CO_2

CO_2 is probably the most important factor in ventilatory control. Control is achieved through the central chemoreceptors. A rise in $PaCO_2$ stimulates ventilation, increasing the rate of CO_2 excretion and reducing $PaCO_2$ to normal values (40 mmHg). $PaCO_2$ is precisely controlled within very fine limits, so that even small changes will trigger a ventilatory response.

- Increasing inspired $[CO_2]$ increases minute ventilation in a linear fashion, by elevating both depth and rate of inspiration.
- Minute ventilation is most sensitive to inspired $[CO_2]$ of 5–10%.
- Elimination of CO_2 becomes difficult when inspired air is similar to alveolar values with respect to PCO_2.
- Inspired $[CO_2] > 10$–15% has little effect on ventilation.

The physical response to this gas is dictated by inspired concentration:

- Low concentrations (>0.3%) cause: increased ventilation.
- Higher concentrations (5–10%) cause:
  - dyspnoea (feeling of difficulty breathing)
  - greatly increased ventilation
  - cerebral vasodilation, causing headache
  - restlessness.
- Very high concentrations (15–20%) cause:
  - respiratory depression
  - coma (CO_2) narcosis
  - rigidity and tremor.
- Grossly high concentrations (>20%) cause: generalized convulsions.

The CO_2 ventilatory response is reduced in certain circumstances (Fig. 6.29):

- Non-REM sleep.
- Trained athletes.
- Narcotics and anaesthetics.
- Increased work of breathing.
- Age.

CO_2 ventilatory response is increased in:

- Metabolic acidosis.
- Hypoxia.

Response to O_2

The ventilatory response to reduced PaO_2 increases significantly below 50–60 mmHg (with normal $PaCO_2$ values). These are much lower than occur normally. Hence, the hypoxic stimulus is usually not a significant contributor to ventilatory drive. Experiments have demonstrated that PaO_2, which relates to dissolved O_2 and not to blood O_2 content (e.g. anaemia), determines this response. The increase in ventilation is brought about via peripheral chemoreceptor afferents (in particular, those from the carotid bodies).

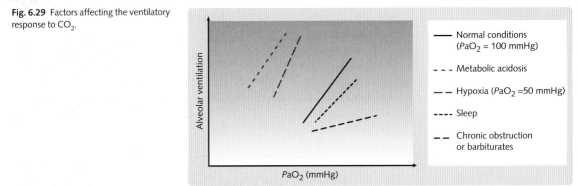

Fig. 6.29 Factors affecting the ventilatory response to CO_2.

The hypoxic response is subject to $PaCO_2$; a greater ventilatory response occurs at any PaO_2 when $PaCO_2$ is greater than normal and in significant hypercapnia, $PaO_2 < 100$ mmHg will actually result in some ventilatory stimulus. As the ventilation response to the two conditions in combination is greater than the sum of their separate effects, hypercapnia is said to interact with hypoxia.

The hypoxic drive becomes the chief ventilatory stimulus in some severe lung diseases, whereby chronic CO_2 retention and resetting of ECF pH has rendered the central chemoreceptors relatively insensitive to their hypercapnia compared with unaffected individuals. Therefore, patients with COPD might have concurrent hypoxaemia on which they rely on to breathe. Attempting to treat this hypoxia by administering high concentrations of O_2 will depress breathing and might stimulate respiratory arrest.

Response to H+

Ventilation is stimulated by decreased pH/raised [H^+], with consequent lowering of PCO_2 as alveolar ventilation increases, which will left-shift the equation and lower H^+:

$$CO_2 + H_2O \leftrightarrow H_2CO_3 \leftrightarrow H^+ + HCO_3^-$$

H^+ concentration is sensed by the peripheral chemoreceptors because the blood–brain barrier is fairly impermeable to ions, which limits its effect on the central chemoreceptors. However, if there is a significant change in blood pH, the permeability of the blood–brain barrier increases somewhat, so H^+ does exert some influence on central chemoreceptor activity.

Lowered pH usually occurs in association with elevated $PaCO_2$, so that it is difficult to separate responses to the two stimuli. However, animal experiments demonstrate increasing ventilation by increasing [H^+] while maintaining PCO_2.

Renal compensation might occur in long-term acidosis. This normalizes pH, thereby decreasing its ventilation stimulus.

Response to exercise

The increased demands of exercise require a prompt ventilatory response that also closely matches elevation in O_2 consumption and CO_2 production. Minute ventilation can rise to a total of 120 L/min – 15 times that at rest. This results from increases in both the tidal volume and respiratory rate. The ventilatory responses to exercise can be considered under the following headings:

- Immediate: at the start of exercise.
- During aerobic: only exercise (< 60% maximum work capacity).
- During strenuous exercise with anaerobic metabolism (> 60% maximum work capacity).

Mechanisms are controversial and their putative contributions relatively unknown.

Immediate response
Neural influences are likely, as they would permit the instantaneous response seen. There is involvement of the following:

- Higher centre:
  - motor cortex neurons send information to skeletal muscles and collaterals to the respiratory centre
  - conditioned learned response to exercise exists.
- Limb proprioceptors: send afferents to the respiratory centre.
- Hypothalamic 'exercise' centre: may contribute to the initial response to exercise.

During aerobic exercise

There is a linear increase in minute ventilation in association with increased O_2 consumption and CO_2 production. Blood gases change very little during this phase:

- PaO_2 may rise slightly.
- $PaCO_2$ is maintained.
- pH is maintained.

Role of CO_2 oscillations   It has been suggested that peripheral chemoreceptors respond to the greater amplitude of oscillations in $PaCO_2$ and PaO_2 during exercise, even though the mean values remain constant.

The periodic nature of ventilation is responsible for fluctuations, which become wider about their mean with increasing tidal volume (as is the case with exercise).

The slow ventilatory increase during this phase of exercise in those who have undergone surgical removal of their carotid bodies, lends support to this hypothesis.

Role of potassium   Exercising muscle cells release K^+ into the bloodstream, elevating its concentration, which can stimulate peripheral chemoreceptors.

Other factors   The increase in temperature that accompanies exercise can also stimulate ventilation.

Pulmonary circulatory receptors that detect elevated mixed venous CO_2 loads to stimulate ventilation have also been suggested in addition to the theory that the central chemoreceptor operates to adjust ventilation to maintain $PaCO_2$.

During strenuous exercise with anaerobic activity

Sixty per cent of an individual's maximum work capacity is the threshold above which anaerobic metabolism will begin and rise. After this point, the rise in ventilation is greater than the rise in O_2 consumption, but remains proportional to CO_2 production. In this phase, the blood gases change:

- PaO_2 may drop.
- $PaCO_2$ increases.
- pH falls due to lactic acid (lactate).

Lactic acid production from anaerobic metabolism lowers the pH. Attempts at buffering $[H^+]$ left-shifts the reaction $CO_2 + H_2O \leftrightarrow H_2CO_3 \leftrightarrow H^+ + HCO_3^-$ to produce more CO_2, which adds to that produced by aerobic metabolism. Hence, an increase in CO_2 production is greater than O_2 consumption, maintaining a linear relationship with increased ventilation.

In addition to probable mechanisms in the previous phase, during this stage there is clearly involvement of peripheral and central chemoreceptors.

Abnormal periodic breathing

Periodic breathing is an abnormal type of ventilation characterized by a continual cycle comprising a short period of deep breathing followed by apnoea or hypoventilation for a comparable length of time.

Cheyne–Stokes breathing is the most common example of periodic breathing. It consists of alternating equal length (20–30 s each) periods of hyperventilation then apnoea.

The tidal volume waxes and wanes with significant variations in PaO_2 and $PaCO_2$. It is commonly seen in disease states and in various conditions:

- Altitude: especially when asleep.
- Congestive cardiac failure: with uraemia.
- Brain disease.

Several factors might have a role:

- Prolonged circulation from lungs to brain: the negative feedback mechanism from lung to brain is delayed.
- Increased sensitivity to $PaCO_2$: apparently due to disruption of the inhibitory pathways and is seen mainly in patients with brain damage where it is a preterminal sign.

Respiratory response to extreme environments

High altitude

As the distance above the Earth increases the composition of the air stays the same, but the barometric pressure decreases almost exponentially. This occurs because of the Earth's gravitational pull on air towards its surface. According to Dalton's law, PO_2 will similarly fall with resultant hypoxia with increased heights:

$$PO_2 = \text{of dry air} = 0.21\,(21\%) \times \text{Barometric pressure}$$

Inspired air contains water vapour, the partial pressure of which, on reaching the alveoli, remains constant at 47 mmHg because it has been warmed to body temperature and is completely humidified:

$$\text{Inspired } PO_2 = 0.21 \times (\text{Barometric pressure} - \text{water vapour pressure})$$

Acute adverse effects

Mainly hypoxic symptoms affecting the nervous system are experienced when rapid ascent to high altitudes occurs, as with sudden loss of cabin pressure when flying. Symptoms include drowsiness,

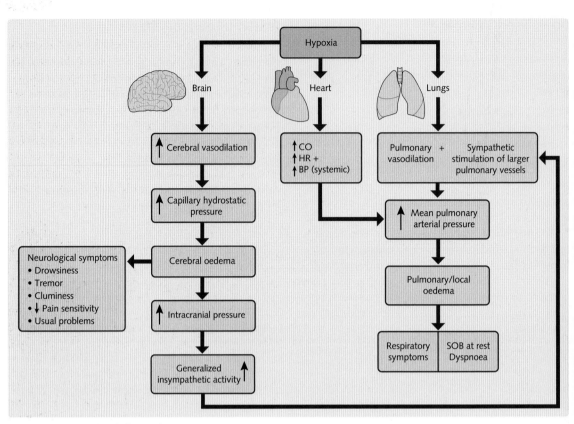

Fig. 6.30 Hypoxia-mediated effects and symptoms.

clumsiness, decreased sensitivity to pain, tremor and visual problems (Fig. 6.30).

Acute mountain sickness   This refers to a syndrome that develops up to 1 day after newcomers ascend to a moderate altitude, lasting for a few days. There are respiratory and neurological symptoms probably caused by cerebral oedema.

A few people may become extremely sick with acute cerebral and pulmonary oedemas, and can die if they are not transferred to lower altitudes or given 100% O_2.

Acclimatization

Many people live at high altitudes despite the hypoxia. The various compensatory mechanisms for low PO_2 can be separated into immediate, early and long-term responses (Fig. 6.31). These include:

- Hyperventilation.
- Increased production of red blood cells and therefore increased haemoglobin concentration.
- Oxygen dissociation curve shifts.

Hyperventilation   Hyperventilation results from hypoxic stimulation of peripheral chemoreceptors. It is the most important response to altitude, as can be deduced from the alveolar gas equation:

Fig. 6.31 Compensatory mechanisms to altitude responses

	Immediate	Early	Long term
Minute ventilation	↑	↑	↑
Arterial pH	↑	↑/→	↑/→
Renal HCO_3^- excretion	−	↑	↑
[Hb] or haematocrit	−	↑	↑
2,3 DPG synthesis	−	↑	↑
Cardiac output	↑	→	→/↓
Cerebral effects with CNS symptoms	↑	−	−

Fig. 6.31 Compensatory mechanisms to altitude responses.

$$P_AO_2 = P_IO_2 - \frac{P_ACO_2}{R}$$

where P_IO_2 = partial pressure of inspired O_2

R = respiratory exchange ratio

If R = 1, (CO_2 evolved = O_2 taken up) the alveolar gas equation can be reduced to an even simple form.

Minimizing P_ACO_2 by hyperventilation therefore increases P_AO_2. The consequences are a low $PaCO_2$ and a respiratory alkalosis, which would normally limit hyperventilation if it were not for early responses permitting its perpetuation:

- Renal compensation of respiratory alkalosis with excretion of HCO_3^- (24 hours).
- Exit of HCO_3^- from CSF, lowering its pH towards normal values.

Additionally, the ventilatory response to CO_2 becomes greater after a few days.

Polycythaemia Hypoxia stimulates renal production of erythropoietin with resultant increase in RBC count observed after a few days. After a few weeks, haemoglobin might be elevated from 15 to 20 g/dL and the haematocrit raised from 40%-60%. Although PaO_2 (dissolved) and haemoglobin saturation are relatively unchanged (low), the arterial content of O_2 is near normal. Mixed venous blood also tends to maintain PCO_2 approximately at normal values.

The main problem with polycythaemia is its increased viscosity with a tendency for thrombus formation and increased ventricular workload.

Oxyhaemoglobin dissociation curve At moderate altitudes, the increased concentration of 2,3-DPG, which arises primarily from respiratory alkalosis and the polycythaemia, shifts the oxyhaemoglobin dissociation curve to the right, aiding with O_2 unloading at the tissues. However, at very high altitudes, where PaO_2 is extremely low, the respiratory alkalosis shifts the curve to the right. This is important physiologically as it allows sufficient loading of O_2 in the pulmonary capillaries.

Other responses
- Increased cardiac output: stimulated by peripheral chemoreceptors and increased lung inflation.
- Increased capillarity: means the number of systemic circulatory capillaries per unit volume in peripheral tissues increases.
- Pulmonary vasoconstriction: occurs as a response to hypoxia with no apparent advantage except for abolishing zone 1 of the lung with more uniform blood flow.

Diving

With every 10 m (33 feet) descended below the surface of the sea, the pressure rises by 1 atm (760 mmHg). Although this tremendous increase in pressure has lesser effects on most of the body's tissues, which are incompressible due to their liquid and solid states, gases in the body are subject to this pressure, and are compressed. Reduction in volume is accompanied by increase in density. In addition, higher pressures increase the partial pressure of gases, according to Dalton's law, more of which will dissolve in the body's liquids (Henry's law).

For a gas, increasing depth causes:
- ↓Volume.
- ↑Density.
- ↑Partial pressure.
- ↑Amount dissolved in solution.

Effect of chest immersion

Immersion up to the neck results in a positive pressure outside the body of approximately 20 cmH$_2$O, with predominantly ventilation and pulmonary blood flow effects (Figs 6.32 and 6.33).

Snorkelling

This describes someone who is completely submerged at a shallow depth (of <1 m), whose airways are communicating with the air via a tube, which adds to the dead space. The maximal inspiratory pressure that can be generated is 80–100 cmH$_2$O, i.e. an intrapleural pressure of 80–100 cmH$_2$O. This is equivalent to the positive pressure of water at a depth

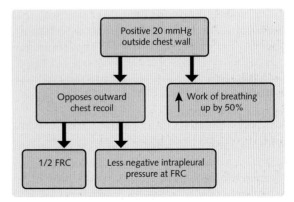

Fig. 6.32 Ventilatory effects of chest immersion.

Fig. 6.33 Effects of chest immersion on pulmonary blood flow.

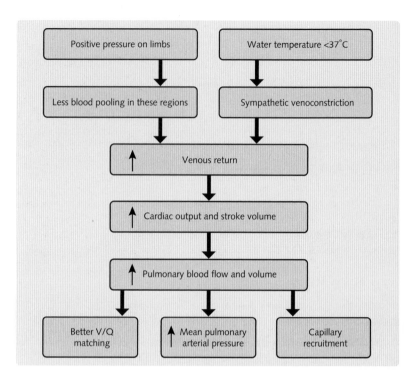

of approximately 1 m, below which breathing by this method would be impossible.

Single-breath dive

This involves a dive made following inhalation of air at the surface. The volume within the thorax wall will decrease and the partial pressure of gases will increase as the subject descends through the water. PCO_2 and PO_2 double at a depth of 10 m. Interestingly, CO_2 exchange at the blood–alveolar interface is reversed, so that CO_2 is retained, thereby significantly elevating its concentration. This has a powerful hypercapnic ventilatory drive through the central chemoreceptor, so that the diver realizes the need to return to the surface.

Divers often hyperventilate before diving to lower $PaCO_2$ and thus prolong their dive. However, this needs to be limited to four breaths, as any more than this can dangerously lower CO_2 such that the person could lose consciousness from hypoxia before the hypercapnic stimulus occurred to remind them of the need to surface.

Scuba diving and conventional underwater apparatus

Self-contained underwater breathing apparatus is self-explanatory. It involves the delivery of compressed gas in a tank on demand for inspiration, and release of expiratory gas into the water. The pressure of inhaled gas within the lung is ambient and so avoids problems relating to lung mechanics. Instead, stresses are consequent upon higher partial pressures and densities of gases that arise at very great depths (> 50 m):

- Higher gas densities increase the work of breathing needed to overcome elevated airways resistance.
- Higher PO_2 and density decrease the respiratory system's sensitivity to CO_2.

If the compressed gas contains a nitrogen component, nitrogen narcosis and decompression sickness are two other possible complications associated with deep-sea diving.

Nitrogen narcosis

At depths below 50 m, very high partial pressures of nitrogen directly affect the CNS. These can result in:

- Feelings of euphoria.
- Lack of coordination.
- Coma (eventually).

The mechanism is unknown but, like anaesthetic agents, nitrogen is highly fat soluble.

Decompression sickness

Nitrogen is normally poorly soluble, with very small amounts in solution present in the body (most of which is in fat). Diving at great depths elevates its partial pressure, driving it into solution in the body's tissues. During a slow ascent, the gradual decrease in ambient pressure results in slow removal of nitrogen from the tissues.

Decompression sickness occurs when there is rapid ascent from a great depth: a rapid decrease in ambient pressure causes nitrogen to come out of solution, forming bubbles in the body's fluids and tissues. These bubbles can enter the bloodstream, or affect the joints of the extremities, causing pain that is known as 'the bends'.

Bubbles are frequently trapped in the pulmonary circulation, occasionally causing symptoms known as 'the chokes', which include substernal chest pain and dyspnoea. More seriously, this might be accompanied by pulmonary hypertension and pulmonary oedema.

Arterial gas embolus is a dangerous consequence of decompression illness, whereby gas bubbles appear in the arterial circulation. These come from ruptured alveoli because the diver has not exhaled in time, thus blocking the escape of expanding gas. Persistence in the CNS circulation can cause paralysis or brain damage.

Treatment for decompression sickness is in two phases:

1. Recompression of the subject in a hyperbaric room, forcing nitrogen back into solution.
2. Careful and slow decompression: this can take a very long time.

Physiology of the gastrointestinal system

Objectives

In this chapter, you will learn to:
- Outline the basic functions of the various components of the gastrointestinal tract
- Describe how feeding is controlled via higher centres and feedback
- Describe the processes involved in digestion in the mouth and the production of saliva
- Outline the phases involved in swallowing and vomiting, and describe how these processes are controlled
- Describe the key components and functions of gastric secretions
- Explain how storage and motility occur within the stomach
- Outline the composition, production, secretion and storage of bile, relating the role of the gall bladder
- Describe the composition and control of pancreatic secretions
- Describe digestion and absorption within the small intestine
- Explain how the microflora are distributed within the gastrointestinal tract
- Outline the different roles of the colonic bacteria
- Explain how transport of electrolytes and water and mucus secretion occurs in the colon
- Describe the various types of movements of the large intestine and relate these to defecation and to the colocolic and gastrocolic reflexes

OVERVIEW OF THE GASTROINTESTINAL SYSTEM

The gastrointestinal tract (also known as the alimentary canal) is a mucosa-lined muscular tube extending from the mouth to the anus. It has an enormous surface area and five primary functions (Fig. 7.1):

1. Motility.
2. Secretion.
3. Digestion.
4. Absorption.
5. Elimination.

In addition, the gastrointestinal system has endocrine and immunological roles.

THE UPPER GASTROINTESTINAL TRACT

Food intake and its control
Definitions
- Hunger = craving for food. Hunger is associated with increased food-seeking behaviour, salivation and rhythmical stomach contractions (hunger cramps).
- Appetite = desire for certain foods that directs choice of food. Can be independent of hunger.
- Satiety = lack of desire to eat. The feeling usually follows ingestion of food and depends on individual energy stores.
- Anorexia = aversion to food ingestion despite sensations of hunger.

Central control
Feeding mechanisms are controlled by brainstem centres. Regulation of food intake occurs via:

- Higher centres: involving prefrontal cortex and amygdala. Implied importance of habit and conditioning to appetite.
- Hypothalamus: appetite is regulated by interaction between two areas within the hypothalamus:

1. Feeding centre (ventrolateral nucleus):
   - controls the emotional search for food by exciting motor drive
   - excites brainstem centres controlling upper gastrointestinal (GI) functions, including salivation, mastication and swallowing

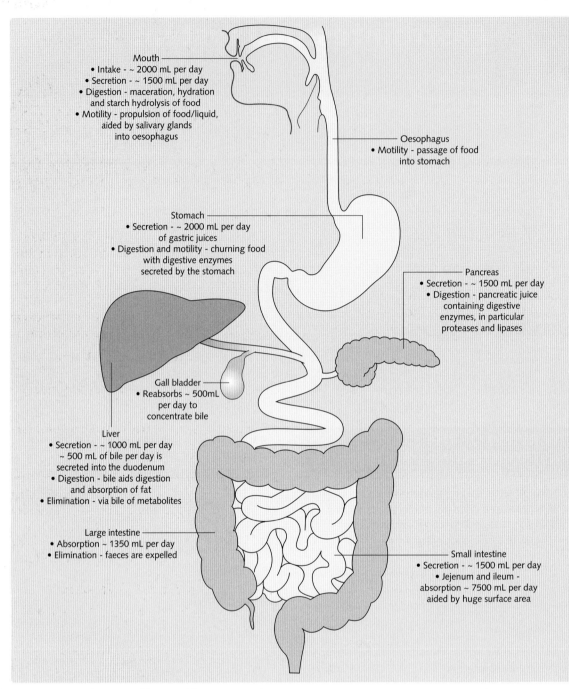

Fig. 7.1 Overview of structure and function of the gastrointestinal system.

- stimulation → hyperphagia (vigorous eating)
- lesions → satiety and weight loss.

2. Satiety centre (ventromedial nucleus):
   - inhibits the feeding centre
   - stimulation → complete satiety
   - lesions → excessive eating.

These centres have a rich supply of receptors, whose neurotransmitters mediate:

- Promotion of feeding: neuropeptide Y and γ-aminobutyric acid (GABA).
- Inhibition of feeding: serotonin and norepinephrine (noradrenaline).

Other hypothalamic regions may also contribute to food intake.

Feedback control

There are two main types of feedback control of appetite:

1. Short-term feedback:
   - involves mechanical feedback from the alimentary tract during feeding
   - limits the quantity of food eaten per meal.
2. Long-term feedback:
   - involves chemical feedback related to the body's nutritional status
   - maintains the body's energy stores.

Short-term or alimentary feedback

- Oral activity: including salivation, chewing, taste and swallowing. This inhibits the feeding centre for approximately 30 minutes.
- Stomach or duodenal distension (as detected by stretch receptors): vagal afferents inhibit the feeding centre.
- Chemical content of food: GI hormones and humoral factors stimulate the release of:
  - cholecystokinin: from fat ingestion
  - insulin and glucagons: from the pancreas.

All the above suppress feeding.

Long-term or nutritional feedback

- Temperature: interaction between the hypothalamus and temperature-regulating systems:
  - cold: stimulates the feeding centre → food ingestion → ↑ metabolic rate (heat) and ↑ fat deposition
  - heat: inhibits the feeding centre.
- Blood concentration of nutrients, including glucose, amino acids and lipids:
  - a low concentration of any of the above stimulates an increase in feeding
  - hypothalamic glucoreceptors monitor glucose; high concentrations stimulate the satiety centre whereas low concentrations stimulate the feeding centre
- The amount of white adipose tissue (stored energy): there is negative feedback via the secretion of leptin, high concentrations of which indicate adiposity. Stimulation of leptin-specific receptors in the hypothalamus results in:
  - increases in: substances inhibiting food intake and metabolic rate through increased sympathetic activity
  - decreases in: appetite stimulating neurotransmitters and pancreatic insulin secretion.

Hypothalamic feeding centres consist of the:
- Ventrolateral nucleus (feeding).
- Ventromedial nucleus (satiety).

Mastication

This involves the cutting and tearing of food by incisor teeth and grinding by molars, stimulating salivary flow, which then lubricates the food. Digestion is further aided by the removal of the indigestible fibrous coat that encases many plant materials.

Mastication involves two muscle groups, which are innervated by the trigeminal nerve motor branches:

1. Jaw-opening muscles: digastric and lateral pterygoid muscles.
2. Jaw-closing muscles: masseters, temporal and medial pterygoid muscles.

The tongue, gums and hard palate are also involved.

Chewing reflex

Brainstem reticular areas elicit and maintain chewing, with probable input from other structures, including the hypothalamus, amygdala and cortical sensory areas. The chewing reflex (Fig. 7.2) lasts about 1 second and consists of coordinated rhythmical jaw movements, the sequence of which is shown below.

Salivation and oral defences

Saliva is important for oral hygiene as well as for digestion. There are two secretions:

1. Serous: contains ptyalin (α-amylase) and enzymes that can digest starch.
2. Mucoid: contains mucin, which has lubrication and oral protective properties.

Composition and production of saliva

Saliva is composed of water, proteins (including enzymes) and electrolytes, whose exact concentrations depend on the production site and rate (Fig. 7.3); it is always hypotonic. The two stages in the production of saliva occur in the salivary compound glands, which are composed of acini and ducts:

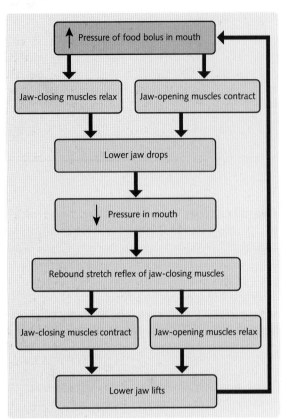

Fig. 7.2 Sequence of chewing reflex muscle movements.

- Stage 1: primary secretion in the acinus – secretion of a solution that resembles extracellular fluid (ECF) plus ptyalin (α-amylase) and/or mucin dependent on the gland – submaxillary and sublingual but not parotid.
- Stage 2: secondary ionic modification in the duct – active transport processes; Na^+ reabsorption is greater than K^+ secretion, generating a negative membrane potential of $-70\,mV$, which favours passive Cl^- reabsorption via Cl^-/HCO_3^- exchanger in the apical membrane.

Increased rate of production (Fig. 7.3) leads to lower $[K^+]$ (basal secretion concentration is 7 × plasma concentration) but higher $[Na^+]$ and $[Cl^-]$ than basal secretion; this may reflect limited capacity of the ductal transport processes. $[HCO_3^-]$ is increased with consequent salivary alkalinization (saliva is slightly acidic at basal secretions).

Nervous supply

Salivary glands receive both sympathetic and parasympathetic innervation, the parasympathetic nerves extending from the salivary nuclei in the upper medulla have more important effects. These nerves:

- Dilate blood vessels supplying the glands, increase the synthesis of ptyalin and mucins and

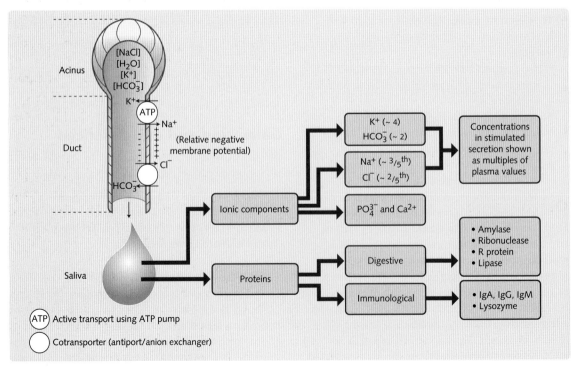

Fig. 7.3 Modification and components of saliva.

enhance epithelial transport processes in the duct, causing hypersalivation.

Receive cortical input: a greater response occurs with favourite foods.

Are involved in stomach and small intestine reflexes: nausea/irritant foods cause hypersalivation.

Respond to tactile stimuli in the mouth and taste: especially to acids (i.e. sour taste).

Sympathetic nerve supply from the superior cervical ganglia causes vasoconstriction and a transient increase in secretion.

Functions of saliva

Oral defence

Rinses away some bacteria and their energy source, i.e. food particles.

Saliva contains immunoglobulins and bactericidal substances, e.g. lysozyme and thiocyanate ions.

Calcium and phosphate promote the mineralization of the teeth.

Proteins provide a protective tooth covering known as an 'acquired pedicle'.

High $[HCO_3^-]$ neutralizes acid that can contribute to the formation of caries.

Saliva also cools hot food and/or liquid that would otherwise damage the upper GI tract.

Digestion

Lubrication:

- Decreases friction of the food bolus, aiding swallowing.
- Moistens the mouth for speech, comfort and breastfeeding.

Hydrolysis

- α-amylase hydrolyses α-1,4 glycoside bonds of polysaccharides.

Oral absorption

Sublingual administration of drugs is not only convenient but allows a smaller dose of drug to be given because it permits rapid absorption directly into the systemic circulation through the rich capillary network of the mouth. Furthermore, unlike oral medications, which pass into the gut, sublingual administration avoids metabolism of the drug by the liver before it has exerted its effect. A common example of a sublingual drug is glyceryl trinitrate, which is used for the relief of angina.

Swallowing

Movement of food from mouth to stomach is a complex event, which is initiated voluntarily but then proceeds through a reflex. As adults swallow about 600 times a day, this process must be closely coordinated with respiration. It consists of three phases:

1. Voluntary phase (closed mouth).
2. Pharyngeal phase.
3. Oesophageal phase.

Voluntary phase

Food is separated into a bolus by the tongue, which then propels it posteriorly and upwards against the hard palate towards the pharynx.

Pharyngeal phase

The food bolus exerts pressure near the opening of the pharynx, stimulating mechanoreceptors. These send afferents to the medullary swallowing centres via the trigeminal (CN V), glossopharyngeal (CN IX) and vagus (CN X) nerves, to elicit the swallowing reflex. The stages in this phase are as follows:

- Soft palate rises → nasopharynx closed off, preventing reflux through this area.
- Palatopharyngeal folds move inwards (medially) → a channel for the food bolus is opened into the posterior pharynx.
- Vocal cords are pulled together, causing the epiglottis to tilt over the larynx → the trachea is closed off to prevent respiration.
- Larynx pulled anteriorly and upwards → entrance to the oesophagus is stretched and enlarged.
- Upper oesophageal sphincter relaxes for 0.5–1 s → food enters the upper oesophagus.

Oesophageal phase

This phase lasts 6–10 s. Immediately after upper oesophageal sphincter relaxation with each swallow, pressure increases, thereby promoting the one-way passage of the bolus of food. Conduction through the oesophagus occurs by two types of peristaltic movement:

1. **Primary peristalsis**
   - Continuation of oropharyngeal swallowing.
   - Consists of a sequence of nerve activation causing a ring of contraction and high pressure:
     - speed: 3–5 cm/s
     - the lower oesophageal sphincter (LOS) is reached 6 seconds later

Fig. 7.4 The stages of swallowing.
(A) The bolus of food (F) is pushed into the pharynx by the tongue (T). (B) The bolus is propelled further back and the soft palate shuts off the nasopharynx. (C) The epiglottis (Ep) closes the opening to the trachea (Tr) and the bolus moves through the upper oesophageal sphincter. (D) Peristalsis now propels the bolus towards the lower oesophageal sphincter and stomach. O, oesophagus.

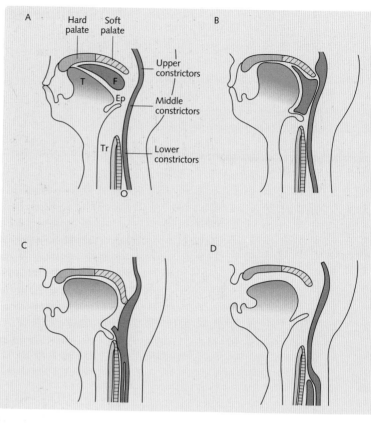

- movement is aided by gravity
- passage is faster with warm liquids than with cold ones.
- Glossopharyngeal and vagus nerves control peristalsis in the pharynx and upper third of oesophagus, which are made up of striated muscle.
- Vagus nerve controls smooth muscle of the lower two-thirds of the oesophagus.

2. Secondary peristalsis

- Occurs in response to oesophageal distension due to:
  - incomplete emptying into the stomach
  - reflux of stomach contents back into the oesophagus.
- Involves intrinsic myenteric circuits and pharyngeal reflexes.
- Peristalsis until emptying into the stomach is complete.
- Has no appreciable perception.

Lower oesophageal sphincter

The LOS consists of a circular ring 3 cm above ʾe gastro-oesophageal junction. It is normally stricted, thereby protecting the oesophagus from the acidic and proteolytic gastric secretions. Th LOS relaxes for 5–10 s when it meets the peristalti wave, permitting passage of the food bolus. Durin, drinking, there is rapid swallowing with a relaxe LOS and no peristalsis until the last swallow.

Vomiting

Vomiting is the forceful expulsion of gastric content (and sometimes duodenal) through the mouth It occurs when the vomiting centres, which are present bilaterally in the lateral medulla, are stimulated by distension, irritation or excitation of any part of the upper GI tract (Fig. 7.5). Impulses are sent here via:

- Higher centres (including cerebral and limbic regions): in response to:
  - psychic stimulation (smells, emotions and disturbing sights)
  - raised intracranial pressure (ICP).
- Chemoreceptor trigger zones (CTZs): these respond to upper GI tract lesions, drugs (e.g. opioids) and metabolic conditions (e.g. uraemia). CTZs are located bilaterally on the fourth ventricle floor near the area postrema.

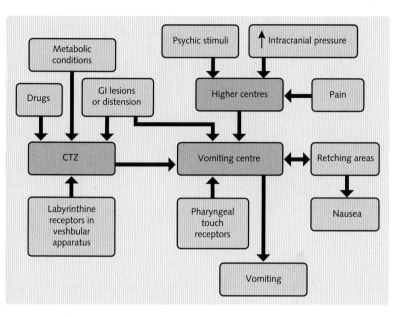

Fig. 7.5 Control of vomiting.

In addition, they respond to prolonged vestibular nuclei stimulation resulting from rapid changing of rhythm or direction of motion.

When there is a gastric lesion, retrograde peristalsis from the small bowel occurs many minutes before vomiting. Consequently, food is unloaded into the stomach and duodenum, which distend and stimulate the vomiting centre.

Stages of vomiting

Vomiting usually includes a prodrome of nausea – an accompaniment of autonomic symptoms that include sweating, hypersalivation and pallor. A sequence of events then follows:

- Deep inspiration.
- Closure of the glottis: this protects the airways and fixes the chest by holding the diaphragm down.
- Raising of larynx and hyoid bone: this extends the opening of the oesophagus (UOS).
- Elevation of soft palate: this seals off the nasopharynx.
- Strong contractions of abdominal skeletal muscles: → ↑intra-abdominal pressure → diaphragm forced up into the thorax → ↑intrathoracic pressure.
- Relaxation of the LOS: this permits the stomach contents to enter the oesophagus.

The oesophagus can empty food back into the stomach, with recycling of the whole process if the UOS remains closed. However, when the UOS opens because of expulsive forces, vomit enters the mouth.

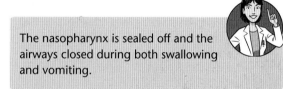

The nasopharynx is sealed off and the airways closed during both swallowing and vomiting.

THE STOMACH

The anatomical and functional regions of the stomach are shown in Figure 7.6.

Storage of food: the fundus

Storage of food is an important motor function that permits controlled release of gastric contents into the duodenum. The fundus, in particular, functions as a reservoir and contributes greatly to the 1–1.5-L capacity of a fully relaxed stomach. This area is particularly suited to its distensible function as:

- The muscular wall is thinner than in other parts of the stomach.
- Intragastric pressure is roughly maintained over a wide range of gastric volumes.
- An additional oblique layer of stomach muscle facilitates expansion.

Storage is facilitated by receptive relaxation (vagovagal reflex) by the stomach. Food entering

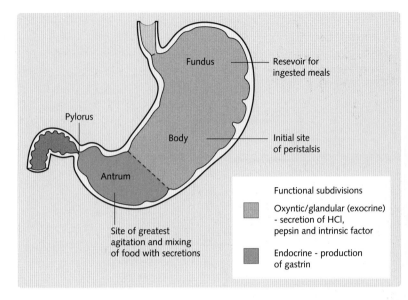

Fig. 7.6 The anatomical and functional regions of the stomach.

Fundus — Reservoir for ingested meals

Pylorus

Body — Initial site of peristalsis

Antrum

Site of greatest agitation and mixing of food with secretions

Functional subdivisions

Oxyntic/glandular (exocrine) - secretion of HCl, pepsin and intrinsic factor

Endocrine - production of gastrin

through the LOS results in decreased muscle tone of the fundus and upper stomach walls, which bulge outwards. Food can stay up to 1 hour in the fundus.

Gastric secretions and their control

There are two main components of the 2 L of stomach secretions produced every day:

1. Non-acid secretion: alkaline material, which is released by mucous cells, that coats the entire mucosal surface.
2. Acid secretion: gastric juice released by parietal and chief cells in the oxyntic region.

Non-acid secretion

The alkaline secretion, which comprises mainly mucus and bicarbonate (HCO_3^-), adheres to the stomach lumen, forming a protective fluid barrier against the acidic gastric juice, which would otherwise cause autodigestion. The viscous secretion is isosmolar to plasma ultrafiltrate but, owing to its high HCO_3^- content, has an alkaline pH (7.7). Acid-resistant trefoil proteins in the GI tract mucosa offer further resistance to tissue damage.

Gastric juice

This is composed of mucus, hydrochloric acid (HCl), intrinsic factor, pepsin and other enzymes. The pH of this fluid is low, with increase in flow rate associated with increase in acidity (reaching pH 1.0).

Mucus

Mucus is secreted principally in the antropyloric region and by surface mucous cells located throughout the stomach. The roles of mucus are:

- Lubrication of food.
- Protection against autodigestion.

Mucus is a gel, which permits HCO_3^- to be trapped within it. Therefore, an alkaline barrier to H^+ diffusion back to the epithelium is created. Hence, a pH gradient is established between the lumen (pH 1.5) and the epithelial surface (pH 6.5).

Prostaglandins and local irritation stimulate mucus production.

Pepsin only modestly degrades mucus, as a low pH is required for optimal pepsin activity.

Hydrochloric acid

Parietal cells actively secrete HCl (Fig. 7.7), the functions of which include:

- Defence against ingested microorganisms.
- Activation of pepsinogen.
- Stimulation of bile and pancreatic juice flow.

Parietal cells are suited to this function as they contain:

- Abundant carbonic anhydrase: this catalyses H^+ and HCO_3^- formation from diffused plasma CO_2. HCO_3^- is then exchanged with Cl^- from interstitial fluid and effluxed into the blood. Hence, parietal cell HCl secretion causes gastric venous blood to have a lower $[CO_2]$ than its arterial counterpart (alkaline tide).

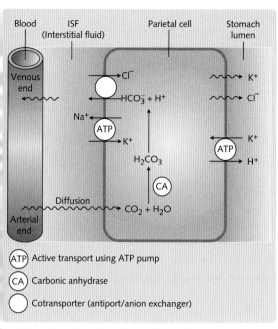

ATP Active transport using ATP pump

CA Carbonic anhydrase

⚪ Cotransporter (antiport/anion exchanger)

Fig. 7.7 Parietal cell secretion of hydrochloric acid.

- Large quantities of H^+/K^+ ATPase molecules: these permit H^+ extrusion against a huge concentration gradient: gastric juice $[H^+]$ can be 1.5×10^{-1} M, whereas plasma $[H^+]$ is 10^{-8} M. These pumps are associated with tubo-vesicular structures that move and fuse with the apical membrane when parietal cells are stimulated.

The rate of formation of HCl is directly related to histamine production by enterochromaffin cells, which in turn are stimulated by the following:

- Gastrin: an antral hormone, the production of which increases with protein consumption.
- Acetylcholine: from vagal nerve endings, which act via M_3 muscarinic receptors.
- Acetylcholine (Ach) and gastrin also have direct effects on parietal cells acting via, respectively, M_3 receptors and gastrin receptors on the surface membrane.
- Signal-transduction pathways for Ach and gastrin involve activation of protein kinase C and release of Ca^{2+} and for histamine (H_2 receptors) adenylyl cyclase, production of cAMP and activation of protein kinase A.

Prostaglandins inhibit HCl secretion.

Intrinsic factor

Parietal cells produce intrinsic factor (IF), a glycoprotein that is essential for vitamin B_{12} (also known as cobalamin) absorption in the terminal ileum. Without IF, pernicious anaemia would develop once liver stores of B_{12} were depleted (after 2–3 years).

Pepsin

Chief cells secrete pepsinogen, the inactive precursor of pepsin. Under acidic conditions (pH < 2), pepsinogen is rapidly converted to pepsin.

Pepsin is a proteolytic enzyme that degrades proteins to peptides. Stimulation by the vagus nerve (containing acetylcholine) or gastric enteric plexus promotes the secretion of pepsinogen. The gastric enteric reflexes involved may in turn be influenced by the amount of HCl present.

Other enzymes

Small quantities of the following are secreted in gastric juice:

- Gastric lipase: principally acts on butter fats.
- Gastric amylase: plays a small role in starch digestion.
- Gelatinize: has a role in liquefying some meat proteoglycans.

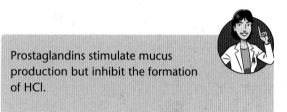

Prostaglandins stimulate mucus production but inhibit the formation of HCl.

Control of gastric secretion

Gastric secretion occurs both continuously at rest (basal secretion) and as a response to food. Stimulation associated with feeding involves neural, hormonal and paracrine mechanisms, and is divided into three phases:

1. Cephalic phase: stimuli involve the brain.
2. Gastric phase: stimulus is food or fluid within the stomach.
3. Intestinal phase: stimulus is food in the small intestine.

Cephalic phase

- Responsible for 20% of the gastric secretion associated with feeding.
- Reflex gastric stimulation occurs with anticipation of eating (e.g. sight, smell and taste).
- Continues up to 30 minutes after ingestion.
- Vagus nerve mediates gastric secretion in preparation for the meal involving CNS centres.

- Anger and hostility increase gastric secretion.
- Fear and depression decrease gastric secretion.

Gastric phase
- Responsible for 70% of gastric secretion associated with feeding.
- Food entering the stomach accelerates gastric secretion in response to:
  - chemical stimuli: luminal amino acids and nutrients stimulate gastrin release, augmenting acid flow
  - distension: mediated via long vagovagal and local short intramural reflexes.
- pH < 3 inhibits gastrin release in response to distension or chemical stimuli via somatostatin release from antral mucosa D cells and secretin from the duodenal mucosa S cells both of which inhibit HCl secretion through inhibition of gastrin release from the antral mucosa and, for somatostatin, by reducing histamine release.

Intestinal phase
- Responsible for < 10% of total gastric acid secretion.
- Response to food (chyme) in the duodenum.
- The release of some duodenal gastrin release can be stimulated by local distension or chemical stimuli.

Fat, acid and hyperosmolar solutes within the duodenum are often inhibitory and mediated by the release of two inhibitory factors: gastric inhibitory peptide (GIP) and cholecystokinin (CCK). Both suppress the secretion of HCl by parietal cells.

Gastric motility and emptying

The motor functions of the stomach are threefold:

1. Storage of food (see above).
2. Mixing of gastric secretion with food to form a mixture known as chyme.
3. Controlled release of chyme into the small intestine.

Gastric motility

The proximal stomach stores food and presses it towards the distal region via low-amplitude, slow peristaltic waves, which occur at a rate of 3–4/min. These contractions are initiated and regulated by pacemaker cells in the middle of the stomach body. Initially, the waves travel at 1 cm/s but gather speed and amplitude towards the antropyloric region. Rostral to caudal propagation is achieved by the staggering of these waves according to the level of the stomach, so that distal segments have a lag phase (Fig. 7.8).

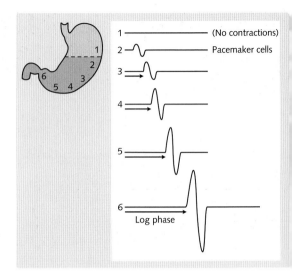

Fig. 7.8 Peristaltic waves through the stomach.

A very high pressure in the antrum is facilitated by a thicker muscle layer than the rest of the stomach, enabling small boluses of food to be ground. During antral peristalsis, pyloric contraction causes reflux of the antral contents to more proximal regions of the stomach. This is known as retropulsion, and has the following consequences:

- Breakdown of solid foods.
- Churning of food with secretions to form the semi-solid chyme.

Gastric emptying
- There is coordinated sequential contraction of the antrum, pylorus then duodenum.
- The partially contracted pylorus both regulates the amount of gastric emptying and prevents the reflux of potentially damaging bile into the stomach.
- Only small amounts of chyme pass through the pylorus into the duodenum, which has a limited capacity.

The rate of gastric emptying depends on factors (Fig. 7.9) relating to:

- Stomach contents.
- Duodenal chyme: generally inhibits emptying, known as the enterogastric reflex.

Protection of the gastric mucosa

In addition to stimulating the secretion of HCO_3^- and mucus, the inhibitory actions of prostaglandins and other factors on acid secretion serve to maintain the gastric mucosa. If the protective barrier is breached,

Fig. 7.9 Factors affecting stomach emptying

	Factor	
	Stomach content	*Duodenal chyme*
Energy content	Carbohydrates empty quickest Proteins empty slower Fats empty slowest	Fats
Bulk	Solid and coarse foods	Distension (duodenum)
Osmolality	Isosmolar ↑ in variation from isosmolar values further ↓ emptying	High osmolality
Temperature	Body temperature Cold or hot substances	
pH		< 3.5

H^+ penetrates the mucosa, causing local inflammation with epithelial cell necrosis and haemorrhage (ulcer). This can occur with the following:

- Non-steroidal anti-inflammatory drugs (NSAIDs): inhibit prostaglandin production, reducing HCO_3^- secretion and mucus.
- *Helicobacter pylori* infection: disrupts the mucosal barrier, causing acute inflammation.
- Zollinger–Ellison syndrome: characterized by multiple gastrinomas with acid hypersecretion.
- Lifestyle:
  - alcohol: increases barrier permeability by interfering with mucosal enzymatic processes
  - smoking: chronic nicotine exposure
  - elevated stress.

THE LIVER AND BILIARY TRACT

Overview of liver metabolism

Liver metabolism (Fig. 7.10) is primarily concerned with:

- Metabolic functions:
  - nutrient metabolism
  - inactivation of drugs, other toxins and steroid hormones.
- Secretion and excretion: involving bile.
- Drug and hormone metabolism.

Liver biotransformation and clearance occur in three stages:

Phase 1 (oxidation):
- Increases polarity and water solubility.
- Enables excretion via urine/bile.

- Involves oxidation and reduction by enzymes (e.g. cytochrome P450).

Phase 2 (conjugation):
- Decreases toxicity.
- Involves conjugation with various groups (e.g. glucuronide).

Phase 3 (elimination):
- Active transport via ATPase pumps into the circulation.

Bile production and function

The liver produces and secretes up to ~ 1000mL of bile daily, which flows equally into either the:

- Small intestine through the ampulla of Vater.
- Gall bladder: stores and concentrates bile until it is stimulated to empty into the intestine by a fatty meal.

Around 90% of the bile in the small intestine is recycled back to the liver via the mesenteric venous drainage through the portal vein – the enterohepatic circulation. The pathway of bile is shown in Figure 7.11.

Bile is an aqueous solution composed of bile salts, inorganic salts (electrolytes), bile pigments, phospholipids and cholesterol. It forms a medium for biotransformed substances and waste products, such as cholesterol and bile pigments (bilirubin and bilverdin), enabling their elimination. In addition, bile facilitates fat handling by:

- Emulsifying fat: helping with digestion.
- Promoting absorption from the small intestine.

Bile salts

These refer to potassium and sodium salts of bile acids, notably cholic and chenodeoxycholic

Fig. 7.10 Overview of nutrient metabolism by the liver

Nutrient	Metabolic process
Carbohydrate	Utilization of major monosaccharides to synthesize other substances: fatty acids, amino acids, glycogen Glycogen breakdown and storage: < 65 g/kg in liver tissue Gluconeogenesis (glucose production)
Lipid	Fatty acid synthesis and storage as triglycerides Ketogenesis from fatty acid oxidation and lipolysis (fat breakdown) Cholesterol synthesis from acetyl CoA involving the enzyme HMGCoA reductase Conversion of cholesterol to bile salts Secretion and metabolism of lipoproteins Metabolism and excretion of steroid hormones
Protein	Synthesis of plasma proteins: albumin and globulins, acute-phase proteins and complement Synthesis of clotting factors (V, VII, IX, X, prothrombin and fibrinogen) Generation of urea from ammonia Amino acid deamination
Vitamins	Storage of fat-soluble vitamins A, D, E and K (depending on bile production) Storage of a 2–3-year supply of vitamin B_{12} Storage of a few months' supply of folate Modification of most vitamins to active coenzyme forms

Fig. 7.11 Pathway of bile in and outside the liver.

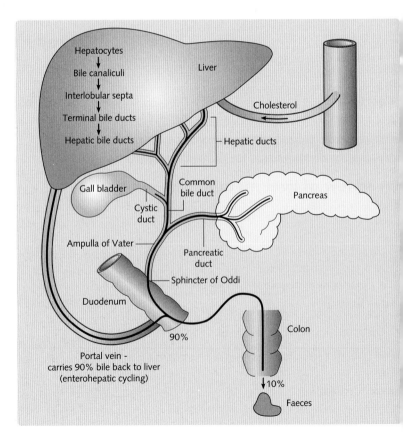

acid, which are synthesized from cholesterol and then conjugated before their secretion into bile. Cholesterol 7-α–hydroxylase is the rate limiting factor in the production of bile salts (Fig. 7.12).

Over a period of 24 hours, 0.2–0.4 g of bile salts are synthesized and recycled 6–8 times via the enterohepatic circulation. Hepatic uptake is by secondary active process involving Na$^+$/bile salt cotransport powered by basolateral Na$^+$/K$^+$-ATPase. Interruption of bile salt circulation – either by inadequate production or exclusion from the small intestine – results in severe fat malabsorption inclusive of fat-soluble vitamins.

Properties of bile salts

Amphipathic properties (hydrophobic and a hydrophilic end) enable bile salts to perform two special actions:

1. Detergent function: surface tension is reduced, permitting lipid emulsification with the aid of lecithin (a phospholipid).
2. Micelle formation: micelles are cylindrical aggregates with a hydrophilic shell and a hydrophobic core. Bile salts can form mixed micelles by incorporating phospholipids, to transport lipids, which are normally very water insoluble. Cholesterol, fatty acids and monoglycerides can thus be maintained in solution, and their absorption greatly assisted.

In addition, bile salts stimulate the secretion of water and electrolytes via the:

- Stimulation of bile: when secreted by the liver.
- Production of diarrhoea: if present in the colon.

Bile pigments

The primary bile pigment is bilirubin. It is derived from the breakdown of haem, which is formed principally from haemoglobin (85%), as well as others (e.g. myoglobin). Bilirubin is bound to albumin unconjugated and is therefore relatively insoluble.

Most bilirubin dissociates into its free form in the liver. It is then taken up by the liver cells (hepatocytes) in exchange for Cl$^-$ via an anion transporter. In the hepatocytes, cytoplasmic proteins (ligandins) bind to bilirubin, which is then conjugated with glucuronate via glucuronyl transferase (UDP glucuronosyltransferase) → bilirubin diglucuronide (conjugated bilirubin).

Conjugated bilirubin has a higher water solubility than free bilirubin and is actively transported into the bile canaliculi. A small amount escapes into the blood, where it is lightly bound to albumin and excreted in the urine. Most of the conjugated bilirubin is excreted, in the bile, into the small intestine, the wall of which is permeable to free bilirubin and urobilinogen but impermeable to conjugated bilirubin. Here intestinal bacteria convert conjugated bilirubin into urobilinogen and a little free bilirubin, some of which passes through the bowel wall before circulating back to the liver (via the enterohepatic circulation). However, most conjugated bilirubin remains in the bowel.

Urobilinogen in the gut is partially reoxidized to stercobilinogen, which is then lost via faeces.

Free/unconjugated bilirubin is relatively insoluble and is not generally excreted. Conjugated bilirubin is more soluble and is easily converted to excretable forms, which are lost in urine and faeces.

Fig. 7.12 Synthetic pathway of bile salts.

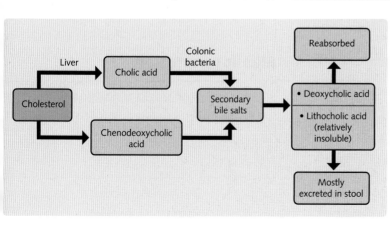

Hepatic bile production

The liver secretes bile by various mechanisms:

- Bile-acid dependent.
- Bile-acid independent.
- Intrahepatic bile-duct dependent.

Bile-acid dependent mechanisms

- Carrier-mediated active bile salt transport into the canaliculi (minor ducts).
- Passive water and electrolyte flow in response to above.

Bile-acid independent mechanisms

- Active transport of electrolytes and passive water flow.
- Insignificant in man.

Intrahepatic bile-duct dependent

- Epithelial cells lining bile ducts alkalinize the bile with HCO_3^- via a Cl^-/HCO_3^- exchanger.
- Greatly increases total volume of bile.
- Important contributor to bile flow in man.
- Strongly stimulated by secretin.

Control of bile synthesis and secretion

Both bile acids concentration and hormones affect the secretion and flow of bile.

Concentration of bile acids in the portal vein

- ↓ to low concentrations: stimulates bile acid secretion; inhibits bile acid synthesis.
- ↑ to high concentrations: inhibits bile acid secretion; stimulates bile acid synthesis.

Hormones

- Secretin: stimulates bile duct-dependent secretion of bile.
- Cholecystokinin:
  - ↑ bile acid secretion during digestion (intestinal phase)
  - ↑ flow of bile into the duodenum.

Role of the gall bladder

Between digestive periods, there is increased resistance to bile flow due to constriction of the sphincter of Oddi. Meanwhile, bile is preferentially diverted towards the cystic duct and relaxed gall bladder. Storage within the gall bladder serves to concentrate bile to up to 20 fold that of hepatic bile:

- The epithelium actively transports Na^+ out of bile.
- Secondary absorption of Cl^-, water and other solutes, including HCO_3^-, follows.
- Bile pH falls consequent to reabsorbed HCO_3^-.

As there is a large volume reduction of bile, the gall bladder can store 12 hours' worth of bile secretions despite its capacity of only 50–60 mL.

Gall bladder emptying

Contraction of the gall bladder occurs:

- Intermittently: interdigestive contraction results in intermittent release of bile into the duodenum.
- In response to a meal:
  - during the cephalic phase of digestion: resistance by the sphincter of Oddi decreases; the gall bladder begins to empty with the early digestion of food
  - in the stomach: antral distension and gastrin secretion stimulate gall bladder contraction via nerve reflexes
  - fatty acids (and amino acids) in the duodenum: stimulates the release of cholecystokinin, which acts to greatly increase gall bladder emptying.

Gall bladder relaxation results from the following:

- The presence of bile acids.
- Vasoactive intestinal peptide (VIP).
- Pancreatic polypeptide.
- Sympathetic stimulation of the gall bladder.

THE PANCREAS

Pancreatic secretions

Exocrine pancreatic secretion has two components:

1. Aqueous alkaline fluid: secreted by duct cells.
2. Organic, e.g. inactive digestive enzymes: secreted by acinar cells.

When it enters the duodenum, the pancreatic fluid regulates the pH and helps with the digestion of fats, proteins and nucleic acids.

Alkaline secretion

Alkaline fluid is isosmolar at all rates of secretion. Its main component is aqueous bicarbonate (HCO_3^-); smaller quantities of other solutes are also present. HCO_3^- secretion by epithelial cells occurs via intracellular hydration of CO_2 (Fig. 7.13). High efflux of Cl^- into the duct lumen is crucial to HCO_3^- secretion; basolateral Na^+/K^+-ATPase provide a favourable electrochemical gradient for Cl^- secretion and osmosis of water intraluminally:

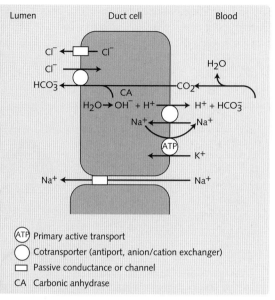

Fig. 7.13 Secretion of bicarbonate by the pancreatic cells.

- CO_2 from the blood diffuses into the cell.
- Intracellular carbonic anhydrase (CA) serves to:
  - catalyse $CO_2 \rightarrow H^+ + HCO_3^-$
  - maintain a chemical concentration gradient for CO_2 between blood and cell.
- Intracellular HCO_3^- is exchanged for luminal Cl^-, facilitated by cystic fibrosis transmembrane regulator (CFTR) Cl^- channels.
- Intracellular H^+ passes into the blood via Na^+/H^+ exchange where it reacts with blood HCO_3^- to form CO_2.

Concentrations of bicarbonate can rise to five times plasma value with high rates of pancreatic fluid secretion.

Enzyme secretion

Acinar cells secrete the following groups of enzymes:

- Peptidases:
  - digest proteins to form peptides
  - form the most abundant of enzyme secretion
  - are secreted in precursor form to prevent pancreatic autodigestion
  - examples: trypsin, chymotrypsin and carboxypeptidase.
- Amylases:
  - hydrolyse carbohydrates to mainly disaccharides
  - main example: α-amylase.

- Lipases:
  - digest fat to release monoglycerides and fatty acids
  - examples: lipase, phospholipase and cholesterol esterase.
- Nucleases:
  - Examples: ribonuclease and deoxyribonuclease which act on RNA and DNA, respectively.

Peptidase precursors are maintained in their inactive form when stored within zymogen granules in acinar cells, until they are secreted into the duodenum. This is achieved by:

- Secretion of a specific inhibitor to trypsin: kazal inhibitor. Trypsin would otherwise initiate the chain reaction of autodigestion by activating the other peptidases.
- An acidic zymogen pH that is beyond the optimal range for proteolytic enzymes.

When it enters the duodenum, trypsinogen is converted to trypsin by brush-border enterokinase, thus activating the other peptidases (Fig. 7.14).

Control of pancreatic secretion

Pancreatic secretion is subject to both neural and hormonal effects:

Hormonal control

The duodenal mucosa secretes the following two hormones in response to the presence of local nutrients:

- Secretin: the most powerful stimulant of ductal cell HCO_3^- secretion, acts via 2nd messenger

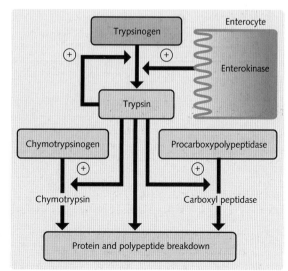

Fig. 7.14 Peptidase activation within the duodenum.

cAMP, protein kinase A and phosphorylation of CFTRCl⁻ channels.
- Ach: acting via 2nd messenger intracellular Ca^{2+}, also increases ductal HCO_3^- secretion.
- CCK (and Ach): stimulate acinar cell release of pancreatic enzymes via activation of intracellular Ca^{2+} and PKC.

Somatostatin potently inhibits all pancreatic secretions, via direct and indirect mechanisms involving the above hormones.

Neural control

- Sympathetic stimulation: globally inhibits secretion, mainly through gland arteriolar vasoconstriction.
- Parasympathetic stimulation: via vagal acetylcholine, increases both enzyme and alkaline fluid secretion.

The phases of secretion occurring in response to a meal can be divided into:

- Cephalic.
- Gastric.
- Intestinal.

Cephalic secretion

- Stimulation occurs with anticipation of feeding.
- Accounts for 15% of total pancreatic secretion.
- Associated with increases in alkaline fluid (10%) and enzymes (25%).
- Mediated by parasympathetic release of acetylcholine.
- Inhibited by sympathetic activity.

Gastric secretion

- Stimulation occurs with either antral distension or food entering the stomach.
- Associated with increases in predominantly enzyme secretion (5–10%).
- Mediated by vagovagal stomach reflexes and gastrin release.

Intestinal secretion

- Stimulation begins with the entrance of gastric chyme into the duodenum.
- Accounts for 80% total pancreatic secretion.
- Mediated by:
  - secretin:
    - responsible for large amounts of alkaline fluid secretion
    - present in duodenal and jejunal mucosal S cells

    - released in response to acidic chyme.
- CCK:
  - responsible for copious enzyme secretion
  - present in duodenal and jejunal mucosal I cells
  - released in response especially to peptones (protein breakdown products) and fatty acids.

Active forms of pancreatic peptidases – chiefly trypsin – are capable of autodigestion and are therefore secreted in their inactive forms.

THE SMALL INTESTINE

Overview of digestion

Food enters the GI tract as complex chemicals: fats, proteins and carbohydrates. Digestion represents the breakdown of these substances into simpler subunits, involving hydrolysis.

Digestion begins in the mouth – by physical (mastication) and enzymatic actions (secreted salivary amylase). Food is then acted on in an orderly fashion as it travels along the alimentary tract. The small intestine is the principal site for digestion, which occurs by enzymatic activity, mainly in the duodenum:

- Intraluminally: by secreted enzymes.
- At the cell (enterocyte) surface: by brush-border enzymes.
- Within the cytoplasm of intestinal mucosal cells (enterocytes).

The products of digestion are then transported across the wall of the small intestine by various processes. The jejunum and ileum have mainly absorptive functions, facilitated by their huge surface areas.

Intestinal circulation

The intestines receive 10% of cardiac output via the splanchnic circulation, most of which supplies the mucosa (Fig. 7.15):

- Coeliac trunk supplies the proximal duodenal half.
- Superior mesenteric artery (extensive branching network) supplies the remainder of the small intestines.

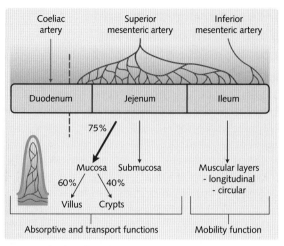

Fig. 7.15 Blood supply and distribution of the small intestines.

- Inferior mesenteric artery contributes a small amount of blood to the terminal ileum.

Variation in intestinal circulation

The autonomic nervous system is one of the factors affecting the blood supply:

- Decreases in splanchnic circulation can result from: sympathetic nervous stimulation → shunting of blood away from the intestines, e.g. during exercise or shock. Mediators include catecholamines, angiotensin II and antidiuretic hormone.
- Increases in splanchnic circulation can result from: response to a meal or functional hyperaemia, involving:
  - peptide hormones: CCK, VIP, gastrin and secretin
  - kinins: released by GI glands into the gut wall.
  - digestive products within chyme: glucose and long-chain fatty acids cause localized intestinal hyperaemia. Bile salts cause direct vasodilation in the terminal ileum
  - decreased intestinal wall O_2 tension: secondary to increased gut metabolic activity, with subsequent rise in adenosine concentrations.

Local activity dictates the distribution of blood flow. The mucosal blood supply is greater with increased absorption or secretion. The blood supply to the muscular layers is greater during increased motility.

There are two main functions of the intestinal circulation:

1. It provides a readily available blood reservoir that can be mobilized on demand.
2. It provides a countercurrent system within the villus.

Circulation within the villus

The villus represents the absorptive unit within which arterial and venous blood lie in close proximity but flow in opposite directions (Fig. 7.16). This anatomical arrangement facilitates diffusion between arterioles and venules. O_2 diffuses directly between the two, with a significant amount bypassing the capillaries supplying the villus. PaO_2 is higher where the arteriole enters the base of the villus than the tip of the ascending limb, with no harmful effects under normal conditions. Ischaemia, leading to necrosis of the villus tip (extending to the whole villus in cases), can interfere with the absorptive capacity of the intestine.

As with the countercurrent system in the kidneys, the tip of the hairpin loop (villus tip) is hyperosmolar. There is net diffusion of substances between arteriole and venule such that nutrient-rich venous blood leaves the gut via the portal vein to the liver.

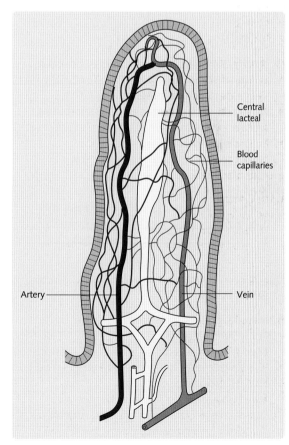

Fig. 7.16 The countercurrent mechanism in the villi of the small intestine. This enables absorption of electrolytes and water.

Although the villus receives a significant portion of intestinal blood supply, it is prone to necrosis, particularly of the tip, since oxygen diffusion largely bypasses the supplying capillary.

Factors that control absorption

Absorptive mechanisms involve:

- Diffusion: ~80% of 9 L of water entering the small intestine per day is absorbed by diffusion.
- Solvent drag: the flow of water (solvent) results in secondary 'dragging' of solute with it.
- Active transport: large amounts of nutrients are absorbed via membrane carrier proteins.

Absorption and rate of transport are facilitated by:

- Increased surface area: villi greatly multiply the surface area of the gut available for transport.
- Vascular supply: the countercurrent system permits the rapid uptake of fluid, electrolytes and nutrients. Rates of blood flow are adequate so as to not limit absorption.
- High cell turnover: increased food intake itself stimulates mucosal growth, which in turn improves substance absorption.
- Carrier protein density: nutrient absorption relies on carrier molecules and plentiful brush border enzymes.
- Permeability gradient: mucosal permeability decreases along the small bowel; absorption is high in the duodenum and low in the ileum.

Handling of water and electrolytes is affected in particular by:

- Osmotic gradient: this depends on the intracellular osmolality.
- Tight junction selectivity: monovalent cations (Na^+) pass more readily than anions (Cl^-) or divalent cations (Mg^{2+}).
- Mineralocorticoids: aldosterone increases intestinal Na^+ absorption and K^+ secretion.
- Autonomic control: absorption is increased and decreased by sympathetic and parasympathetic stimulation, respectively.

Intestinal digestion and absorption of different components of the diet
Fats

The arrival of lipid (triglycerides) in the duodenum serves as a stimulus for the secretion of bile. Bile contains phospholipids and bile salts, which form micelles. A minimum bile acid concentration is required for micelle formation (critical micelle formation occurs at 5 mmol/L).

Micelles function as detergents by breaking the fat globules into tiny emulsified droplets, thus presenting a greater surface for hydrolysis by pancreatic lipase, cholesterol esterase and phospholipase (Fig. 7.17). The breakdown products – fatty acids and 2-monoglycerides – are incorporated into mixed micelles, which have longer fatty acids at their centre.

Micelles transport lipids through an unstirred layer of fluid at the enterocyte surface. Lipid then passively diffuses across the intestinal cell wall whereupon it is taken up by smooth endoplasmic reticulum. Here rapid re-esterification mostly occurs, reforming triglycerides. Triglycerides, phospholipids and cholesterol are then combined with lipoprotein (in particular apolipoprotein B) to form chylomicrons, which consist of:

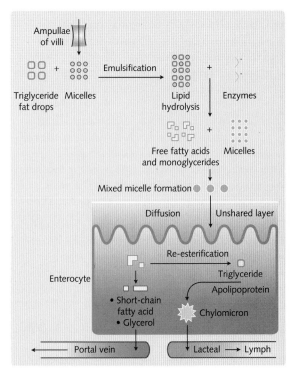

Fig. 7.17 Overview of intestinal fat digestion and absorption.

- Outer coat: composed of phospholipids and apolipoprotein.
- Inner core: composed of fatty acids, cholesterol and cholesterol ester.

Chylomicrons exocytose from the epithelial basolateral surface and enter the villus lacteal to be transported to venous blood via the lymph system. A minority of lipid digestion products, including short-chain fatty acids and glycerol, pass into the portal system to the liver.

The digestion of the fat-soluble vitamins A (retinol), D, E (via the lymph system) and K follow the same principles.

- Micelles have a hydrophilic outer shell and a hydrophobic core.
- Chylomicron structure differs from that of micelles due to the addition of lipoproteins.

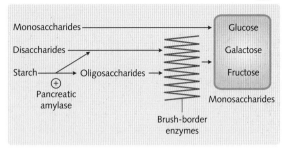

Fig. 7.18 Overview of intestinal digestion of carbohydrates.

Carbohydrates

Digestion of carbohydrates begins in the mouth, where starch is broken down by the action of salivary amylase (also known as ptyalin), which is subsequently destroyed by the low pH in the stomach. The α-1,4 linkages in up to 50% of the ingested starch are hydrolysed in the stomach to produce maltose. Thus, carbohydrate enters the duodenum in a variety of forms:

- Complex carbohydrates (starch).
- Disaccharides (sucrose, lactose and maltose).
- Monosaccharides (glucose, galactose and fructose).

Digestion continues with the release pancreatic α-amylase, which, like its salivary form, requires a neutral or mild alkaline pH and chloride for optimum activity (Fig. 7.18). Breakdown starch products are:

- Maltriose.
- Maltose.
- Short oligosaccharides (alpha-dextrins). Amylases have no reaction site (specificity) for some glucose–glucose branching points, thus cannot fully hydrolyse starch.

The appropriate oligo- and disaccharidases (e.g. lactase) in the intestinal epithelial brush border act

to release monosaccharides. Free concentrations of these subunits at the mucosal surface are likely to be adequate for the passive process of facilitated absorption. As concentrations fall, however, other transport processes are necessary. Apical membrane-associated transporters powered by basolateral Na$^+$/K$^+$-ATPase are employed - Na$^+$/glucose cotransporters (SGLT1) that also bind galactose. Entry of fructose, independent of Na$^+$ movement, and much slower than glucose and galactose, is by facilitated diffusion utilising GLUT-5, a member of another group of protein transporters (GLUT 1-5).

Monosaccharides move out of the epithelial cells by facilitated diffusion utilising the GLUT 2 transporters in the basolateral membrane and enter the capillaries of the villi.

Proteins

In the stomach, proteins are denatured by acidic gastric juice before hydrolysis by pepsin to produce polypeptide fragments. These then pass into the duodenum, where pancreatic bond-specific proteolytic enzymes, including chymotrypsin, elastin and carboxypeptidases, continue the hydrolysis of peptide bonds. The resultant

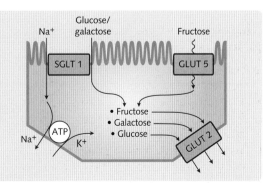

Fig. 7.19 Monosaccharide absorption by the enterocyte. ATP, Na$^+$/K$^+$-ATPase; SGLT, cotransporter (Na/glucose); GLUT, facilitated diffusion (glucose).

mixture of free amino acids and small peptides is then transported into the intestinal cells by a series of carrier systems. Di- and tripeptides are either:

- Absorbed intact by secondary active transport mechanisms with Na^+ (cf SGLT), although not all protein transporters are Na^+ dependent (Fig. 7.20).
- Broken down into free amino acids by the action of peptidases on the microvillus membrane (brush-border enzymes) before absorption by active transport.

Peptides are hydrolysed within the epithelial cells into amino acids, which, along with other amino acids, pass into the capillaries of the villus by three different mechanisms:

- Passive diffusion: relatively hydrophobic amino acids (e.g. tryptophan).
- Facilitated diffusion: L-isomers of amino acids.
- Active transport: as above.

Acidic amino acids serve as energy substrates for enterocytes; they do not appear to exit via carrier proteins.

Water and water-soluble vitamins

The small intestine absorbs about 80% of the 9 L of water that enters it per day. This is achieved by solvent drag through paracellular and transcellular routes

Absorptive mechanisms from lumen to enterocyte:
- Carbohydrates (monosaccharides): passive (fructose) or Na^+-coupled active transport.
- Protein or amino acids: Na^+-independent or dependent active transport.
- Fat: passive diffusion from micelles.

consequent upon sodium uptake (in association with carbohydrate and amino acid absorption). Passive diffusion accounts for a significant amount of water-soluble vitamin uptake, although there are other mechanisms.

Vitamin C (ascorbic acid)
Secondary active transport mechanisms employing Na^+ cotransport in the proximal ileum.

Folate
Carrier-mediated facilitated diffusion.

Vitamin B 'complex'
- Biotin, riboflavin and choline are absorbed by facilitated diffusion.
- Thiamin, inositol and nicotinic acid rely on Na^+-dependent active transport.

Fig. 7.20 Absorption of peptides and amino acids.

Fig. 7.20 Carrier systems for the transport of amino acids		
Transport system	**Substrates**	**Dependence on Na^+ gradient**
Brush-border membrane		
Neutral	All neutral aromatic and aliphatic AA	Yes
PHE	Phenylalanine and methionine	Yes
Acidic	Glutamate, aspartate	Yes
Imino	Proline, hydroxyproline	Yes
y^+	Basic AA	No
L	Neutral AA with hydrophobic side chains	No
Basolateral membrane		
A	Small neutral AA	Yes
ASC	Three-and four-carbon neutral AA	Yes
L	Neutral AA with hydrophobic side chains	No
y^+	Basic AA	No
AA, amino acids		

Vitamin B$_{12}$ (cobalamin)

On entering the duodenum, B$_{12}$-R protein complexes are degraded. B$_{12}$ then binds to intrinsic factor (IF), which is secreted by the gastric parietal cells in the stomach, enabling absorption of vitamin B$_{12}$ in the terminal ileum. The ileal epithelial brush border contains receptors that bind the IF–B$_{12}$ complexes, enabling cobalamin uptake.

Vitamin B$_{12}$ crosses the basolateral border and enters the portal blood by an unknown mechanism. It is then bound to a globulin – transcobalamin II – as a complex before being taken up by the liver and other tissues. Passive diffusion throughout the small intestine accounts for 1–2% of vitamin B$_{12}$ absorption.

Ions and minerals

Sodium

The distal small bowel and colon are largely responsible for absorbing 95% of the sodium load entering the gut. The four mechanisms implicated are powered by basolateral Na$^+$/K$^+$-ATPase enabling Na$^+$ diffusion into the cells from the lumen via passive and coupled means.

1. *Nutrient-coupled* Na$^+$ absorption is the primary postprandial process: the main mechanism in the jejunum with some in the ileum
2. *Na$^+$/H$^+$ exchange* in the apical membrane: is stimulated by alkaline, HCO$_3$$^-$ rich luminal environment in the duodenum and jejunum and is important in the postprandial period.
3. *NaCl coupled* transport resulting from the parallel Na$^+$/H$^+$ and Cl$^-$/HCO$_3$$^-$ exchange in the apical membrane of the ileum and proximal colon. This is the important transport process in the interdigestive period.
4. *Passive* Na$^+$ reabsorption via specific Na$^+$ channels (ENaC) occurs primarily in the apical membrane of the proximal colon; it is important in salt conservation.

Active basolateral pumping of Na$^+$ lowers luminal [Na$^+$], enabling diffusion into the cells from the lumen via passive and coupled means.

Chloride

Rapid diffusion in the proximal small bowel and distal colon occurs through Cl$^-$ channels via both transcellular and paracellular routes powered by the electrochemical gradient secondary to Na$^+$ absorption. Cl$^-$ absorption also occurs via apical Cl$^-$/HCO$_3$$^-$ exchange.

Bicarbonate

The absorption of HCO$_3$$^-$ is important, because large quantities are secreted by the pancreas and in bile: significant loss in faeces will lead to a metabolic acidosis. The mechanism of absorption is the same as that in some renal tubules: luminal HCO$_3$$^-$ combines with H$^+$ exchanged for Na$^+$ to form H$_2$CO$_3$. This then dissociates into H$_2$O and CO$_2$; the latter is absorbed into the blood and contribute to plasma HCO$_3$$^-$ homeostasis.

Calcium

Absorption occurs:

- Passively: high chyme [Ca^{2+}] via the paracellular pathway.
- Actively: low chyme [Ca^{2+}] and is rate limited.

Active absorption occurs throughout the small intestine in balance with the body's needs. It is subject to:

- Parathyroid hormone: activates vitamin D, which in turn enhances Ca^{2+} absorption.
- Vitamin D: stimulates the synthesis of basolateral Ca^{2+}-ATPase pumps and binds to enterocyte nuclear receptors to stimulate synthesis of brush-border and cytosolic calcium-binding proteins.

Diffusion of luminal Ca^{2+} into the enterocyte through Ca^{2+}-channels is facilitated by a reduction in the intracellular [Ca^{2+}] that is brought about by:

- Basolateral transport proteins:
  - the Ca^{2+}-ATPase active transport protein is more effective at low extracellular [Ca^{2+}].
  - the Na$^+$/Ca^{2+} exchanger is more effective at high extracellular [Ca^{2+}].
- Calbindin: binds two calcium ions, which enter the cell, enabling large amounts of enterocytic absorption of Ca^{2+}.

Iron

Haem represents the majority of iron consumed by non-vegetarians. Iron exists in two forms:

1. Fe^{2+} (ferrous): present in haem and more easily absorbed as it has a lower tendency to form insoluble complexes.
2. Fe^{3+} (ferric): reduced by ascorbic acid or HCl, both secreted by the stomach, to Fe^{2+}. It is less well absorbed, more readily forming insoluble complexes with tannins and phytins.

Only the proximal 10 cm of the small intestine absorbs haem and Fe^{2+}:

- Haem: uptake is via facilitated diffusion. Intracellular xanthine oxidase releases Fe^{2+} ions.

- Fe^{2+}: duodenal and jejunal epithelial cells secrete transferrin, each of which binds with two Fe^{2+} ions. Brush-border transferrin receptors bind to the complex and are then endocytosed. Fe^{2+} ions are released inside the cell; transferrin and the receptor are returned to the lumen.

Within the enterocyte, iron is either:

- Stored: in combination with apoferritin, known as ferritin (majority).
- Transported to transferrin receptors in the basolateral membrane bound to an intracellular protein, mobilin; these receptors transfer Fe^{2+} to transferrin bound at the ECF surface; the complex is released into the blood.

Cells with a large ferrous store take up this ion less readily, discouraging iron overload.

The intestinal flora

Bacteria colonize most of the GI tract, exceptions being the stomach, gall bladder and salivary glands (Fig. 7.21). The majority of these microflora are commensal (do not have adverse effects) and differ according to site. Although there are skin flora and bacteria in the mouth, the harsh acidic conditions result in a low bacterial count in the duodenum.

The bacterial count increases distally from the duodenum and there are progressively fewer aerobic bacteria, which favour the more aerated, better-oxygenated conditions that predominate in the proximal small bowel.

Duodenum

- Contains few microflora.
- Species include Gram-positive cocci and rods.

Jejunum

- Contains more bacteria.
- Examples include *Enterococcus faecalis*, lactobacilli and diphtheroids.

Ileum

- Contains a large number of flora.
- Species include *Bacteroides*, *Clostridia* and *Bifidobacterium*.

Colon

- Contains more bacteria than the small intestines by a factor of ~10^3.
- >400 species, including *Bacteroides* and coliforms (e.g. *Escherichia coli*, *Entercoccus faecalis*).

- Large microflora maintained by relatively low peristalsis.
- Bacterial residue contribute a third of the dried weight of faeces.

Functions of colonic bacteria

- Immunity: commensal bacteria compete with pathological types for food and space, thus keeping them under control.
- Bilirubin metabolism: the bacteria in the small bowel assist in the enterohepatic cycling of bilirubin by converting conjugated bilirubin back into its free form. They also form urobilinogen, which can be reabsorbed, excreted in urine, or further acted upon to produce stercobilinogen (which is excreted in the faeces).
- Production of vitamins: bacterial activity forms B vitamins (riboflavin and thiamin). Anaerobes are particularly important for the production of vitamin K, which would be insufficient for coagulation if relying on ingested amounts alone.

- Small intestine has fewer microflora, with a higher aerobic proportion, than the large intestine.
- Large intestine contains large microflora with mainly anaerobes, which function in vitamin K and B formation.

THE LARGE INTESTINE

Transport and secretion in the large intestine

The colon represents a 1.5-m reservoir that has important functions in absorption, secretion, motility and elimination of faeces.

Water and electrolyte transport

Approximately 90% of water is absorbed from the 1.5 L of fluid chyme that enters the colon. Most of this occurs in the proximal colon, leaving ~150 mL per day to be excreted in faeces.

Water absorption is by solvent drag, following the passage of electrolytes which themselves employ active transport.

As in the small bowel, basolateral Na^+/K^+-ATPase is key to permitting Na^+ influx by secondary active

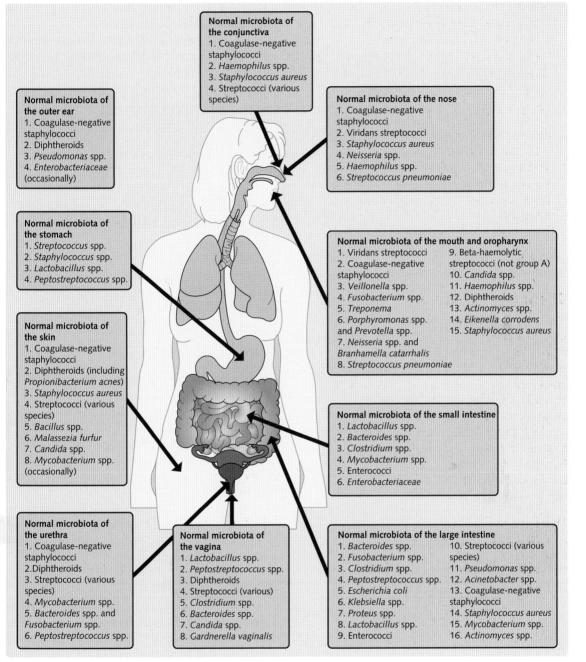

Fig. 7.21 The gut flora. Adapted from Prescott et al (2005) Microbiology, 6th edn. New York, McGraw-Hill Company Inc, with permission.

transport. However, a luminal electrogenic Na⁺ channel results in a potential difference of 30 mV across the mucosa (the lumen is relatively negative). This favours K⁺ secretion through tight junctions intraluminally. Another effect is Cl⁻ movement into the intercellular space, taking water with it (Fig. 7.22).

Although the mechanisms for water absorption are similar to those operating to concentrate bile in the gall bladder, there are some important differences. The colonic mucosa has relatively small pores compared to the small bowel and offers high resistance to the passage of water. Hence, passive

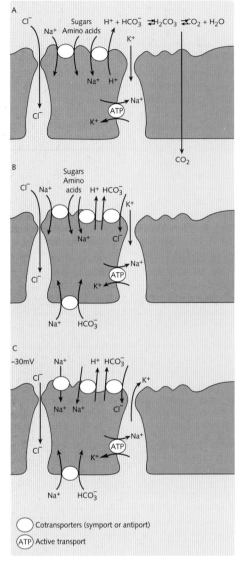

Fig. 7.22 Electrolyte absorption. (A) The jejunum, (B) the ileum, (C) the colon.

water movement into the intercellular spaces is much slower than electrolyte transport.

Water absorption is promoted by the following:

- Aldosterone: stimulates electrogenic Na channels and increases basolateral Na^+/K^+-ATPase pumps.
- Glucocorticoid: stimulates electroneutral NaCl absorption.
- Angiotensin II: increases Na^+ and water absorption.

Mucus secretion

Thick mucus, rich in HCO_3^- and K^+, is secreted by the abundant goblet cells of colonic mucosa. Movement of these ions in response to the mucosal

potential difference discussed above is thought to be responsible for their contributions.

Mucus lubricates the colon and serves to minimize mechanical trauma from the passage of faeces. Secretion is promoted by:

- Distension or mechanical irritation: probably involving VIP and acetylcholine.
- Parasympathetic stimulation.

Secretion of mucus is inhibited by:

- Sympathetic stimulation: mediated by adrenaline and somatostatin.

The motility of the large intestine

Colonic movements resemble a reduced and slow version of the small intestine. These are adequate for its main functions of absorption in the proximal half and storage in the distal half. The large intestine has both mixing and propulsive contractions:

- Mixing: haustration: no net movement of chyme.
- Propulsive:
  - segmental propulsion: movement of chyme from one segment to another
  - peristalsis: propagated wave of contraction behind a relaxed area
  - mass movement: contraction of a large length of bowel at once.

Haustration

Contraction of the circular muscles in one portion of the colon causes distension of the distal segment. Hence, mixing of the contents with mucosal contact is enabled, facilitating absorption.

Segmental propulsion

Contraction of the three longitudinal muscle bands in the wall of the colon (known as the taeniae coli) results in sequential haustration. This slowly moves the intestinal contents (8–15 hours from ileocaecal valve to transverse colon) towards the anus. Distension of the longitudinal bands of the muscularis externa itself stimulates contraction. Movement is initiated by acetycholine and release of substance P.

Peristalsis

Contractions are less frequent than in the small bowel and hence are slower, although they do propel chyme further.

Mass movement

This refers to the intense contraction that occurs between the transverse colon and the sigmoid colon between one and three times a day, shortly after ingestion of a meal. The colonic contents are pushed towards the anus and the urge to defecate is experienced when there are faeces in the rectum, usually during a period following a large meal. This is known as the gastrocolic reflex, and involves the coordination of physical, hormonal and neuromyoelectric mechanisms of bowel movement:

- Parasympathetic stimulation (vagal and pelvic nerves): increases proximal colonic contraction.
- Sympathetic stimulation (inferior mesenteric and hypogastric plexi): decreases colonic movement and is responsible for reflex relaxation of bowel adjacent to a contracting portion: colo-colonic reflex.

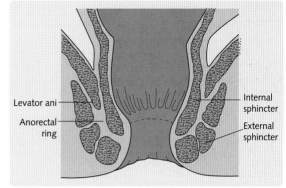

Fig. 7.23 Rectum and anal canal.

Defecation

Continence is maintained through the action of both internal and external sphincters (Fig. 7.23):

- Internal sphincter: thickened continuation of circular muscle inside the anus, which is under involuntary control.

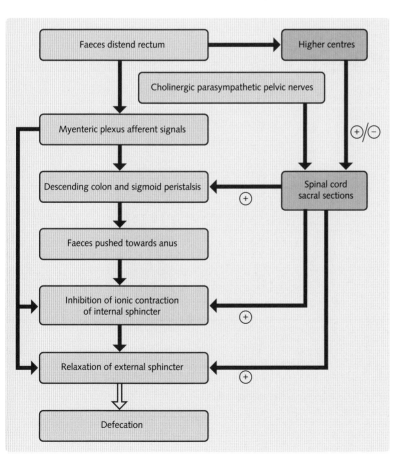

Fig. 7.24 Sequences of events in defecation.

- External sphincter: striated voluntary muscle extending more distally than the internal sphincter and surrounding it.

Defecation occurs as an intrinsic reflex in response to faeces distending the rectum. However, it is subject to input from higher centres, so that defecation occurs when socially acceptable (Fig. 7.24).

Autonomic reflexes involving the parasympathetic system augment this intrinsic pathway to further increase distal colon motility.

Physiology of the kidneys and the urinary tract

Objectives

In this chapter, you will learn to:

- Outline the basic structure and function of the kidney, including hormone production
- Describe the various components of the nephron
- Relate the function of the blood supply to vascular organization within the kidney
- Describe the components of the glomerulus and explain how they facilitate the production of an ultrafiltrate
- Define glomerular filtration rate, explaining what governing forces and factors are involved and how feedback is achieved
- Describe control of renal blood flow and glomerular filtration rate, explaining how measurements are made
- Outline the various mechanisms of transport, including sodium handling
- Describe the transport functions of the proximal tubule, relating this to its structure
- Describe the countercurrent mechanism and reabsorptive functions of the loop of Henle, explaining how these are achieved through its design
- Explain how osmoreceptors and baroreceptors maintain osmolality of the body's fluid
- Describe how the renin–angiotensin system and renal handling of sodium influence regulation of body fluid volume
- Explain how acid–base balance is achieved, describing the factors involved and the buffering systems of the body
- Outline the various acid–base disorders and explain their methods of correction involving compensation
- Outline how the kidney regulates ions and minerals, with reference to normal values and derangements
- Describe the regulation of erythropoiesis

ORGANIZATION OF THE KIDNEYS

Morphology and internal structure

The kidneys have the following dimensions:

- Length: 10–12 cm.
- Width: 5–7 cm.
- Breadth: ~3 cm.
- Weight: ~140 g.

They lie retroperitoneally between the levels of T12 and L3, with the right kidney positioned slightly lower than the left (by ~12 mm) owing to the position of the liver.

The hilus of the kidney lies near the centre of each concave medial border and constitutes a deep vertical fissure. Here, the ureter, nerves and blood vessels enter and exit the kidney.

Three layers of tissue surround each kidney. From superficial to deep these are:

- Renal fascia.
- Adipose tissue.
- Renal capsule.

All have mainly protective and anchoring properties.

Two distinct regions within the kidney form the parenchymal – functional – portion of the kidney (Fig. 8.1):

1. Outer renal cortex: comprising tubules and filtering capsules.
2. Inner renal medulla: consisting of 8–18 cone-shaped pyramids of straight tubules with the apex or renal papilla pointing towards the hilum.

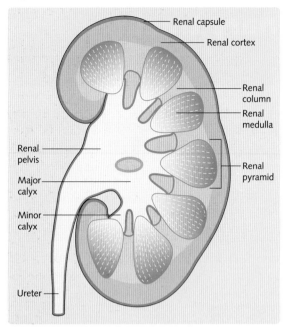

Fig. 8.1 Gross anatomical features of the kidney.

The strips of cortex that extend between renal pyramids are referred to as renal columns.

The nephron

The nephron (Fig. 8.2) is the functional unit of the kidney. Approximately 10^6 nephrons are present within the parenchyma of each kidney. Each nephron consists of:

- Renal corpuscle, which acts as a filter and comprises the:
  - glomerulus
  - Bowman's (i.e. the glomerular) capsule.
- Renal tubule, which acts as the receptacle for the filtrate. It is a luminal structure lined with epithelium and has several parts:
  - proximal tubule
  - loop of Henle
  - distal tubule
  - collecting duct.

Plasma is filtered at the glomerular capillaries into the tubules by a process known as glomerular filtration. In the tubules, water and solutes are removed from the filtrate – tubular reabsorption – and solutes can be added – tubular secretion. The effect is to reduce the volume and change the composition of the fluid, forming urine as it passes through the tubule and down the collecting duct. Urine enters the renal pelvis via papillary ducts, which drain into the minor calyces. These then deliver urine to the major calyces (of which there are two or three per kidney), which drain via the pelvis into the ureters.

Fig. 8.2 Structure of the nephron with cell types.

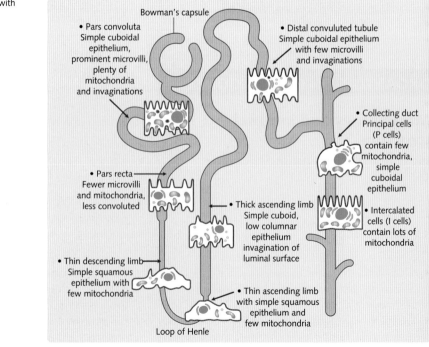

Between 80 and 85% of nephrons are cortical, penetrating only into superficial regions of the medulla. The remainder are juxtamedullary nephrons whose long loops of Henle extend into the deepest parts of the medulla, enabling the production of very dilute or very concentrated urine.

Glomerulus and Bowman's capsule

The glomerulus is a ball of interconnected capillaries that arise from the afferent arteriole and are drained by the efferent arteriole. The glomerulus projects into the dilated, hollow, cup-like end of the nephron known as Bowman's capsule. This structure is approximately 200 μm in diameter.

The glomerulus acts like a molecular sieve. It has three membranes through which a protein-free filtrate is received into Bowman's space (between the inner surface contacted by glomerulus and the outer surface) (see Fig. 8.7, p. 166).

Proximal tubule

From Bowman's capsule, fluid flows into the proximal tubule (length ~15 mm; diameter ~55 μm). This segment of the nephron is the primary site for reabsorption of salt, water and organic solutes.

The cells making up the first portion of the proximal tubule – pars convoluta – have the following features:

- Tight junctions at their luminal surfaces.
- Extensive interdigitations.
- Magnified absorptive surface area due to the microvillus-dense brush border.
- Extracellular space between bases of cells (the lateral intercellular spaces).
- Large numbers of mitochondria.

The second, straight, part of the proximal tubule is the pars recta, which has fewer microvilli and mitochondria. It is continuous with the first part of the loop of Henle.

Loop of Henle

The loop of Henle is a hairpin structure with the following parts:

- Thin descending limb: there are a few interdigitations between cells of this section, which ends at the tip of the hairpin loop.
- Thin ascending limb: is of variable length (20 μm diameter). It differs from the preceding part with respect to permeability and has widespread interdigitations.
- Thick ascending limb: this is about 12 mm long, with an abrupt transition from the thin ascending limb. Its start position is determined by the length of the loop.

The loop (up to about 14 mm long) can superficially enter the medulla or extend as far down as the renal pelvis in the case of some juxtamedullary nephrons. The loop of Henle works with the high solute/ionic concentration of the medulla to drastically concentrate urine at the bottom of the hairpin, and then dilute it as it ascends.

Distal tubule

The distal tubule is a continuation of the loop of Henle in the cortex of the kidney; it is approximately ~5 mm long. The cells forming this segment are concerned with hormone-sensitive water balance. The terminal portions of eight to ten distal tubules coalesce to form the collecting ducts.

Collecting ducts

In the cortex, the collecting duct is about ~20 mm long and composed of predominantly principal (P) cells and intercalated (I) cells:

- P cells: have few organelles and are involved in Na^+ and antidiuretic hormone (ADH)-mediated water reabsorption.
- I cells: contain more organelles, including mitochondria, than P cells; they secrete H^+ and are involved with the transport of HCO_3^-.

The cortical collecting duct becomes progressively larger as it passes downwards through the medulla to form the medullary collecting duct. Here, pairs of distal tubules join, forming the ducts of Bellini. The duct then empties into the pelvis of the kidney, which is lined with transitional epithelium via the renal papillae and calyces.

Blood supply and vascular structure

The kidneys are supplied with between 20 and 25% of resting cardiac output (1100 mL/min) via the right and left renal arteries (Fig. 8.3). On entering the hilum, the renal artery branches into segmental arteries. These divide to form the interlobar arteries, which enter the parenchymal tissue and pass through the renal columns.

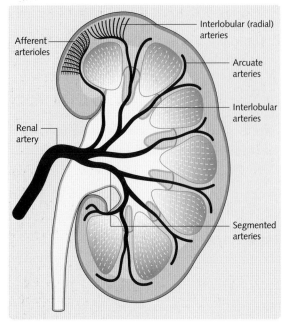

Fig. 8.3 Blood vessel arrangements.

Further branching continues and, at the junction between cortex and medulla, the arteries become known as the arcuate arteries. These arch around the bases of the renal pyramids, giving off vessels at 90° into the cortex – the radial or interlobular arteries. The interlobular arteries finally divide to form the afferent arterioles, one of which supplies each nephron, dividing to form a tuft of capillaries – the glomerular capillary. It branches and anastomoses before coalescing and draining as the efferent arteriole.

The efferent arterioles form two capillary networks that are unique to the renal circulation and dependent on the type of nephron they are draining:

- Peritubular network: this is formed from efferent arterioles in the outer two-thirds of the cortex. It supplies all parts of the nephron located in the cortex.
- Vasa recta: this is a hairpin loop of capillaries. It is formed in addition to the peritubular network from the efferent arterioles of juxtamedullary nephrons in the inner one-third of the cortex. These vessels are closely associated with the loop of Henle and collecting duct as they follow their courses down into the medulla.

Function of the renal blood supply

The enormous blood flow to the kidney dictates a high filtration rate and is vital for maintaining homoeostatic mechanisms: the kidney can alter blood flow in response to physical demand. In the face of altered blood pressure, this is maintained by autoregulation. However, blood flow far exceeds metabolic requirements. Consequently, the arteriovenous O_2 difference is only 1–2%, $\approx$ 15 mL/L, compared to 62 mL/L and 114 mL/L in the brain and heart, respectively.

Interestingly, the loop arrangement of the vasa recta results in inefficient O_2 and CO_2 exchange, such that O_2 supply to medullary cells is no more than adequate despite a rich blood supply. The vasa recta returns reabsorbed substances to the circulation from filtrate in the tubular portion of the nephron.

The juxtaglomerular apparatus

As the thick ascending loop of Henle re-enters the cortex to become the distal tubule, it passes very close to its own Bowman's capsule. Part of the distal tubule lies adjacent to the capsule and comes into contact with the arterioles from its own glomerulus. The term 'juxtaglomerular apparatus' (JGA) refers to this area of association (Fig. 8.4). There are three special cell types within the JGA:

1. Granular cells: specialized thickened cells in the tunica media of the afferent arteriole wall that secrete the enzyme renin, which is vital for the

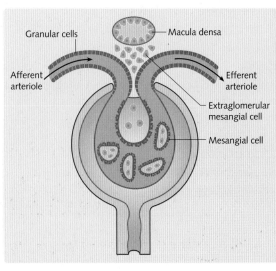

Fig. 8.4 The juxtaglomerular apparatus and important cell types.

control of blood pressure and renal function. Additionally, the granular cells deform with changes in afferent arteriolar pressure, function as detectors.

2. Macula densa cells: specialized epithelial cells found at the distal end of the thick ascending loop of Henle. These cells monitor the luminal content of the tubule and contribute to the control of glomerular filtration rate (GFR) and renin secretion.

Renal hormones

Renin

Renin is a proteolytic enzyme that is important in the renin–angiotensin pathway (see Fig. 5.18). Its only known function is to cleave the liver α_2-globulin, angiotensinogen, to form angiotensin I, the precursor of angiotensin II – a potent vasoconstrictor. The effect of angiotensin II on regulating blood pressure is additionally mediated by:

- Tubular reabsorption of Na^+.
- Aldosterone secretion from the zona glomerulosa of the adrenal cortex.

Renin release is stimulated by the following mechanisms, which can occur together (Fig. 8.5):

- ↓Na load delivered to the distal tubules.
- ↓in afferent arteriolar pressure.
- Sympathetic stimulation of granular cells via β-adrenergic receptors.

A significant drop in blood pressure, e.g. following haemorrhage, will cause all three mechanisms above to occur in unison.

Erythropoietin

Cells of the inner cortex and peritubular interstitium of the kidney produce >80% of the body's erythropoietin (EPO). Small amounts are also made in the liver, which is the major site of EPO production and erythropoiesis during fetal and neonatal life.

The synthesis of EPO, which is mediated by prostaglandins, occurs in response to a reduction in PO_2 in the kidneys: hypoxia: most importantly, hypoxic anaemia and renal ischaemia. It acts on erythrocyte stem cells of the bone marrow, stimulating increased production of red blood cells (RBC) and thus improving the oxygen-carrying capacity of the blood.

EPO has a half life of 5 hours. The rate of secretion of EPO can be high or low, depending on the type of renal disease:

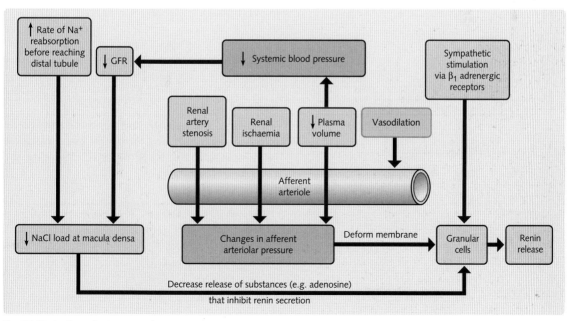

Fig. 8.5 Mechanisms affecting renin release.

- High levels: resulting in polycythaemia (high RBC), feature in polycystic kidney disease and renal cell carcinoma.
- Inappropriately low levels: are seen in chronic renal failure. In addition 90% of people who have had their kidneys removed or are suffering end-stage renal disease and are on haemodialysis develop anaemia.

The decrease in EPO production by the kidney cannot be compensated by the rest of the body and so, in chronic renal failure, intravenous or subcutaneous recombinant EPO will treat the anaemia.

Vitamin D

The last two biochemical transformations involved in the synthesis of vitamin D take place in the kidneys (Fig. 8.6). The active metabolite 1,25-dihydroxycholecalciferol ($1,25 (OH)_2 D_3$) is produced from the liver metabolite 25 hydroxycholecalciferol ($25 (OH) D_3$) by hydroxylation in the proximal tubule, a process that is mediated by the enzyme 1-α-hydroxylase. The rate of synthesis of $1,25 (OH)_2 D_3$ is regulated by parathyroid hormone and negative feedback.

Vitamin D plays an important role in calcium (Ca^{2+}) and phosphate (PO_4^{3-}) balance, promoting absorption of these two substances from the gut and stimulating their reabsorption by the renal tubules so that less is excreted. It has a significant function in the mineralization of bone. Patients with renal failure commonly have hypocalcaemia:

- In excessive amounts, vitamin D favours bone resorption.
- In smaller amounts, it promotes the mineralization of bone.

THE GLOMERULUS

Structure of the glomerular filter

The glomerular capillary wall acts a three-layered sieve (Fig. 8.7), allowing a virtually protein-free filtrate containing inorganic ions and low molecular weight (LMW) organic solutes, to pass through into Bowman's capsule. The properties of this filtration barrier are accounted for by its three layers:

1. Glomerular capillary endothelium.
2. Basement membrane (connective tissue).
3. Bowman's capsule epithelial cells (podocytes).

Glomerular capillary endothelium

The high filtration rate is facilitated in part by thousands of small holes or fenestrae perforating the thin, flat endothelial cells and occupying approximately 10% of the endothelium. Their diameter is 60 nm, which allows only small molecules to pass through and confines blood cells and platelets to the vascular space. In addition, the endothelial cells are extensively equipped with fixed negative charges, further discouraging the passage of anionic plasma proteins.

Basement membrane

This collagen and proteoglycan meshwork is particularly effective at limiting any further passage of plasma proteins due to its strong negative charge. At the

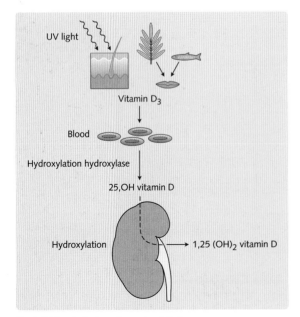

Fig. 8.6 Vitamin D absorption and metabolism.

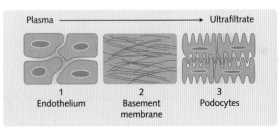

Fig. 8.7 The three layers of the glomerular filter.

same time, it permits large amounts of water and small solutes through its large spaces.

Bowman's capsule epithelial lining

This final layer comprises negatively charged cells called podocytes, providing a further barrier to plasma proteins. They are so named because of the long, foot-like processes embedded in the basement membrane, encircling the capillary. At the end of these extensions are small 'toe-like' processes, called pedicels, which interdigitate with the pedicels of neighbouring podocytes. The gaps formed are called slit pores, and the glomerular filtrate passes through these into Bowman's capsule. These cells are also involved in macromolecular phagocytosis.

The mesangium

Mesangial matrix and mesangial cells – the latter of which lie between the glomerular capillaries – are also part of the renal corpuscle. Mesangial cells have a number of properties. They:

- Act as phagocytes, preventing the build-up in the basement membrane of escaped macromolecules from the capillaries.
- Have a supportive function in maintaining the delicate structure of the glomerulus.
- Can alter the surface area of the glomerular capillaries available for filtration using their large amounts of myofilaments to contract in response to various stimuli.
- Can secrete extracellular matrix and prostaglandins.

Process of glomerular filtration

Glomerular filtration is the first step in the formation of urine. It involves the passive flow of solvent through a molecular sieve. The filtrate formed is known as glomerular ultrafiltrate (because the filter operates at the molecular level) and is composed of only LMW substances from plasma at the same concentrations as in plasma.

Ultrafiltrate contains:
- Virtually no protein.
- H_2O, low molecular weight organic solutes and inorganic ions

Glomerular filtration rate

All the kidney's copious blood flow passes through the glomeruli. The volume of filtrate formed (from blood in the glomeruli) per unit time is known as the glomerular filtration rate (GFR):

Normal GFR = 90 to 125 mL/min

This value correlates with body surface area, so 125 mL/min is an appropriate value for an average man. In females the GFR is always 10% less than the male value even after correcting for surface area.

The total volume filtered is 180 L/day.

GFR is used to indicate renal function: very low values are a cardinal sign of renal failure.

Normal urine volume is 1-2 L/day, hence most filtrate is reabsorbed. The advantage of a high GFR is therefore to permit rapid extraction of waste from the body and also to repeatedly filter and process all the body fluids each day. Hence, the kidney can tightly and swiftly control both the composition and the volume of body fluids.

Molecular size

The glomerular filter acts selectively, discriminating mainly by molecular size and, to a lesser degree, by the charge of the solutes. Although all three layers of the filter have a negative charge, they exert an inhibitory effect on macromolecules, not negative small ions or LMW solutes. Smaller positive molecules can also remain in the capillaries if they are protein bound. In terms of molecular weight, molecules less than 70 kDa are generally filtered freely. Shape can also be a factor, with filtration affected by radius:

- Radius < 20 Å (Angstroms): freely filtered.
- Radius > 36 Å: generally not filtered.
- 20 Å < radius < 40 Å: filtered to various degrees depending on molecular weight and charge.

Albumin – the smallest of the plasma proteins (radius = 35.5 Å) – is filtered but only in an amount equivalent to ~0.01% of that which passes through the glomerular capillaries. However, almost none appears in the urine as it is avidly reabsorbed by the proximal tubule. Significant amounts of protein in the urine (proteinuria) indicate kidney disease. Many diseases can cause proteinuria by affecting the negative charges of the layers of the glomerular filter, in particular that of the basement membrane.

In general, a total of ~30 g per day protein enters the lymphatic channels within the kidney and is returned to the vascular space.

Forces governing tissue fluid formation and glomerular filtration rate

Filtration across the glomerular capillaries are determined by the same factors that determine filtration across all capillaries – Starling forces:

- P: hydrostatic pressure (from water): favours filtration out of that space.
- π: oncotic (colloid osmotic) pressure (exerted by proteins): opposes filtration out of that space.

The arteriolar end of each systemic capillary is narrower than the opposite end. This results in increased resistance to flow: P>π and fluid is forced out (filtration) through the highly permeable cells. Fluid continues to move out of the capillaries until P has decreased to a value lower than π. When π>P, fluid moves back into the vascular space. This usually happens at the venule end of the capillary.

In the nephron, the formation of filtrate depends on the hydrostatic and colloidal pressures of both the glomerular capillary and Bowman's capsule (Fig. 8.8).

Changes in these forces will change the GFR because:

GFR ∝ forces favouring filtration – forces opposing filtration

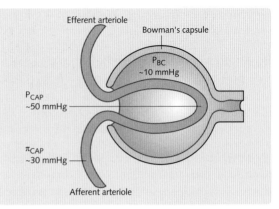

Fig. 8.8 Starling forces acting across a glomerular capillary.

As the filtrate in Bowman's capsule is virtually protein free, there is a negligible oncotic pressure in capsule (π_{BC}), which can therefore be disregarded when calculating net glomerular filtration force:

$$GFR = k_f(P_{CAP} - P_{BC} - \pi_{CAP})$$

where GFR = glomerular filtration rate, k_f = coefficient of filtration, P_{CAP} = glomerular capillary hydrostatic pressure, P_{BC} = Bowman's capsule hydrostatic pressure and π_{CAP} = glomerular capillary oncotic pressure (Fig. 8.9).

The surface area of the glomerular capillaries is relatively large when compared to normal systemic capillary beds such as the peritubular capillaries. However P_{CAP} does not decline along the capillary as in systemic capillaries; it is maintained essentially constant because the efferent arteriole acts as a secondary resistance vessel.

Imagine the glomerular capillary is a hosepipe, with afferent being connected to the tap and efferent being the other end. A slight kink at the tap end of the hosepipe causes a resistance to flow and produces a good hydrostatic pressure. If this were the only kink, then the rest of the pipe would be relatively wider. Hence the pressure of water as it flowed along the hosepipe would slacken because it would be easier for the water, i.e. there would be less resistance to flow. A second kink at this distal (efferent) end would prevent the movement of water from getting any easier and the pressure would have to remain high as the water travelled along the hosepipe. Hence, if the efferent arteriole provides some resistance to flow, hydrostatic pressure can be more or less maintained in the glomerular capillaries.

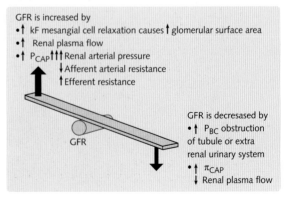

Fig. 8.9 The effect of Starling forces on glomerular filtration rate.

From the above equation, it follows that net filtration will occur until the sum of $P_{BC} + \pi_{CAP} = P_{CAP}$ so that there is an equilibrium. If this occurs in the capillary it will be consequent on the rise in π_{CAP} as fluid leaves the glomerular capillary (~20% capillary plasma fluid). If there is filtration pressure disequilibrium at the end of the glomerular capillary, equilibrium will occur along the efferent arteriole as P_{CAP} declines (point X in Fig. 8.10). When fluid in the efferent arteriole reaches the peritubular capillaries $P_{CAP} < \pi_{CAP} + P_{ISF}$ (P_{ISF} = hydrostatic pressure in renal interstitial fluid) hence fluid moves into the capillary with consequent dilution of π_{CAP}, and a reduction of the decline in P_{CAP} in the peritubular capillaries (Fig. 8.10).

Feedback control of glomerular filtration

Although GFR can be altered by various mechanisms, renal blood flow (RBF) is a clearly a major determinant. Both RBF and GFR are maintained through a large arterial pressure range: 90–180 mmHg. This autoregulation occurs by adjusting vascular resistance – namely that of the afferent arteriole. Two main mechanisms are responsible for this:

1. Myogenic mechanism: affected by arterial pressure changes.
2. Tubuloglomerular feedback: affected by changes in the flow rate of tubular fluid.

Myogenic mechanism

This relies on the property of smooth muscle to contract in response to stretching:

$$\text{Flow} = \frac{\text{Change in pressure } (\Delta P)}{\text{Resistance (R)}}$$

RBF and hence GFR are therefore kept constant if ($\Delta P/R$) is the same. For a change in arterial pressure, the vascular resistance must be altered.

Tubuloglomerular feedback

This feedback from the tubules to the glomerulus relies on the relationship between the filtration rate and the amount of Na^+ escaping reabsorption in the proximal tubule and loop of Henle. The more Na^+ entering the capsule and proximal tubule, the more Na^+ remains in the tubular fluid and enters the distal tubule, where its levels are detected by the macula densa of the JGA. The macula densa acts as both detector (of osmolality, Na^+ and/or Cl^- load, or tubular flow) and effector:

- ↑GFR causes higher Na^+ and Cl^- loads in the distal tubular fluid.
- This change in load increases cellular NaCl transport which is sensed by macula densa cells.
- Macula densa cells release signals (perhaps adenosine, ADP, ATP or thromboxane) into the interstitial space.
- These transmitters decrease filtration by causing vasoconstriction of the afferent arteriole thus hydrostatic pressure in the glomerular capillaries decreases.
- Signals also result in mesangial cell contraction, ↓kf.

Thus, GFR is lowered.

Renal blood flow and the glomerular filtration rate

Clearance

The GFR can be calculated by measuring the clearance of a substance (C):

$$C = \frac{U \times V}{P}$$

where C = volume of plasma cleared of a particular substance in a unit time, U = urine concentration of substance (mg/mL), V = urine flow rate (mL/min), P = plasma concentration of substance (mg/mL).

For example, for substance S, plasma contains 60 mg/mL (P), urine flow rate 1 mL/min (V) and urinary concentration 7500 mg/mL (U), then:

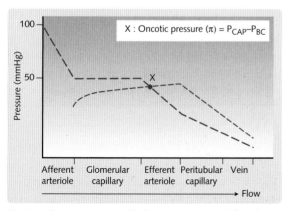

Fig. 8.10 Changing oncotic and hydrostatic pressures with flow along renal vessels.

$$\text{Clearance} = \frac{7500}{60} = 125 \text{ mL} / \text{min}$$

However, if clearance of a compound is to provide an accurate estimate of GFR, the compound must be completely eliminated by the kidney. For this the compound must:

- Be freely filtered.
- Be unaffected by tubular processing, i.e. it is not reabsorbed or secreted.
- Not be synthesized or metabolized by the kidney.

Inulin, a polysaccharide of fructose found in Jerusalem artichokes and dahlia tubers and molecular weight of 5500, is such a substance. After intravenous introduction, its urinary clearance can be used to calculate GFR. GFR is a measure of glomerular function – especially in disease. However, although normal inulin clearance is equal to GFR, the method involved is invasive and seldom used. Instead, the clearance of creatinine, a product of muscle metabolism, is used therefore to estimate GFR in clinical practice:

Phosphocreatinine + ADP

$$\xrightarrow{\text{Creatinephosphokinase}} \text{Creatine} + \text{ATP}$$

$$\text{Creatine} + H_2O \longrightarrow \text{Creatinine}$$

A little creatinine is secreted by the tubules, therefore GFR is slightly overestimated, although overestimation of true plasma concentration by the methods used is thought to negate this.

Since plasma creatinine concentrations are reciprocally related to GFR, plasma creatinine concentrations are commonly used to reflect renal function. Values depend on:

- Renal function.
- Metabolism.
- Muscle mass: this dictates the exact value and so size, age and sex must also be taken into account.

Clearance ratio

Inulin is the gold standard for which clearance best reflects GFR. The clearance of other substances can be measured (C_{Sub}) and then compared with that of inulin (C_{In}) to see how the body deals with these substances in comparison:

- $C_{Sub} : C_{In} = > 1$ there is additional tubular secretion (p-aminohippuric acid).
- $C_{Sub} : C_{In} = < 1$: either the substance is not freely filtered or the substance is partially reabsorbed from the tubule.

Measurement of renal blood flow

Renal blood flow (RBF) is calculated as the amount of substance excreted by the arteriovenous difference in concentration of the substance. The Fick principle applies, whereby as the plasma is filtered by the kidney, the amount of substance filtered by the kidney per unit time is equal to that appearing in the urine.

$$\text{Renal plasma flow (RPF)} = \frac{V \times U}{\text{Art} - \text{Ven}}$$

Where Art = arterial concentration of substance, Ven = venous concentration of substance, U = urinary concentration of substance and V = urinary flow rate.

Infusion of para-aminohippuric acid (PAH) can be used to calculate RPF. As 90% PAH in arterial blood is removed after a single passage through the kidneys, the extraction ratio (or arteriovenous concentration difference) is said to be sufficiently high for the venous concentration value to be ignored. Hence, the calculation can now be based on PAH clearance as the arterial concentration by itself is the same as plasma value in the clearance equation:

$$C = \frac{U \times V}{P}$$

It is important to use a low concentration of PAH because secretion by the proximal tubules has a transport maximum such that at concentrations below 0.1 mg/mL, extraction approaches 100%.

Normal RPF = 600 mL/min

Renal plasma flow (RPF) can be converted to renal blood flow (RBF) using the haematocrit value (total percentage of total blood volume composed of erythrocytes). Therefore, if the haematocrit is 45% then 55% must be plasma.

$$\text{RBF} = \frac{\text{RPF} \times 100}{55} = \frac{600 \times 100}{55} = 1100 \text{ mL}$$

The PAH estimates only cortical plasma flow and does not include the blood flowing to the capsule, perirenal fat, medulla, tubules and glomeruli. Hence the RPF calculated this way is referred to as the effective RPF (ERPF). The following significantly decrease PAH excretion and therefore render the RPF deduced less reliable:

- [PAH] > 0.1 mg/mL (exceeds transport maximum).
- Chronic renal failure.
- Tubular dysfunction.

RBF can also be measured using the following methods in animals:

- Inert gas 'washout'.
- Isotope uptake.

Both the above methods are invasive, requiring intravenous injections of radioactive isotopes; furthermore, removal of the kidneys is required (experimental studies on animals) in the latter procedure.

Filtration fraction

$$\text{Filtration fraction} = \frac{\text{GFR}}{\text{RPF}}$$

The filtration fraction shows how much of the plasma passing through the glomerulus is filtered. Normal values are about 20%.

Fractional excretion

Fractional excretion (FE) is used to determine whether a freely filtered substance undergoes net reabsorption or secretion, by comparing the amount excreted in the urine (U × V) with the amount filtered (GFR × P).

- Net secretion: mass excreted > mass filtered (>100%).
- Net reabsorption: mass excreted < mass filtered (<100%).

Regulation of RBF and GFR

Autoregulation of RBF and GFR by blood flow essentially involve changes in tone of the afferent and efferent arterioles. Normal values are:

- RBF: 1100 mL/min.
- RPF: 600 mL/min.
- GFR: 120 mL/min.

As the perfusion pressure increases over the autoregulatory range 90–180 mmHg, so too does resistance to flow, because blood flow is independent of perfusion pressure at these values. Additionally, a high-protein diet leads to increases in renal blood flow, glomerular capillary pressure and glomerular filtration rate.

Vasoactive substances (Fig. 8.11) are particularly important in maintaining renal perfusion, as in cases of acute haemorrhage, where increased sympathetic activity leads to vasoconstriction and decreased blood flow. In this situation, intrarenal prostaglandin production prevents excessive vasoconstriction, preserving blood flow.

It is the afferent arteriole diameter with respect to the efferent, that controls the RBF and GFR separately (Fig. 8.12).

- RBF: changes reflect changes in total arteriolar resistance.
- GFR: increases by relative constriction of the efferent arteriole compared to the afferent.

Age-related changes in RBF and GFR

RBF
- 5% cardiac output in newborn.
- Rises progressively during the first year.
- 20% cardiac output in the adult.

Renal plasma flow cannot be measured reliably using PAH clearance in infants, as its extraction is even lower than that in the adult owing to immature tubular secretory mechanisms.

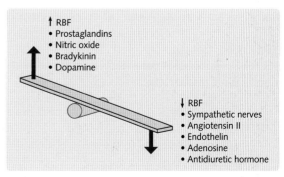

Fig. 8.11 The effects of various vasoactive substances on renal blood flow.

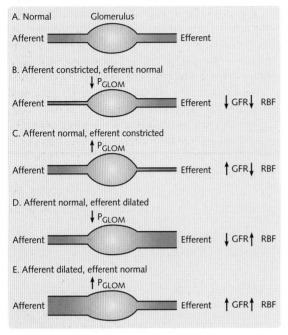

A. Normal
Afferent — Glomerulus — Efferent

B. Afferent constricted, efferent normal
$\downarrow P_{GLOM}$
Afferent — Efferent $\downarrow$ GFR $\downarrow$ RBF

C. Afferent normal, efferent constricted
$\uparrow P_{GLOM}$
Afferent — Efferent $\uparrow$ GFR $\downarrow$ RBF

D. Afferent normal, efferent dilated
$\downarrow P_{GLOM}$
Afferent — Efferent $\downarrow$ GFR $\uparrow$ RBF

E. Afferent dilated, efferent normal
$\uparrow P_{GLOM}$
Afferent — Efferent $\uparrow$ GFR $\uparrow$ RBF

Fig. 8.12 The effects of afferent and efferent arteriolar constriction and dilatation on renal blood flow and glomerular filtration rate.

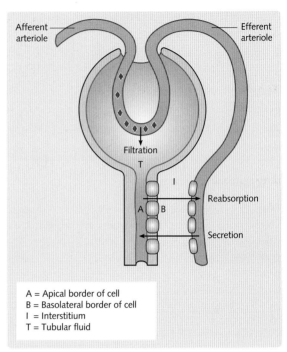
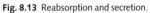
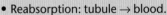

A = Apical border of cell
B = Basolateral border of cell
I = Interstitium
T = Tubular fluid

Fig. 8.13 Reabsorption and secretion.

GFR

- Filtration of fluid and urine production: contributing to amniotic fluid, starts in week 10 of gestation.
- 25 mL/min in the newborn.
- Progressive increase in GFR from 1 month.
- Adult values reached by 1 year of age.
- Progressive decline with old age.

TRANSPORT PROCESSES IN THE RENAL TUBULE

Once the ultrafiltrate enters the renal tubule, its contents are modified by reabsorption and secretion as it flows through the proximal tubule to the successive portions before reaching the collecting ducts as final urine (Fig. 8.13).

Definitions

- Reabsorption: removal of a substance from the tubular fluid back into circulation.
- Secretion: addition of a substance via the tubular cells from blood into the tubular fluid.

- Excretion: waste product removal from the blood following net processing (filtration, secretion and reabsorption).

Note: reabsorption and secretion are defined according to the direction of movement:
- Reabsorption: tubule → blood.
- Secretion: blood → tubule.

Solute can cross the epithelial lining of the tubule by passing either between or through the cells:

- Paracellular movement (around cells): through the matrix of tight junctions, driven by concentration, electrical and osmotic gradients.
- Transcellular movement (through cells): osmotic and relies on two factors:
  - solute crosses the apical membrane (in contact with filtrate), then the basolateral wall (facing the capillary at an angle), followed by water
  - polarization of epithelial cells (this is crucial to transcellular movement): proteins differ

between the apical and basolateral membranes. Hence net influx of Na^+ from tubular fluid to the cell is permitted – a process crucial to the transport of most other substances (see later).

Transport mechanisms

Diffusion

This is the passive simple movement of a substance down its electrochemical gradient. Substances that are lipid soluble, such as blood gases and steroids, diffuse through the membrane.

Facilitated diffusion

This process is faster than simple diffusion and the passage of substances down their concentration gradients requires a transporter (a specific carrier protein located within the cell membrane); e.g. GLUT-2 for glucose movement out of the enterocytes, (p. 153). Examples for the proximal convoluted tubule are GLUT-1 and GLUT-2 in the basolateral membrane (cell to interstitial fluid).

Primary active transport

Primary active transport refers to the direct coupling of ATP hydrolysis with a transport process concerned with movement of a substance against its electrochemical and concentration gradients. It uses energy derived from hydrolysis of the terminal phosphate bond of the ATP molecule, which itself forms part of the protein structure of the transporter.

In the nephron the key example is the $3\,Na^+/2\,K^+$-ATPase pump located on the basal and basolateral membranes of the tubular epithelium: for every molecule of ATP hydrolysed, three Na^+ are moved out of the cell while two K^+ are simultaneously pumped into the cell from ISF, both against their electrochemical gradients. This allows reabsorption of over 99% filtered Na^+ as low $[Na^+]$, in particular, and high $[K^+]$ are maintained within the cells. Hence, ionic gradients across the nephron cell membrane are established, which act as driving forces for reabsorption or secretion.

Other important primary active transport systems are:

- H^+-ATPase: moves protons out of cells into tubular fluid.
- Ca^{2+}-ATPase: moves calcium out of cells.
- H^+/K^+-ATPase; H^+ into lumen/K^+ into cell.

Secondary active transport

Transporters that require energy but do not hydrolyse ATP are called secondary active transporters. Instead, of utilizing ATP, they rely on the ionic gradient established by the ATPases. This is best illustrated by the basolateral Na^+/K^+-ATPase, which causes the cell to have a low $[Na^+]$ relative to the lumen (Fig. 8.14). Na^+ will now diffuse along its electrochemical gradient into the cell from the tubular fluid. The energy produced by this passive movement can be used to power other solutes against their gradients if the energy required for this active component can be met by that produced from the Na^+ diffusion. The processes involved are:

Symport (cotransport)

Movement of solute by protein carriers down the Na^+ gradient or in the same direction as Na^+. Examples include:

- Proximal tubule:
  - Na^+/glucose: (SGLT-2; lumen to cell): SGLT 1–4: a family of protein transporters.
  - $Na^+/3\,HCO_3^-$ (cell to interstitial fluid).
- Thick ascending limb of Henle's loop: $Na^+/K^+/2Cl^-$ (triple transporter; lumen to cell).

Antiport (exchange/counter transport)

Protein carrier effected movement of solute in the opposite direction to Na^+. Examples include:

- Na^+ (into cell)/H^+ (into lumen).
- Na^+/Ca^{2+} exchange.

Ion channels

Within the epithelial membrane are pored proteins constituting ion channels. They permit much faster transport (10^6–10^8 ions/s) than both the relatively abundant ATPases and transporter molecules (100 ions/s). However, ion channels are greatly outnumbered by the active transporters (100 versus 10^7 per cell). In addition to the Na^+/K^+-ATPase pump in the basolateral membranes, specific Na^+, Cl^- and K^+ ion channels are located in the apical membranes of all parts of the nephrons.

Handling of sodium by the kidney

Free filtration in Bowman's capsule means that filtrate $[Na^+]$ = plasma $[Na^+]$. Modification of filtrate as it passes through the successive segments of the nephron can be illustrated using Na^+ ions, 99% of those filtered are reabsorbed into the circulation. Na^+ transport occurs via both paracellular and transcellular routes. The transcellular path consists of movement via Na^+ ion channels and symports, which rely on the ATPase pump keeping the intracellular environment relatively negative.

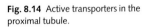
Fig. 8.14 Active transporters in the proximal tubule.

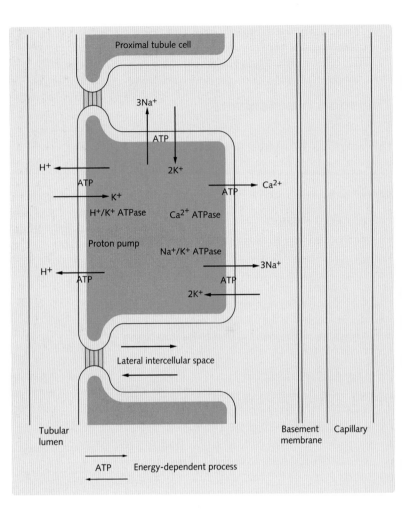

In addition to conservation of Na^+ in the body, the concentration of other solutes whose transport processes are dependent on Na^+ reabsorption are also maintained (Fig. 8.15):

- Glucose.
- Amino acids.
- Lactate.
- Cl^-.
- HCO_3^-.
- PO_4^{3-}.

THE PROXIMAL TUBULE

Microstructure of the proximal tubule

The proximal tubule (PT) is the site where most reabsorption takes place. It is divided into two parts:

1. Pars convoluta: convoluted portion immediately distal to the glomerulus. This is the primary site of active transport in the PT.
2. Pars recta: straight portion connecting to the loop of Henle.

Cells in the PT are particularly adapted to suit its function:

- The large numbers of mitochondria: support the highly metabolic and active processes of the PT.
- Tight junctions: connect cells at the luminal side, leaving gaps (lateral intercellular spaces) at the peritubular sides. These serve to increase the absorptive surface area greatly.
- Brush border at the luminal surface: adds to the effects of above (see p. 162).

Fig. 8.15 Sodium reabsorption from different nephron segments

Nephron segment	% Na⁺ reabsorbed	Route and mechanism of apical Na⁺ entry	Hormone involved
Proximal tubule	67	Paracellular	Angiotensin II
		Transcellular	
		Na^+/H^+ exchange	
		Cotransport with glucose, amino acids, PO_4^{3-}	
		HCO_3^-/Cl^- anion exchange	
Thick ascending limb of the loop of Henle	25	Paracellular	Aldosterone
		Transcellular	
		$Na^+/K^+/2Cl^-$ symport	
Early distal tubule	4	Na^+/Cl^- symport	
Late distal tubule and collecting ducts	3	Na^+ channels	Aldosterone
			Atrial natriuretic peptide

Transport of sodium and chloride

Movement of sodium into the lateral space

The linchpin in PT reabsorption is the Na^+/K^+-ATPase pump on the basolateral membrane, which actively transports Na^+ into the interstitial space. Most of the ≈ 70% reabsorption of Na^+ from the filtrate occurs in the early portion of the PT. However, cell junctions are leaky and unable to maintain a secure concentration gradient, as the PT is highly permeable to Na^+ in both directions. However, the more distal parts of the PT can establish a better concentration gradient because the cell junctions are tight, even though less reabsorption takes place here.

Sodium entry into the cell

The ATPase pump has already established:

- Low [Na^+] (<30 mmol/L) within the PT tubular cells.
- A relatively negative transmembrane potential (–70 mV) compared to the lumen.

Both serve to drive Na^+ intracellularly along its electrochemical gradient from the tubular fluid (Fig. 8.16). In the first half of the PT this is coupled with movement of other solutes:

- Symport with glucose, amino acids, phosphate, lactate, etc.
- Antiport with H^+ (which is linked to HCO_3^- reabsorption).

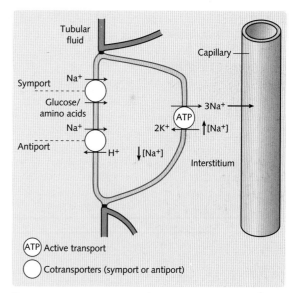

Fig. 8.16 Some Na^+ transporters in the proximal tubule. Open circle - cotransporter (antiport, anion/cation exchanger).

Little of the glucose or other solutes symported with Na^+ remains in the later portion of the PT. Chloride ions are the main method for Na^+ reabsorption in late PT. Here there is a relatively high [Cl^-] (from preferential HCO_3^- reabsorption in early PT) and different Na^+ transport mechanisms compared with the early portion of the PT.

Chloride reabsorption

In the early PT, the relatively negative transmembrane potential opposes intracellular Cl^- entry from the tubular lumen. Coupling of HCO_3^- with Na^+ reabsorption maintains osmolality in this section although $[Cl^-]$ rises. A concentration gradient that favours Cl^- diffusion across tight junctions or directly into the cell is thus established in the middle to late portions of the PT.

- 60% of Cl^- reabsorption occurs in the middle to late PT.
- Most Cl^- diffusion is paracellular (followed by Na^+).

This is handled differently in superficial and juxtamedullary nephrons. In the superficial, late PT, Cl^- permeability of cells is greater than that for other ions, whereas in juxtamedullary nephrons there is no permeability preference of anions.

Smaller amounts of Cl^- enter by Cl^-/HCO_3^- and $Cl^-/HCOO^-$ (formate) antiport (Fig. 8.17). Both HCO_3^- and $HCOO^-$ are continuously generated in the cell by dissociation of H_2CO_3 and $HCOOH$, respectively, which recombine with protons secreted into the tubular fluid (antiport with Na^+) and then freely diffuse back into the cell where they dissociate again.

Water reabsorption

In the PT there is tight coupling between reabsorption of Na^+ and H_2O. Hence, 70% of the filtered water passes back into the circulation: approximately the same value as for Na^+. In addition, there is a high H_2O permeability within this section of the nephron due to the presence of aquaporins (water pores), which permit small changes in osmolality to have an effect on H_2O reabsorption.

Solute reabsorption (including Na^+, or else dependent on Na^+ transport) into the intercellular spaces (the equivalent to ISF) lowers the osmolality of the tubular fluid while raising the osmolality of the ISF. This results in an osmotic gradient causing osmosis from the lumen across the plasma membrane or tight junction into the ISF.

Note that glomerular filtration causes a high plasma protein concentration within the peritubular capillaries. This equates to a high oncotic pressure and this Starling force causes H_2O from the ISF to be reabsorbed into these capillaries.

Although the main function of the PT is reabsorption, it does not alter the concentration: fluid leaves this section of the nephron having virtually the same osmolality with which it entered.

Transport of other solutes

Glucose

In general, almost all glucose is reabsorbed in the PT, maintaining plasma glucose levels of approximately 5 mmol/L (90 mg/dL). Only a few milligrams at most may be excreted in the urine per 24 hours. The transport systems within the PT for glucose are by secondary active transport (SGLT 1 and SGLT 2):

- 1 Na^+/1 glucose symport (SGLT 2) in the pars convoluta.
- 1 Na^+/1 glucose symport (SGLT 1) in the pars recta.

The amount of glucose reabsorbed is proportional to the amount filtered (plasma glucose × GFR). Plasma glucose concentrations transiently rise after meals and so the amount of glucose filtered will rise. However, glucose reabsorption is a T_m system, which means there is a point of saturation for the carrier. Hence, if the plasma glucose is above a certain concentration, such that the amount filtered exceeds the reabsorptive capacity of the SGLT of the PT cells, glucose will be excreted in the urine (glycosuria) (Fig. 8.18).

The reabsorptive rate limit differs between nephrons, ranging from thresholds at plasma glucose levels of 10 mmol/L to 20 mmol/L; known as the lack of nephron heterogeneity and accounts for the splay on Fig. 8.18. Therefore at plasma glucose concentrations of:

- 10 mmol/L: T_m will be exceeded for a few nephrons and glucose will start to appear in urine.
- 20 mmol/L: No nephrons will be able to reabsorb their entire filtered glucose load.

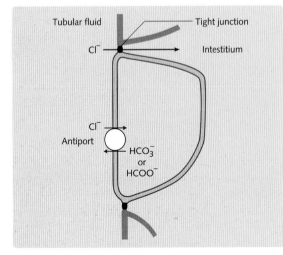

Fig. 8.17 Cl^- transport within the proximal tubule.

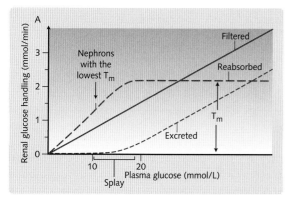

Fig. 8.18 The effect of increasing plasma glucose concentration on filtration, reabsorption and excretion of glucose (glucose titration curves).

Glycosuria (Fig. 8.20) occurs if:

- The renal threshold is exceeded by the filtered glucose load: this is seen with the elevated plasma glucose concentration of uncontrolled diabetes mellitus.
- T_m is lower than normal: this occurs in certain inherited tubular diseases and in pregnancy. Hence, glycosuria will appear with plasma glucose concentration < 10 mmol/L.

In both cases, a glucose tolerance test is useful to distinguish diabetes from a low T_m, especially as diabetes can occur during pregnancy (gestational diabetes).

Amino acids

Plasma amino acids are in a constant state of dynamic equilibrium. They are continuously being absorbed from the gut, while at the same time their steady removal for restructuring/building maintains a concentration of 2.5 mmol/L. Their small size means that they undergo free filtration by the glomerulus. Most amino acids are reabsorbed, primarily in the PT by symport with Na^+ (i.e. a secondary active process). A negligible amount is excreted in the urine, as there is an effective T_m-limited reabsorptive process for amino acids. Amino aciduria can occur if this mechanism is saturated, or in Fanconi syndrome (Fig. 8.20) where the mechanism is faulty.

In the PT, there are at least five different systems, depending on the character of the amino acids:

1. Glycine.
2. Imino acids.
3. Glutamate and aspartate.
4. Neutral amino acids.
5. Cystine and basic amino acids.

Phosphate

Phosphate is essential for mineralized tissue structure (bones and teeth) and 80% of the body PO_4^{3-} content is found in the skeleton. The remaining 20% is distributed mainly in the intracellular fluid (ICF) and least in the plasma, whose concentration is approximately 1 mmol/L.

PO_4^{3-} is freely filtered, the amount filtered being proportional to plasma values. 80% of filtered PO_4^{3-} is reabsorbed in the PT with Na^+ by the $2\,Na^+/1\,PO_4^{3-}$ symporter in the apical membrane. The rate of PO_4^{3-} uptake is hormonally regulated, which primarily affects excretion (Figs. 8.19 and 8.20).

Normally, < 20% filtered PO_4^{3-} is excreted, functioning as an important urinary buffer for H^+. Excretion is increased when plasma concentration exceeds 1.2 mmol/L. A fall in GFR will also cause plasma values to rise. Importantly, the normal amount of filtered PO_4^{3-} (dictated by plasma concentration) is very close to T_m. Hence, plasma $[PO_4^{3-}]$ is regulated by the reabsorptive process.

Urea

Urea is continuously produced by the liver, with a normal plasma concentration of 2.5–7.5 mmol/L. Production is increased in high-protein diets and decreased during starvation. As it is waste, and the end product of protein metabolism, 50–60% is excreted.

Urea is a small molecule and is therefore freely filtered; 40–50% is reabsorbed by passive diffusion. Its concentration increases in the filtrate passing down the PT because of H_2O reabsorption.

In addition to its role as a waste product, urea is useful in the control of water balance. Its presence

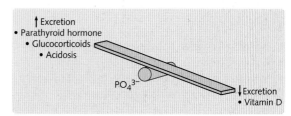

Fig. 8.19 Hormones affecting PO_4^{3-} excretion.

Fig. 8.20 Summary of organic solute and phosphate transport in the proximal tubule and factors affecting excretion

Substance	Normal plasma values (mmol/L)	% Reabsorbed (normal)	% Excreted (normal)	Factors affecting excretion
Glucose	5	100	0	Low T_m
				Diabetes mellitus
Amino acids	2.5	>99	Negligible	High plasma concentration
				Fanconi syndrome
PO_4^{3-}	1	80	21	Plasma >1.2
				Hypoparathyroidism
				Acidosis
				Glucuronidases
				Vitamin D
Urea	2.5–7.5	40–50	50–60	Antidiuretic hormone

in the later segments of the nephron is important in concentrating the urine, especially when under the influence of ADH (Table 8.2). ADH stimulates UT1 (a member of a family of urea transporters (UT1–4) found in renal tissue) in the apical membrane of inner medullary collecting ducts transporting urea into the cell; exit from the cell across the basolateral membrane is via a different transport protein, UT4. The distal tubule and outer medullary collecting ducts are impermeable to urea.

Bicarbonate

HCO_3^- plays a key role in acid–base balance and, by regulating this substance, the kidneys contribute to the regulation of body pH. Normal plasma values are maintained at 20–30 mmol/L.

HCO_3^- is freely filtered and it is essential that virtually all is reabsorbed, otherwise the body fluids would become very acidic. The PT reabsorbs about 80%. HCO_3^- reabsorption is a secondary active process, coupled to either Na^+/H^+ exchange or H^+-ATPase in the apical membrane; transporters deliver H^+ into the lumen:

- Na^+/K^+-ATPase sets up an ionic gradient driving Na^+ into cells.
- Most Na^+ entry is coupled with H^+ secretion into the lumen via an Na^+/H^+ antiporter in the apical membrane.
- Filtered HCO_3^- combines with this secreted H^+ in the lumen, forming H_2CO_3 (carbonic acid).

- On the brush border, the enzyme carbonic anhydrase (CA) catalyses dissociation of H_2CO_3 → CO_2 and H_2O.
- Both the CO_2 and H_2O diffuse readily into the cell, where intracellular CA catalyses reformation of H_2CO_3.
- Intracellular H_2CO_3 then dissociates into:
  - H^+: which is secreted by antiport into the lumen and starts the cycle again
  - HCO_3^-: some of which moves across the basolateral membrane into the interstitial fluid either in exchange for Cl^- or by cotransport with Na^+. Some intracellular fluid HCO_3^- exchanges with luminal Cl^-.

When there is an increased load of HCO_3^-, the PT automatically reabsorbs more. HCO_3^- reabsorption behaves in a T_m manner, although it can be increased somewhat by H^+ – driven by Na^+.

The remainder of HCO_3^- is reabsorbed in the thick ascending limb of Henle, the distal tubule and the collecting duct. In addition to the primary active H^+-ATPase in all H^+-secreting distal tubule segments, type A intercalated cells have a primary active H^+/K^+-ATPase that moves H^+ into the lumen while reabsorbing K^+.

Sulphate

Sulphate reabsorption is an active T_m-limited process maintaining plasma levels at 1–1.5 mmol/L.

Potassium

Normal plasma values are 3.5–5 mmol/L. K^+ is freely filtered with the PT reabsorbing 80–90%. This occurs predominantly by passive diffusion through the tight junctions (paracellular route), although there appears to be a small active component as the PT can reabsorb K$^+$ against a concentration gradient. In the nephron K$^+$ can be:

- Reabsorbed: in the thick, ascending limb of loop of Henle (cotransported with Na$^+$ and Cl$^-$) and the distal tubule (only in severe dietary depletion).
- Secreted: in the thin limb of loop of Henle/distal tubule (generally, Na$^+$ is reabsorbed and K$^+$ secreted). K$^+$ secretion is proportional to the rate of tubular fluid flow and delivery of Na$^+$ to distal parts of the nephron. Aldosterone, secreted in response to increased plasma K$^+$, in enhancing Na$^+$ reabsorption from the cortical collecting duct increases K$^+$ secretion.

In healthy people, the K$^+$ balance is maintained so that secretion = K$^+$ intake. Excretion of filtered K$^+$ can vary from 1 to 110%. It will be decreased when small amounts of Na$^+$ reach the distal tubule and when the amounts of H$^+$ are increased. Hence the following affect the amount of K$^+$ excreted:

- Dietary K$^+$.
- Acid–base balance.
- Delivery of Na$^+$ to the cortical collecting duct.
- Aldosterone.

Secretion by the proximal tubule

The PT actively secretes a large number of substances (Fig. 8.21), including many drugs in addition to its reabsorptive capacity. Some of these substances are filtered freely whereas others are bound to plasma proteins and therefore have a limited filtration, so secretion is vital for their excretion. Substances can be secreted both passively and actively, which can be T$_m$ limited (carrier saturation) or gradient time limited (e.g. PT secretion of H$^+$). Three T$_m$-limited PT secretory mechanisms exist for:

1. Strong organic acids: secreted by the pars recta, as substances move out from peritubular capillaries (e.g. PAH).
2. Strong organic bases: secreted in the pars convoluta (e.g. histamine, thiamine).
3. EDTA (ethylene diamine tetra-acetic acid).

Gradient-time-limited mechanisms handle H$^+$ and K$^+$ secretion:

- H$^+$: Na$^+$/H$^+$ antiporter on the apical membrane is responsible for H secretion- which also depends on HCO$_3^-$ reabsorption.
- K$^+$: secretion is variable and dependent on a number of factors (tubular fluid flow, diet, acid–base balance, aldosterone).

THE LOOP OF HENLE

Role of the loop of Henle

The loop of Henle (LoH) functions primarily as a site for reabsorption and as a countercurrent multiplier, which serves to increase the osmolality of the medullary interstitial fluid.

Reabsorptive function

Proportionally more Na$^+$ and Cl$^-$ are reabsorbed than water: 25% versus 15% of the filtered loads. There is a differential permeability to substances of the ascending and descending limbs:

- Descending limb: reabsorbs H$_2$O not NaCl (very H$_2$O permeable).
- Ascending limb: reabsorbs NaCl not H$_2$O (H$_2$O impermeable).

Countercurrent multiplier

The actions of the above set up an osmotic pressure gradient between the ascending and descending limbs:

- Descending limb: functions as a concentrator, increasing tubular fluid concentration as it flows down towards the hairpin.
- Ascending limb: this functions as a dilutor, progressively removing solutes only as the fluid moves away from the tip of the hairpin, thus becoming more dilute.

	Acidic	Basic
Drugs	Penicillin Salicyclates Furosemide Sulfonomide	Atropine Cimetidine Morphine Quinine
Endogenous substances	Bile salts Fatty acids Prostoglandins Urate	Acetyleholine Creatinine Dopamine Epinephrine

Fig. 8.21 Acidic and basic substances secreted by the proximal tubule.

In addition to the concentration gradient, the effects are multiplied because the flow of tubular fluid occurs in different directions (down the descending limb, up the ascending limbs). Thus solute extrusion out of the ascending limb creates and maintains the high ionic concentration of the medulla.

Fluid entering the LoH is isosmolar to plasma but during its passage more solute than water is absorbed, so it leaves hypo-osmotic (dilute) relative to plasma. This is important for enabling urine concentration in the collecting duct using the least amount of energy; there is osmosis of H_2O into the concentrated salty renal medullary interstitium (set up by LoH).

Microstructure of the loop

The LoH consists of three functionally distinct parts:

1. Thin descending limb.
2. Thin ascending limb: only present in very long loops.
3. Thick ascending limb.

Thin segments, as their name suggests have thin, quite plain epithelia.

Thin descending limb

- Thin, flat epithelia without brush borders, minimal metabolic activity and hence few mitochondria.
- Highly permeable to H_2O.
- Relatively low permeability to Na^+, Cl^- and urea.
- Main function is simple diffusion through its epithelia:
  - H_2O diffuses by osmosis down its concentration gradient created by hypertonic medullary interstitium. This accounts for $\approx 15\%$ filtered H_2O reabsorption and is the primary reason for osmotic equilibration between luminal fluid and medullary interstitial fluid.
  - Some Na^+, Cl^- and urea (UT2 transporter) can enter the lumen but do so in limited amounts.

This continues until osmotic equilibrium in between tubular fluid and medullary interstitium is reached (at the tip of hairpin).

Thin ascending limb

- Epithelial cells are similar to those of the thin descending limb.
- Quite impermeable to H_2O, but exhibits some permeability to Na^+, Cl^- and urea .
- Previous H_2O reabsorption in the descending limb concentrates luminal $Na^+ \rightarrow$ small amounts of mainly passive Na^+ and Cl^- reabsorption via paracellular route plus urea (UT2 transporter) entry.

Thick ascending limb

- Start position is dictated by length of the loop – beginning at the tip of the hairpin in shorter LoH to about half way up the ascending limb in nephrons with very long loops.
- Large cells with plenty of mitochondria, which suit the high metabolic activity required for active transport.
- Reabsorption of $\approx 25\%$ filtered Na^+, Cl^- and K^+ loads. This occurs via the apical $Na^+/K^+/2\,Cl^-$ transporter, which:
  - In association with K^+ leaking back through apical K^+ channels and Cl^- efflux across the basolateral membrane through Cl^- channels generates a transepithelial, lumen positive, pd of ~8 mv.
  - Depends on the Na^+ gradient established by the basolateral Na^+/K^+-ATPase (primary active transporter).
  - The lumen positive pd is the driving force for paracellular movement of monovalent and divalent cations. In fact ~50% Na^+ reabsorption in this nephron segment occurs via this route (Fig. 8.15).

The H_2O impermeability of the thick ascending limb of the LoH means that H_2O cannot move down the osmotic gradient generated by the accumulation of mainly NaCl in the medullary interstitial fluid (Fig. 8.22).

Countercurrent multiplication

The osmolality of fluid as it enters the LoH is 300 mosmol/kg H_2O, isosmolar to plasma. When it leaves, it is 100 mosmol/kg H_2O and therefore more dilute than plasma (Fig. 8.23).

A gradient of osmolality (pooling of NaCl and urea in the interstitial fluid) is set up along the renal pyramids of the medulla by the LoH acting as a countercurrent multiplier. The LoH itself has a concentration gradient along its course. This is highest at the hairpin loop, and is in equilibrium with the medullary interstitium (maximum 1400 mosmol/kg H_2O).

As substance movement in the descending limb is passive whereas it is active in the ascending limb, we can consider the ascending limb to initiate the osmolality gradient/countercurrent multiplier. In addition, the cells in the wall of this segment are able to sustain an osmotic pressure difference of 200 mosmol/kg H_2O between the intraluminal contents and the interstitium at any point. In summary the major factors contributing to high solute contribution in the medullary interstitium are:

- Active Na^+, K^+, and Cl^- movement and passive movement of Na^+ and Cl^- out of, respectively,

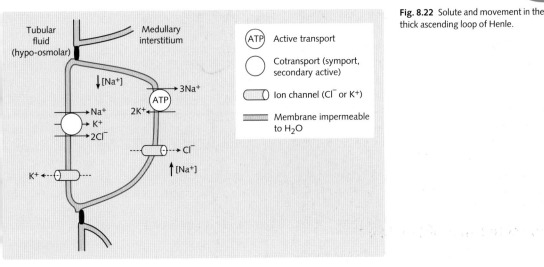

Fig. 8.22 Solute and movement in the thick ascending loop of Henle.

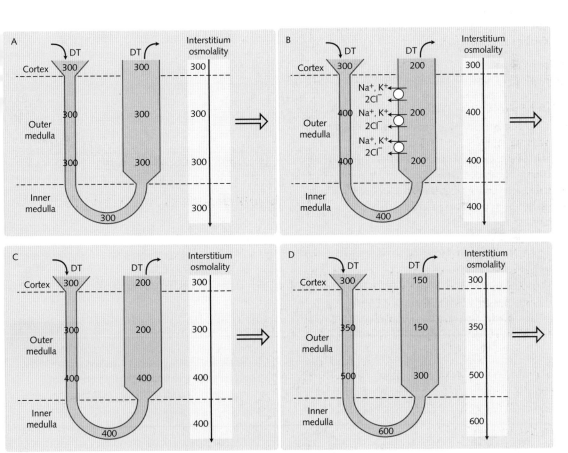

Fig. 8.23 The establishment of the countercurrent multiplier in the loop of Henle. (A) The situation as it is before the countercurrent multiplier system is established, i.e. no movement of substances in the loop of Henle. Fluid stays isosmolar to plasma throughout. (B) Secondary active Na+/K+/2Cl− transport from the thick ascending limb and passive Na+ and Cl− movement (not shown) from the thin ascending limb into the medullary interstitium and establishes a gradient of 200 mosmol/kg H$_2$O. This is followed by osmotic movement of H$_2$O out of the descending limb and to a lesser extent, passive movement of ions and urea into the descending limb. (C) Additional flow from the proximal tubule pushes more concentrated fluid further down the descending limb; this ends up in the ascending limb. (D) Additional ions move out from the ascending limb and maintain the gradient of 200 mosmol/kg H$_2$O. H$_2$O and solute movement across the descending limb reaches osmotic equilibrium with the interstitial fluid. Tubular fluid osmolality reaches its maximum value at the tip of the loop Henle. *Continued*

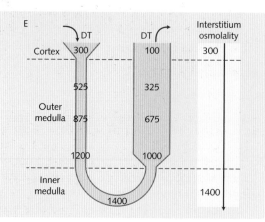

Fig. 8.23 Cont'd (E) Further cycles of stage D result in an increase in the osmolality of the fluid in the thin descending limb as it comes into equilibrium with the interstitial fluid. More and more solute is added from the ascending limb to the medullary interstitial fluid. In time, these solutes become trapped within the medulla and added H_2O is removed from the medulla by the action of the vasa recta.

the thick ascending limb and the thin ascending limb of the LoH.
- Active transport of ions from collecting ducts into the interstitium.
- Passive diffusion of large amounts of urea out of the inner medullary collecting ducts in the interstitial fluid.
- The solutes added become trapped within the medulla and the additional H_2O is removed from the medulla by the action of the vasa recta acting as countercurrent exchangers.

Role of the vasa recta

The vasa recta are derived from the efferent arterioles in juxtamedullary nephrons. Their hairpin arrangement and close association with the LoH enables them to function as a countercurrent exchanger. Hence, the descending and ascending vasa recta run in parallel near each other. The results of this are:

- Medullary osmotic gradient generated is not dissipated.
- H_2O and waste products are removed (aided by high permeability to these substances).

Examples of a countercurrent mechanism are vasa recta and LoH. The inflow runs in:
- Parallel.
- Opposite direction to the outflow.
- Close proximity.

These blood vessels have a relatively low blood flow (1–2% total renal blood flow), which, although not required for the creation of a hyperosmolar medulla, is important in maintaining it. It means that solute can accumulate in the medullary interstitium such that it is grossly higher than that in the entering blood.

Like other capillaries, the vasa recta are permeable to solute and water. Therefore, as the vessels descend progressively further into the medulla, solutes (NaCl and urea) enter the vasa recta while more H_2O diffuses out into the interstitium.

At the tip of the loop, the plasma osmolality approaches that of the interstitium (very high). In addition, the blood is very viscous, with a high plasma protein concentration (high oncotic pressure) secondary to H_2O loss. Therefore, in the ascending vasa recta, H_2O diffuses back with loss of most of the solutes, as it tends to equilibrate with the decreasing medullary interstitial fluid osmolality.

The collecting ducts pass through the cortex and medulla; a functional difference (arising from the differential permeability to urea) exists between the portions in the two different parts of the parenchyma. The vasa recta also reabsorb 5% of the filtered H_2O load from the parts of the collecting ducts and thin descending limb of LoH within the medulla, preventing dilution of the medullary interstitium. This is crucial to the concentration of urine.

Role of urea

Urea contributes to the interstitial osmotic pressure in the medullary pyramids:

- Half of freely filtered urea is reabsorbed in the PT.
- Urea is secreted in the thin segment of the LoH, diffusing into the tubule and restoring the urea to ~100% of that which was filtered.
- When tubular fluid reaches the inner medullary collecting ducts, the luminal [urea] has risen to 500 mmol/L or greater (100 times plasma concentration). This is due to the differential permeability of the cortical collecting duct to urea and H_2O; permeable to H_2O (ADH present) but is not permeable to urea (presence or absence of ADH).
- This high [urea] in the medullary collecting duct lumen results in diffusion of urea out into the interstitium via specialized urea transporters (UT1 and UT4).
- There is low blood flow in the medulla, so the urea that has diffused into the interstitium is essentially trapped but the luminal [urea] is always higher than that in the interstitial fluid because of concomitant H_2O reabsorption.

- Approximately half of the filtered load stays in the lumen, which is then excreted.

This recycling of urea into the interstitium greatly contributes to its hyperosmolality.

ADH is required for the permeability of medullary collecting tubules to urea and hence for its diffusion into the medulla, with consequent diffusion into the thin limbs of the LoH. It then follows that ↑ADH results in ↑medullary osmolality. In addition, ADH acts to increase H_2O permeability in medullary collecting tubules, causing H_2O reabsorption from the cortical components, both of which result in a more concentrated urine.

Maximum urine osmolality is influenced by dietary protein: low protein diets result in a reduced capacity to elaborate concentrated urine because of the reduced availability of urea for sequestration in the medullary interstitium, i.e. reduced medullary urea, reduced medullary osmolality, reduced urine osmolality.

Regulation of urine concentration

Urine volume can range from 400 mL to 23 L, with an average value of 1–1.5 L (normal range 400 mL to 2–3 L). Osmolality can similarly vary from 60 to 1200–1400 mmol/kg H_2O, with an average value of 600 mmol/kg H_2O.

The obligatory volume of urine excreted per day is dictated by the maximal urine osmolality and the amount of solute to be excreted over that period (see page 186). Thus if the maximum urine osmolality is 1400 mosmol/kg H_2O and the solute load to be excreted is 800 mosmole/24 h, the obligatory urine volume is 800/1400 ~ 570 ml/24 h:

$$\frac{\text{Osmolality of urine}}{\text{Max. osmolality of urine}} = \text{Litres of urine produced}$$

Both the concentration and volume of urine are determined by plasma ADH concentration. High [ADH] conserve body water and increase concentration of urine by increasing water permeability of the distal tubules and collecting ducts. As mentioned earlier, the high osmolality of the medullary interstitium provides the gradient for H_2O reabsorption, in the presence of ↑ADH.

Aldosterone – an adrenal steroid – acts on the cortical collecting tubule via the P cells, increasing absorption of NaCl, and also water, which follows by diffusion. This produces a smaller volume of fluid of unchanged osmolality for delivery to inner medullary collecting ducts where water reabsorption (hence urine osmolality) is determined by plasma [ADH] and the magnitude of the transepithelial osmotic gradient.

Body fluid osmolality

Concepts of osmolality

Body fluid volume, manifested as body weight, remains relatively stable on a daily basis. Therefore, intake of water over 24 hours (via thirst mechanisms) is balanced by kidney excretion (output). A minimal volume 400 mL/day of urine permits the maintenance of body fluid homoeostasis.

Osmolality refers to the ratio of solute to water, essentially:

Body Na^+ content + anion : water content

Normal plasma osmolality is strictly maintained at levels of 285–295 mosmol/kg H_2O. If a change in either H_2O balance or that of Na^+ causes osmolality to vary by 3 mosmol/kg H_2O, the body's regulatory mechanisms will be stimulated. These are:

- Osmoreceptors: triggered by change in osmolality.
- Baroreceptors: triggered by ↓ in plasma volume.

If both plasma volume and osmolality are reduced then osmoreceptors exert more influence. However, a large drop in plasma volume takes priority (plasma becomes hypo-osmolar).

Osmoreceptors

Osmoreceptors in the supraoptic and paraventricular areas of the anterior hypothalamus, supplied by the internal carotid artery, are responsive to changes in plasma osmolality. Their function is to regulate H_2O excretion and thirst:

- ↑Plasma osmolality → Stimulation of osmoreceptors → ↑rate of ADH secretion → ↓renal H_2O excretion → ↓plasma osmolality.
- Conversely, ↓plasma osmolality causes ↓ADH secretion.
- ↑Plasma osmolality also stimulates thirst centres in the lateral preoptic area which, together with the osmoreceptors, cause the person to feel thirsty and drink H_2O.

Synaptic input to ADH secreting cells from many other areas in the brain also exist, such that to a much lesser degree, pain, fear and nausea can alter ADH secretion.

Effect of other solutes

Plasma osmolality is dictated mainly by total body Na^+ with its associated anions. However, other solutes can alter osmolality without altering water balance. These solutes vary in their effectiveness to stimulate osmoreceptors, depending on their ability to cause cellular dehydration, i.e. more effective stimulants will be substances which have more difficulty crossing the membrane.

Antidiuretic hormone

ADH is synthesized as part of a large precursor molecule by cells of the supraoptic paraventricular nuclei in the hypothalamus (Fig. 8.24). Synthesis is completed after transportation to the posterior lobe of the pituitary, where it is thought to undergo progressive cleavage as it moves down the axons. ADH is stored in nerve terminals in the posterior pituitary, associated with neurophysin, until its release is triggered by a rise in plasma osmolality.

Membrane depolarization by action potentials in the neurons → Ca^{2+} influx → secretory membrane fusion with cell membrane → release of ADH and neurophysin into the bloodstream.

Cellular actions

All three ADH or V receptors are G-protein coupled. Their functions are as follows:

- V1 receptors: stimulate vasoconstriction by increasing intracellular $[Ca^{2+}]$ in smooth muscle in the walls of blood vessels.
- V2 receptors: reduce kidney excretion of H_2O. This only occurs if ADH is present on the peritubular (non-luminal) side of the cell, as V2 receptors are located in the basal membrane of renal tubules.
- V3 receptors: appear to mediate the effect of ADH on the pituitary gland, facilitating the release of ACTH.

ADH enhances H_2O reabsorption in the cortical collecting tubule by causing fusion of aquaporins (water channels; AQP2) with the luminal membrane.

Fig. 8.24 Role of antidiuretic hormone in maintaining plasma osmolality.

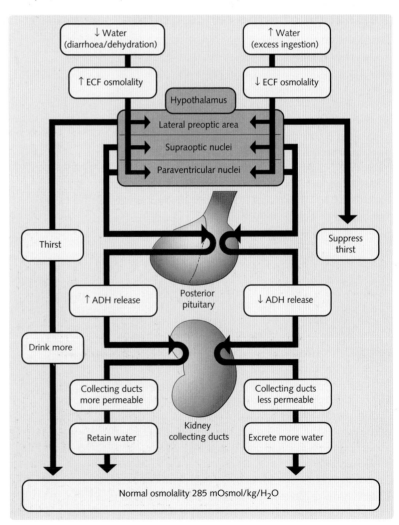

This is particularly important in cortical and outer medullary portion of the collecting duct (CD) as they would have a negligible H_2O permeability in the absence of ADH.

Plasma ADH concentrations dictate the permeability of the CD system. Therefore, with very low [ADH], the hypo-osmotic urine from the ascending LoH will pass through the distal tubule (DT) and proximal parts of the CD. When it enters the inner medullary portion, there is a massive osmotic gradient between tubular fluid and interstitium, and so there is some H_2O reabsorption by osmosis, the amount being limited by the finite H_2O permeability of the tubule cells. Hence, the high tubular volume containing a lot of water (therefore hypo-osmotic) will be excreted (diuresis).

Fate of ADH

Certain metabolic features of ADH ensure plasma osmolality is controlled precisely:

- Rapid release.
- Short half-life (time taken for plasma levels to halve): 10–15 min.
- Rapid cessation of release.
- Rapid removal from the blood:
    - 50% removed by the liver and kidney
    - 40% is metabolized
    - 10% is excreted in urine.

Drugs and ADH

Osmoregulation can be disturbed by the actions of certain drugs on ADH release, which can be:

- Increased: with nicotine, morphine or barbiturates.
- Decreased: with alcohol or antiotensin-converting enzyme (ACE) inhibitors (by inhibiting angiotensin II, which is a promoter of ADH secretion).

Diseases due to disruption of ADH regulation

Inappropriate secretion of ADH can occur from the pituitary or elsewhere (Fig. 8.25). ADH secretion fails to be suppressed even with low [Na^+], low plasma osmolality and H_2O overload. Clinical features are:

- Inappropriately ↑urine osmolality: typically 400 mosmol/kg H_2O in the face of ↓plasma osmolality.
- Inappropriately ↑urine [Na^+] > 20 mmol/L: high plasma volume secondary to retained H_2O switches off aldosterone production so more Na^+ is excreted.
- ↓plasma [Na^+] (typically 125 mmol/L with normal values at 135–145 mmol/L) and ↓plasma osmolality ≤ 260 mosmol/kg H_2O.

Hyponatraemia occurs without the following:

- Hypovolaemia.
- Oedema.
- Endocrine dysfunction.
- Renal failure.
- Drugs.

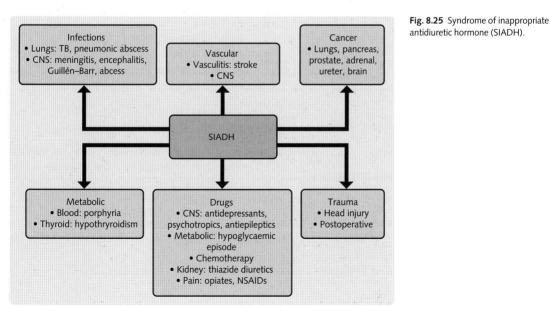

Fig. 8.25 Syndrome of inappropriate antidiuretic hormone (SIADH).

Diabetes insipidus

This syndrome is caused by ADH deficiency or when the kidney fails to respond to ADH (Fig. 8.26). There is continual H_2O diuresis with a urine output of approximately 25 L/day because ADH-dependent H_2O reabsorption in the kidney is about 23 L. The clinical features are:

- Polyuria (production of large amounts of dilute urine).
- Polydipsia (consumption of large quantities of water).
- Hypo-osmolar urine.

Water clearance and reabsorption

Plasma osmolality altered by water balance has an effect on urine production:

- Dehydration: ↑ plasma osmolality →
  ↑ osmotically free H_2O reabsorption by the kidneys resulting in:
  - dilute plasma
  - concentrated urine osmolality higher than that of plasma.
- Too much water: ↓ plasma osmolality →
  ↑ osmotically free H_2O excretion (by the kidneys) → dilute urine (osmolality lower than that of plasma).

The rate at which osmotically active substances are cleared from plasma, is known as the osmotic clearance (C_{OSM}; mL/min). Recall that clearance (C):

$$C = \frac{U \times V}{P}$$

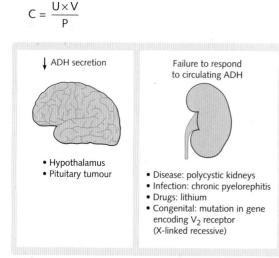

ADH secretion

Failure to respond to circulating ADH

- Hypothalamus
- Pituitary tumour

- Disease: polycystic kidneys
- Infection: chronic pyelorephitis
- Drugs: lithium
- Congenital: mutation in gene encoding V_2 receptor (X-linked recessive)

Fig. 8.26 Control of diabetes insipidus. Two main mechanisms are involved in diabetes insipidus. These relate to the secretion or response to antidiuretic hormone (ADH).

Where U = urinary concentration of a substance, V = urine flow rate and P = plasma concentration of a substance. Then:

$$C_{OSM} = \frac{U_{OSM} \times V}{P_{OSM}}$$

Where U_{OSM} = urine osmolality and P_{OSM} = plasma osmolality.

Urine osmolality = that of plasma (iso-osmolar urine)
The amount of H_2O used to excrete the osmotically active substances of the urine is sufficient to keep these substances at the same osmotic pressure as the plasma.

$$\frac{U_{OSM}}{P_{OSM}} = 1$$

So C_{OSM} = V, i.e. osmotic clearance = urine flow.

Urine osmolality > that of plasma (hyperosmolar urine)
As $U_{OSM} > P_{OSM}$, C_{OSM} = > 1 × V. Therefore $C_{OSM} > V$, i.e. osmotic clearance is greater than urine flow. This is because a volume of osmotically free water is reabsorbed ($T^c{}_{H_2O}$), not excreted, accounting for the difference between clearance and urine flow. Hence $V = C_{OSM} - T^c{}_{H_2O}$.

Urine osmolality < that of plasma (hypo-osmolar urine)

$$U_{OSM} < P_{OSM}, \text{ so } C_{OSM} = < 1 \times V$$

Therefore $C_{OSM} < V$, i.e. osmotic clearance is less than urine flow. An additional volume of free water is excreted (C_{H_2O}) to the urine volume per minute: $V = C_{OSM} + C_{H_2O}$.
$T^c{}_{H_2O}$ and C_{H_2O} provide information on the ability of the kidney to conserve or excrete osmotically free water and the function of the nephron segments involved in the production of a dilute or concentrated urine.

Effect of solute output on urine volume

Limitation of the kidney's concentrating ability restrict the maximum urinary osmolality to about 1400 mosmol/kg H_2O with the volume dependent on the balance between:

- Plasma ADH concentration.
- Amount of excreted solute.

$$\frac{\text{Volume of}}{\text{urine produced}} = \frac{\text{Amount of excreted solute}}{\text{Max. urine osmolality (fixed)}}$$

So $\uparrow$ solute = $\uparrow$ volume, and $\downarrow$ solute = $\downarrow$ volume. Therefore, more excretory solute, even with maximal ADH concentration, will still result in a degree of diuresis. This can be illustrated by the non-reabsorbable solute, mannitol, which decreases the kidney's ability to concentrate urine and is thus termed an osmotic diuretic.

Adrenal steroids and urinary dilution

The normal renal response to H_2O loading, i.e. very dilute urine production, is impaired with adrenal insufficiency and/or $\downarrow$ adrenal steroids. The effects of deficiency of the steroids known as glucocorticoids (e.g. cortisol) and mineralocorticoids (e.g. aldosterone) are:

- $\downarrow$ Glucocorticoids: $\uparrow$ CD H_2O permeability (more H_2O reabsorption).
- $\downarrow$ Glucocorticoids and mineralocorticoids: $\uparrow$ Plasma ADH concentration with the same effect as above (production of concentrated urine).

Body fluid volume

The portion of ECF that perfuses tissue is known as the effective circulating volume (ECV). Na^+ content, which can be altered by changes in either filtration or reabsorption, will affect ECF volume and hence ECV. This is because, in the ECF, Na^+ is the main osmotically active solute and therefore largely dictates osmolality. This affects the osmoregulatory mechanism, resulting in change of the ECF volume and osmolality. However, osmolality is corrected within minutes, whereas disturbances in body volume (with altered Na^+ body content) can take days to correct.

If there is a sufficient drop in ECF volume, GFR can be lowered by tubuloglomerular feedback to prevent further volume loss. Maximal Na^+ reabsorption results in a urinary excretion of < 1 mM per day with possible consequences on acid–base balance through K^+ and H^+ excretion.

Renin and angiotensin

This system is also discussed in the section on renal hormones (see p. 165).

Renin

The enzyme renin is both synthesized and stored in the JGA. Its release is stimulated by:

1. $\uparrow$ Activity of sympathetic nerves: due to $\downarrow$ systemic blood pressure detected by carotid artery baroreceptors.

2. $\downarrow$ Afferent arteriole wall tension: $\downarrow$ ECF volume $\rightarrow$ $\downarrow$ systemic blood pressure $\rightarrow$ $\downarrow$ renal perfusion pressure $\rightarrow$ $\downarrow$ wall tension at granular cells $\rightarrow$ renin release.

3. $\downarrow$ Na^+ detection by macula densa: which leads to inhibition of synthesis of substances inhibitory to renin release *and* secretion of prostaglandins $\rightarrow$ renin release.

In addition, renin secretion is increased by diuretics, haemorrhage, upright posture, heart failure and constriction of the renal artery.

Angiotensin substances

Once it is released from the JGA into the blood, renin acts on angiotensinogen (an alpha$_2$-globulin) produced in the liver to yield the physiologically inactive decapeptide angiotensin I. Angiotensin I undergoes removal of two amino acids by angiotensin-converting enzyme (ACE) in the lungs, to form angiotensin II (octopeptide). Angiotensin II acts via two classes of receptor: A_{T1} and A_{T2} to:

- Release aldosterone: by acting directly on the adrenal cortex zona glomerulosa.
- Directly vasoconstrict renal arterioles: affecting efferent more than afferent arterioles.
- $\uparrow$ Na^+ reabsorption in the PT.
- Release ADH.
- Stimulate thirst.
- Inhibit renin release by negative feedback on JGA cells.

In addition, angiotensin II decreases the baroreceptor reflex sensitivity by acting on the brain. Hence its effects in raising blood pressure can be sustained by this 'faulty' reflex mechanism. Inhibitors of ACE (e.g. ramipril) are used to treat hypertension (high blood pressure) by $\downarrow$ angiotensin II production with consequential:

- $\downarrow$ Vasoconstriction.
- $\downarrow$ Aldosterone secretion (with its subsequent $\downarrow$ ECF volume).

Aldosterone

Aldosterone is an adrenal mineralocorticoid hormone. Release is mediated by mechanisms that either directly stimulate the adrenal cortex or promote the renin–angiotensin–aldosterone (RAA) system.

Direct stimulation of the adrenal gland

- $\uparrow$ Plasma [K^+]: very small rises can cause large increases in aldosterone release resulting in increased DT K^+ secretion, returning plasma [K^+] to normal.
- $\downarrow$ Plasma [Na^+]: low [Na^+] stimulates aldosterone, although this is not an important mechanism.
- Adrenocorticotrophic hormone (ACTH): causes transient stimulation of aldosterone secretion even if high [ACTH] is maintained.

Activation of the RAA system

- Angiotensin II.
- $\downarrow$ ECF.
- $\downarrow$ body Na^+ content stimulates renin release.

Secondary hyperaldosteronism, which can result from heart, liver and kidney failure, can occur by chronic stimulation of the RAA system.

Aldosterone stimulates Na^+ reabsorption by entering P cells. Here, through a series of interactions and steps, it increases transcription of basolateral Na^+/K^+-ATPase and expression of distal nephron Na^+ channels and $Na^+/K^+/2Cl^-$ transporters in the apical membrane to $\uparrow$ Na^+ reabsorption. 2% of filtered Na^+ load is under control of aldosterone, which translates as a large amount (~ 520 mmole/24 h) considering the volume filtered per day (180 L/day).

Its intracellular actions involve increasing Na^+ reabsorption within the kidney as well as other sites: colon, salivary glands, gastric glands and sweat glands. It further encourages the secretion of K^+ and H^+ in the kidney.

Factors affecting Na reabsorption

Starling factors

The two variations of renal function controlling Na^+ excretion are

1. GFR (discussed above).
2. Rate of Na^+ reabsorption.

Na^+ reabsorption in the PT is automatically adjusted to correct disturbances of body Na^+ content and hence that of ECV. This involves altered Starling forces – hydrostatic (P) and oncotic (π) pressures. The rate of reabsorption of NaCl and H_2O in the PT is related to the amount and rate of uptake from the lateral intercellular spaces into the capillaries, which depend on these forces:

Forces favouring capillary uptake

- π_{CAP}: oncotic pressure of the capillary.
- P_{ISF}: interstitial fluid hydrostatic pressure.

Forces opposing capillary uptake

- π_{ISF}: interstitial fluid oncotic pressure.
- P_{CAP}: hydrostatic pressure of the capillary.

For example, $\downarrow$ NaCl intake causes $\downarrow$ ECF volume with resultant $\downarrow$ P_{CAP} but $\uparrow$ π_{CAP}. Hence, PT reabsorption of NaCl by tubule cells will increase.

A rise in P_{CAP} or P_{ISF} results in the leaking back of fluid reabsorbed from interstitial fluid across tight junctions, the usual cause of the former being $\uparrow$ venous pressure.

Sympathetic drive from the renal nerves

Renal sympathetic activity is controlled by arterial baroreceptors, which detect changes in plasma/ECF volume that correspond with blood pressure (BP). Na^+ reabsorption will then be altered to correct values of ECF volume and BP. The effects of renal nerve stimulation are:

- $\uparrow$ Renin secretion: by direct action on β_1-adrenergic receptors of granular cells.
- Vasoconstriction of afferent and efferent arterioles: activating JGA.
- $\uparrow$ Na^+ reabsorption: by nerve endings containing catecholamines directly acting on tubular cells.

Prostaglandins

Prostaglandin (PG) synthesis from arachidonic acid is stimulated by $\downarrow$ ECV. There are at least three sites for PG synthesis; these differ in the type of PG that predominates:

1. Cortex (including glomeruli and arterioles): prostacyclin or PGI_2 (vasodilator).
2. Medullary interstitial cells: PGE_2 (vasodilator).
3. Collecting duct epithelial cells.

In addition to PGI_2 and PGE_2, other forms of prostaglandin exist and have important renal effects, some of which are manifest in only pathological conditions e.g. haemorrhage.

PGI_2

- Most important vasodilator: useful in $\downarrow$ renal perfusion pressure.
  Mediates renin release.

PGE_2

- Vasodilator.
- Mainly affects the CD:
  - natriuretic: limits stimulation of ATP-dependent Na^+ reabsorption and is therefore useful in the prevention of medullary tubule cell anoxia with hypovolaemia ($\downarrow$ ECF) by decreasing the energy requirements of the cell

- diuretic: impairs ADH action, promoting H_2O excretion.

Thromboxane A_2 Thromboxane A_2 (TXA_2) is a vasoconstrictor: small amounts are produced after renal damage, which typically results from obstruction of the ureter. The consequence is that less blood is available for a kidney that is not filtering properly.

In healthy people, NSAIDs, such as aspirin, ibuprofen and indometacin, inhibit renal PG synthesis without altering RBF or GFR. However, where the vasodilator effects of PGs are needed, NSAIDs can cause drastic impairment of RBF and GFR.

Atrial natriuretic peptide

Cardiac atrial cells produce the hormone atrial natriuretic peptide (ANP) when atrial fibres stretch in response to ↑venous return as a consequence of ↑ECV. Granules containing ANP precursor in the cardiocytes are modified before release into the plasma. The various vascular and tubular effects of ANP act to promote Na^+ excretion, which is mediated through cell-specific receptors to which the ANP binds. This causes an increase in intracellular cyclic guanine monophosphate (cGMP), with the following actions:

- Afferent arterial vasodilation: promoting ↑GFR.
- Inhibition of granular cell release of renin: which will itself ↓secretion of aldosterone while ↑dopamine release.
- Inhibition of Na^+/K^+-ATPase with additional closure of Na^+ channels: acting on various tubular sites, especially the inner medullary CD. Na^+ reabsorption is ↓here and somewhat ↓in the PT, with ↑Na^+ and H_2O excretion.

Dopamine

In the kidneys, mostly PT cells synthesize dopamine with the following actions:

- ↓Tubular Na^+ transport by:
  - inhibition of Na^+/K^+-ATPase
  - ↓Na^+/H^+ antiport activity.
- Natriuresis (Na^+ excretion): even when RBF and GFR are unchanged.
- Vasodilation.

Kinins

Kinins are vasodilator proteins that are cleaved off kininogens by the enzyme kallikrein. Renal-produced kinins have similar functions to PGs:

- Vasodilatation.
- Natriuresis.
- Inhibition of ADH release.

Natriuretic hormone

The hypothalamus probably produces this hormone in response to ↑ECV, ↑ECF or ↓ in the quantity of functioning nephrons. Their role is to increase the fraction of filtered Na^+ to be excreted by inhibiting Na^+/K^+-ATPase.

Regulation of body fluid pH

Precise involvement of H^+ balance, which is the basis of pH, is imperative because it influences almost all the body's enzymes' activity. The normal arterial blood pH is 7.4, although variations up to 0.05 pH units have no associated negative effects. The range is therefore 7.35–7.45, with a corresponding [H^+] of 45–35 nmol/L ([H^+] which is inversely related to pH). A range of pH 7.0–7.8 is compatible with life; if pH decreases to 6.8, life can be sustained for only a few hours.

Key to the function of acid–base balance are:

- Principles of balance applied to input and output of acids and bases.
- Control of substances involved in minimizing disturbances of [H^+]. These are known as physiological buffer systems.

Input and output

- Output of acids and bases normally matches input on a daily basis. It occurs in the form of respiratory elimination and by kidney excretion.
- Input involves the addition by many physiological processes:
  - metabolic generation
  - addition by gastrointestinal activity
  - ingestion or products of ingested materials.

Buffer systems

The purpose of a buffer solution is to minimize pH change with the addition of acid or base to it. It contains either:

- A weak acid and its conjugate base:

$$HA \quad \leftrightarrow \quad H^+ \quad + \quad A^-$$

  Weak acid   Proton   Conjugate base

- A weak base and its conjugate acid:

$$BH \quad \leftrightarrow \quad H^+ \quad + \quad B^-$$

  Conjugate acid   Proton   Weak base

Addition of H^+ will cause the reaction to swing to the left of the equation; combining with the conjugate base or weak base. Buffers do not eliminate or create

H^+ de novo, but instead lock them away or release them (if H^+ is removed or a strong alkali added).

Importantly, amphoteric substances such as amino acids and PO_4^{3-} can both donate and accept H^+.

Equilibrium and pH

Different reactions have different equilibrium values (K) depending on whether the reaction is shifted more to the left or the right.

$$HA \quad \leftrightarrow \quad H^+ \quad + \quad A^-$$

$$K \quad = \quad \frac{[H^+][A^-]}{[HA]} \quad = \quad \frac{[H^+][base]}{[acid]}$$

Rearrangement in terms of $[H^+]$ gives:

$$[H^+] = \frac{K[acid]}{[base]}$$

Owing to the small value of K, the equation can be re-expressed using the term pK, which is the log of the inverse of K:

$$pK = log\left[\frac{1}{K}\right]$$

Similarly:

$$pH = log\left[\frac{1}{[H^+]}\right]$$

Combining these equations results in the Henderson–Hasselbalch equation:

$$pH = pK + log_{10}\left[\frac{[base]}{[acid]}\right]$$

Physiological buffers

Several buffer systems operate within the body fluid compartments:

- Blood: HCO_3^-, PO_4^{3-} and plasma proteins.
- ICF: HCO_3^-, PO_4^{3-} and proteins.
- ECF: HCO_3^-, PO_4^{3-} and proteins.

Bicarbonate buffer system

This is the most important system in all compartments. The reaction concerned is:

$$CO_2 + H_2O \leftrightarrow H_2CO_3 \leftrightarrow H^+ + HCO_3^-$$

Where H_2CO_3 = carbonic acid, a weak acid.

$[CO_2]$ and $[HCO_3^-]$ are tightly controlled by the lungs and kidneys, respectively, both of which therefore regulate pH. It is no great surprise, then, that carbonic anhydrase (the catalyst in the above reaction, and without which the reaction would be very slow) is richly distributed in the lung alveoli walls and renal tubule epithelium.

The Henderson–Hasselbalch equation for this system can be written as:

$$pH = pK + log\left(\frac{[HCO_3^-]}{[H_2CO_3]}\right)$$

As $[H_2CO_3]$ is determined by the amount of CO_2 dissolved per unit plasma, then:

$$pH = pK + \frac{log[HCO_3^-]}{[CO_2]}$$

The solubility coefficient of CO_2 at $37°C = 0.23$. Hence:

$$[CO_2] = 0.23 \times PCO_2$$

So:

$$pH = pK + log\left(\frac{[HCO_3^-]}{0.23 \times PCO_2}\right)$$

Normal values for this system are: $[HCO_3] = 20–30\,mmol/L$, $PCO_2 = 4.4–5.3\,kPa$ and $pK = 6.1$. Substituting these values into this final equation gives a calculated pH of 7.4.

Renal handling of acids and bases

HCO_3^- handling is essential to acid–base balance. The kidney can produce urine with a pH ranging from 4.5 to 8.0, depending on the pH of the ECF and the consequent balance between the renal excretion of H^+ and HCO_3^-. PO_4^{3-} and ammonium (NH_4^+) processing by the kidney further aids its maintenance of body pH.

HCO_3^- reabsorption

Metabolic processes produce H^+, which reacts with HCO_3^- to form CO_2. This is exhaled by the lungs, and effectively removes HCO_3^- from the body:

$$H^+ + HCO_3^- \leftrightarrow H_2CO_3 \leftrightarrow CO_2 + H_2O$$

The kidney is involved in conservational HCO_3^- handling – primarily its reabsorption. HCO_3^- is freely filtered and 25 mmol/L enters the PT, where approximately 80% is reabsorbed:

A T_m-dependent mechanism applies, whereby the T_m value lies very close to normal plasma values, thus maintaining them. T_m can, however, be changed, primarily, according to the rate of H^+ secretion.

The process of HCO_3^- reabsorption is secondary active, involving tubular secretion of H^+. There is no direct transport of HCO_3^- but of conversion products, which reform within epithelial cells; reclamation of filtered HCO_3^- to preserve buffer stores:

- H^+: is secreted into the lumen of the PT by Na^+/H^+ exchanger and/or H^+-ATPase. It reacts with luminal HCO_3^-, forming H_2CO_3.
- Carbonic anhydrase (CA): catalyses the breakdown of $H_2CO_3 \rightarrow CO_2$ and H_2O.
- CO_2 and H_2O diffuse into epithelial cells.
- Intracellular CA catalyses the reformation of H_2CO_3.
- H_2CO_3 dissociates into:
    - H^+: secreted back into the lumen
    - HCO_3^-: most of which enters plasma on a $3Na^+/HCO_3^-$ cotransporter in the basolateral membrane; a small amount is secreted back into the lumen on the Cl^-/HCO_3^- exchanger.

H^+ secretion is also vital for the reabsorption of 10% of the HCO_3^- in the thick ascending limb of LoH, and for the reabsorption of a similar amount in DT. In the DT:

- Intercalated cells: are involved specifically in this reabsorption; they secrete H^+ by H^+-ATPase and H^+/K^+-ATPase.
- CA: is limited so less CO_2 and H_2O are formed.

Although filtered HCO_3^- buffers H^+ in the PT, other mechanisms overcome the problem of effective H^+ secretion.

Conversion of alkaline PO_4^{3-} to acidic PO_4^{3-}

Acidic and basic forms of inorganic PO_4^{3-} exist in plasma, with a predominance of the basic type ($H_2PO_4^-$).

$$H^+ + HPO_4^{2-} \leftrightarrow H_2PO_4^-$$
$$\text{Base} \qquad \text{Acid}$$

The PO_4^{3-} are filtered through to the PT, where acidic conversion of basic PO_4^{3-} occurs as a result of H^+ secretion by this portion of the nephron – formation of titratable acid: defined as the amount of base needed to titrate the pH of the urine back to that of plasma. This conversion occurs at all nephron segments where H^+ is secreted. H^+ excretion associated with PO_4^{3-} is ~40 mmol/day. Equally important is that for each H^+ secreted (from the dissociation of H_2CO_3 in the renal tubular cells)

$1 HCO_3^-$ is formed and returned to the blood; new HCO_3^- to replenish plasma buffer stores.

Ammonium (NH_4^+) secretion

The liver generates NH_4^+ and HCO_3^- from the products of protein catabolism. The NH_4^+ and HCO_3^- are further processed to urea or glutamine, the latter of which is taken up by PT cells, which convert glutamine back into HCO_3^- and NH_4^+ (dissociates into NH_3 and H^+).

Cell membranes are freely permeable to NH_3; diffusion into the lumen occurs where it forms NH_4^+ with secreted H^+ (diffusion trapping/non-ionic diffusion). In addition NH_4^+ can masquerade as other ions, some is secreted into the lumen of the PT by the Na^+/H^+ exchanger (replaces H^+). Approximately 50% of NH_4^+ delivered to the LoH is reabsorbed in the thick ascending limb by the $Na^+/K^+/2Cl^-$ transporter (replaces K^+). Dissociation of the NH_4^+ to NH_3 and H^+ in the medullary interstitial fluid occurs; NH_3 diffuses into the CD lumen where it combines with secreted H^+ (from H_2CO_3; apical H^+/K^+-ATPase) and is excreted as NH_4^+ (diffusion trapping; Fig. 8.27).

Hence, the purpose of NH_4^+ secretion is to facilitate the regeneration of HCO_3^-. Excretion of NH_4^+ removes 1 H^+ and returns 1 HCO_3^- to the blood due to overall glutamine handling by the nephron.

NH_4^+ excretion is increased by acidosis, as $\downarrow$ECF pH increases the generation of glutamine by the liver. In the kidney, $\downarrow$ECF pH also stimulates oxidation of glutamine in the PT cells (the opposite happens with $\uparrow$ECF pH). Hence, renal synthesis and excretion of NH_4^+ are massively increased and more HCO_3^- is produced, returning ECF pH to normal.

Hence the value of NH_3/NH_4 secretion/excretion is:

- Toxic NH_3 is removed.
- Excretion of H^+.
- Generation of HCO_3^-.

Acid–base disturbances

Arterial pH change activates the buffering systems to prevent body pH disturbance, which would have negative consequences for most cellular processes. However, disturbances can occur. They can be categorized according to:

- pH change:
    - acidosis if pH < 7.35
    - alkalosis if pH > 7.45.
- System responsible:
    - respiratory: if PCO_2 is affected
    - metabolic: if HCO_3^- is affected.

Fig. 8.27 Renal handling of ammonium.

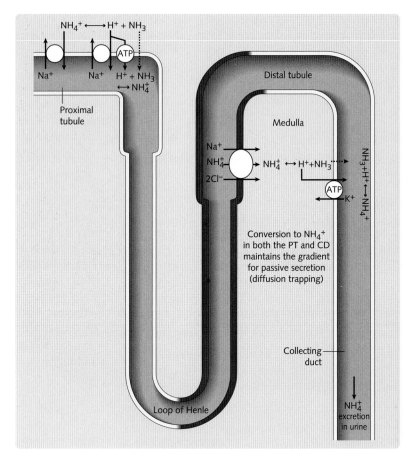

The four types of disturbance are therefore recognized, as classified using the Davenport diagram (Fig. 8.28):

1. Respiratory acidosis.
2. Respiratory alkalosis.
3. Metabolic acidosis.
4. Metabolic alkalosis.

Knowing the value of two of the variables enables the third to be calculated:

$$pH \; \alpha \frac{\left[HCO_3^-\right]}{PCO_2}$$

Arterial blood gases (ABG), which measure pH, PCO_2 and PO_2, are useful for such calculations.

Compensation

In compensation, there is restoration of pH but $[HCO_3^-]$ and/or $[PCO_2]$ are deranged. It occurs by the other system's response. The changes in $[HCO_3^-]$ and PCO_2 will mirror each other – both high or both low.

Correction

In correction, there is restoration of pH and of $[HCO_3^-]$ and PCO_2 values. Correction involves rectification by the system at fault.

Examples of acid–base disturbances

Respiratory acidosis

Anything resulting in inadequate removal of CO_2 from the body by the respiratory system will cause respiratory acidosis, i.e. PCO_2 increases. ABG shows: ↓pH and $PCO_2 > 6$ kPa. Causes include:

Factors affecting the respiratory centres
- Drugs: general anaesthesia, morphine, barbiturates (these depress the respiratory centre).
- Infection of the brainstem.
- Injury to the respiratory centre, which is located in the brainstem.

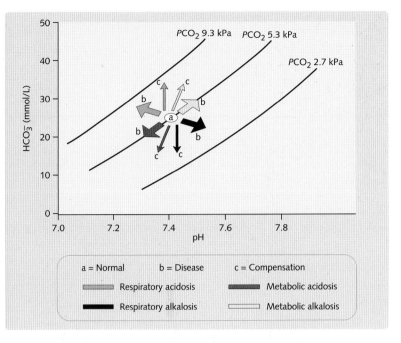

Fig. 8.28 The Davenport diagram illustrating changes in PCO_2 and plasma HCO_3^- resulting from different disturbances in acid/base balance and their compensation.

Conditions decreasing elimination of CO_2 by the lungs

- Conditions affecting work of breathing (inertia and elastic forces):
  - burns (restrict chest expansion and chest wall movement)
  - obesity (if severe enough, often acting in conjunction with another factor)
  - mechanical chest injuries
  - muscular dystrophies: Eaton–Lambert syndrome and Duchenne disease
  - infections: Guillain–Barré and poliomyelitis
  - chest wall deformities
  - spinal defects: scoliosis.
- Conditions affecting work of breathing (airway resistance):
  - severe asthma (smaller airways constitute 20% total airway resistance)
  - chronic bronchitis (large airways constitute 40% total airway resistance)
    - obstruction of airway
    - foreign body
    - mucus plug
    - tumour.
- Conditions affecting the gas exchange:
  - increased diffusion distance: interstitial pneumonia.

Insufficient pulmonary ventilation results in $\uparrow PCO_2 \rightarrow \downarrow pH$

$$CO_2 + H_2O \longleftrightarrow H_2CO_3 \longleftrightarrow H^+ + HCO_3^-$$

and

$$pH = pK + \log \frac{\left[HCO_3^-\right]}{PCO_2}$$

$$\left(\text{subsequently abbreviated as } pH \, \alpha \frac{\left[HCO_3^-\right]}{PCO_2}\right)$$

$\uparrow PCO_2$ will lead to $\downarrow pH$ and the above reaction will shift to the right. Although acid and base are generated, HCO_3^- buffer cannot be used since this would necessitate displacing the reaction to the left, i.e. in a direction of raised PCO_2. The $\uparrow H^+$ will be offset by non-bicarbonate buffers (Bf^-). The small increase in HCO_3^- offsets the fall in pH but is insufficient to return pH to normal. Compensation by $\downarrow PCO_2$ through $\uparrow$ventilation cannot occur until the cause of the disturbance is removed.

The kidney responds to the $\uparrow PCO_2$ in body fluids by increasing its secretion of H^+ and by $\uparrow HCO_3^-$ reabsorption and generation. This renal compensatory response normally corrects pH, although biochemical markers ($[HCO_3^-]$ and PCO_2) are elevated/deranged. Only when the primary disturbance is removed will full correction be achieved by hyperventilation leading to $\downarrow PCO_2$; $[HCO_3^-]$ will follow.

Respiratory alkalosis

This is caused by lung overventilation and the consequent loss of CO_2; it is rarely due to pathological conditions. An arterial blood gas sample (ABG) shows: $\uparrow pH$ and $PCO_2 < 4.5\,kPa$. The stimulation of respiratory centres can be the result of:

- Hypoxia: hypoxic respiratory drive induces hyperventilation:
  - high altitude
  - pulmonary conditions: pulmonary embolus/infarct.
- Higher centres: hysteria, anxiety.
- Brainstem damage.

$$CO_2 + H_2O \longleftrightarrow H_2CO_3 \longleftrightarrow H^+ + HCO_3^-$$

$$pH \, \alpha \, \frac{\left[HCO_3^-\right]}{PCO_2}$$

$\downarrow PCO_2$ will lead to $\uparrow pH$ and the above reaction will shift to the left. Non-bicarbonate buffers (BfH) can supply protons to lower HCO_3^-. HCO_3^- buffer cannot be used since this would necessitate displacing the reaction to the right. The small $\downarrow HCO_3^-$ will offset the $\uparrow pH$ but is insufficient to return pH to normal. Compensation by $\uparrow PCO_2$ through $\downarrow$ ventilation cannot occur until the cause of the disturbance is removed.

Renal compensation involves $\downarrow H^+$ secretion with $\downarrow HCO_3^-$ reabsorption and consequent $\uparrow HCO_3^-$ excretion in alkaline urine.

As with all forms of respiratory acid–base disturbance, correction comes from the respiratory system itself.

Metabolic acidosis

This term refers to all other acidoses not due to $\uparrow PCO_2$. ABG shows: $\downarrow pH$, $\downarrow [HCO_3^-]$ and normal PCO_2. Causes include:

H+ addition

- Ingestion of acid.
- Excessive metabolic acid production: diabetic ketoacidosis (DKA) and lactic acidosis.
- Drugs: methyl alcohol and aspirin (which initially causes respiratory alkalosis).
- Tubular acidosis: insufficient H^+ secreted, if any at all.

HCO3– loss

- Loss of alkaline intestinal fluid:
  - vomiting of intestinal contents

- severe diarrhoea
- fistula drainage.
- Proximal renal tubular acidosis: PT is primary site for HCO_3^- reabsorption.

$$CO_2 + H_2O \longleftrightarrow H_2CO_3 \longleftrightarrow H^+ + HCO_3^-$$

$$pH \, \alpha \, \frac{\left[HCO_3^-\right]}{PCO_2}$$

$[H^+]$ is elevated so the reaction (above) shifts to the left; HCO_3^- is lowered as is pH. Bf$^-$ accepting protons will offset the fall in HCO_3^- but compensation is still required.

Consequently, the respiratory system compensates with hyperventilation and hence $\downarrow PCO_2$.

Resolution of pH permits a further decrease in $[HCO_3^-]$. The kidneys respond to increasingly acidic plasma by increasing HCO_3^- reabsorption and excreting H^+ as titratable acid and NH_4^+ to generate new HCO_3^-. If plasma pH is normal, these renal mechanisms are not switched on.

Anion gap and metabolic acidosis

Changes in the anion gap are important in ascertaining the causes of metabolic acidosis. The anion gap = plasma cations – plasma anions:

$$Anion\,gap = (Na^+ + K^+) - (Cl^- + HCO_3^-)$$

The anion gap is increased in DKA and lactic acidosis.

Metabolic alkalosis

This refers to an alkalosis that is not due to a disturbance in PCO_2. ABG shows: $\uparrow pH$, $\uparrow [HCO_3^-]$ and normal PCO_2. Causes include:

- Loss of H^+: vomiting of gastric contents:
  - infection: gastritis
  - outflow obstruction: pyloric stenosis, peptic ulcer disease, gastric cancer.
- Ingestion of alkali:
  - excessive antacid ingestion
  - antacid taken with milk: milk–alkali syndrome
  - citrate from transfused blood.
- ECF depletion (strong stimulation to reabsorb Na^+):
  - shock
  - diuretics ($\uparrow Na^+$ delivery to distal nephron $\rightarrow$ $\uparrow H^+$ secretion).
- Excessive mineralocorticoid activity:
  - Cushing's disease, primary and secondary hyperaldosteronism, Bartter's syndrome.
- K^+ depletion in tubular fluid:
  - Na^+ reabsorption will occur by exchange preferentially with H^+ rather than K^+ in cortical CD

- decreased dietary intake
- gastrointestinal: vomiting, diarrhoea and intestinal obstruction
- amphotericin – damages tubule, mediated by mineralocorticoid receptor
- Liddle's syndrome.

$$CO_2 + H_2O \longleftrightarrow H_2CO_3 \longleftrightarrow H^+ + HCO_3^-$$

$$pH \: \alpha \frac{\left[HCO_3^-\right]}{PCO_2}$$

$\uparrow HCO_3^-$ leads to $\uparrow pH$. The $\uparrow HCO_3^-$ will be reduced but not prevented by BfH supplying protons. Respiratory compensation, hypoventilation, to $\uparrow PCO_2$ is limited in its efficacy; it further $\uparrow[HCO_3^-]$. Again respiratory compensation of metabolic acid-base derangements causes further worsening of $[HCO_3^-]$ as normalization of pH hinders renal correction of the problem (Fig 8.29).

Regulation of calcium and phosphate

Calcium

Ca is the most abundant cation in the body and is essential for many bodily functions. Of the body's Ca 99% is stored with PO_4^{3-} as complex salts in bone, which functions as a large reservoir or buffer system. Approximately 1% Ca is in the ECF (normal plasma values 2.2–2.6 mmol/L). Plasma $[Ca^{2+}]$ regulates the threshold for excitation in the nerves and muscles. The difference between the resting membrane potential and the threshold potential, which is inversely proportional to plasma $[Ca^{2+}]$, dictates the excitability of these cells' membranes. Therefore, concentration should be kept within a fine limit.

Fig. 8.29 Summary of features of acid–base disorders

	PCO₂ = Respiratory	[HCO₃⁻] = Metabolic
pH ↓	↑ PCO₂ (cause: COPD)	↓ [HCO₃⁻] (cause: diabetic ketoacidosis)
Acidosis	Compensation: ↑ [HCO₃⁻]	Compensation: ↓ PCO₂
pH ↑	↓ PCO₂ (cause: hyperventilation)	↑ [HCO₃⁻] (cause: vomiting)
Alkalosis	Compensation: ↓ [HCO₃⁻]	Compensation: ↑ PCO₂

Calcium is present in three forms in the plasma:

1. $<50\%$ ionized Ca^{2+} (normal plasma value 1.25 mmol/L), which is the only biologically active form.
2. 40% reversibly bound to plasma proteins (especially albumin).
3. $\approx 10\%$ complexed to relatively LMW anions (PO_4^{3-} and citrate).

Approximately 0.1% of the body's Ca^{2+} is found in the ICF (normal concentration 1×10^{-4} mmol/L). Ca^{2+} is:

- Sequestered in smooth endoplastic reticulum and mitochondria.
- Complexed with specific Ca^{2+}-binding proteins such as calmodulin.

Low ICF $[Ca^{2+}]$ is maintained by active transport mechanisms.

Plasma pH is an important regulator of the degree of Ca^{2+} binding to nerve membranes. H^+ and Ca^{2+} compete for binding sites on albumin. $\uparrow pH$ lowers free $[Ca^{2+}]$, which explains the presence of features of hypocalcaemia, such as tetany, in alkalotic patients (Fig. 8.30).

Calcium handling by the kidney

Normally urine loss per day = net intestinal absorption for homeostasis of body Ca^{2+} in health:

- Approximately 60% plasma Ca^{2+} is filtered (i.e. not those bound to proteins).
- In the PT:
  - 65% reabsorption occurs mostly in the pars convoluta
  - membrane Ca^{2+} permeability is very low
  - Ca^{2+} transport is largely passive and paracellular, driven by the small lumen positive pd.
  - Small component of transcellular movement involves passive entry via Ca^{2+} specific channels in the apical membrane and exit across the basolateral membrane via Na^+/Ca^{2+} exchanger and Ca^{2+}-ATPase
- In the thick ascending LoH:
  - 20–25% reabsorption occurs by both paracellular movement (50%) driven by the lumen positive pd and transcellular movement by mechanisms described for the PT.
  - furosemide – a diuretic inhibiting NaCl transport in the LoH – inhibiting $Na^+/K^+/2Cl^-$ transporters reduces lumen positive pd in the LoH – inhibits paracellular Ca^{2+} reabsorption.

195

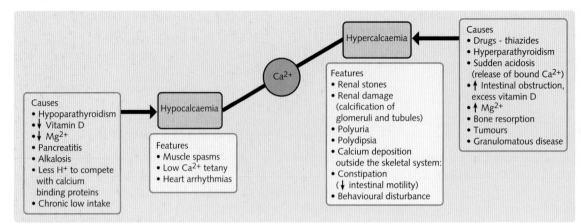

Fig. 8.30 Causes and features of disturbances in calcium homeostasis.

- In the distal tubule (DT):
  - 5–10% reabsorption via transcellular pathway by mechanisms described for the PT
  - renal Ca^{2+} handling is controlled
  - thiazide diuretics inhibit apical NaCl symport, promoting Ca^{2+} reabsorption, possibly by $\uparrow$ the Na^+ gradient across the basolateral membrane and enhancing Na^+–Ca^{2+} exchange.
- In the collecting duct (CD): <0.5% Ca^{2+} is reabsorbed against an electrochemical gradient.
- Ca^{2+} reabsorption through apical channels is controlled by parathyroid hormone in PT, LoH and DT

Proportionally less Ca^{2+} will enter the blood than is consumed. Hence, changing dietary input has little effect on renal excretion of Ca^{2+}.

Phosphate

Inorganic PO_4^{3-} exists in plasma in acid and alkali forms:

- $H_2PO_4^-$: acid.
- HPO_4^{2-}: alkali.

The relative proportions are controlled by pH:

$$H_2PO_4^- \leftrightarrow H^+ + HPO_4^{2-}$$

At plasma pH 7.4, these two exist in the ratio 4 HPO_4^{2-} : 1 $H_2PO_4^-$.

Organic PO_4^{3-} is also present inside the cell, incorporated into organic molecules such as cyclic AMP, ADP and ATP. 90–95% PO_4^{3-} is filtered ($\approx$ 5% is protein bound) with the same ratio of inorganic PO_4^{3-} as in the plasma. In the DT and the PT, H^+ secretion converts alkali PO_4^{3-} into the acid form.

Reabsorption mainly occurs in the PT, and $\approx$ 75% occurs by symport with Na^+.

Renal tubules have a T_m-limited mechanism for PO_4^{3-} reabsorption of 0.1 mM/min. If more than this is present in glomerular filtrate, the excess is excreted. This is normal, as many people consume large amounts of PO_4^{3-} in milk products and meat.

Renal excretion of PO_4^{3-} is:

- Increased by PTH, calcitonin and glucagon.
- Decreased by insulin.

PTH is the only hormone known to regulate tubular PO_4^{3-} transport.

Calcium and phosphate homoeostasis

Calcium and phosphate enter the ECF orally via the intestines, and from bones. They leave via the kidneys and are taken up again by the bones (Fig. 8.31):

$$[Ca^{2+}] \times [PO_4^{3-}] = constant$$

Increasing one will decrease the other; therefore, their concentrations are inversely proportional. However, small changes in plasma $[Ca^{2+}]$ can be mimicked by changes in plasma $[PO_4^{3-}]$. Regulation of these two substances is by:

- PTH: dissolves bone and mobilizes Ca^{2+}.
- Active vitamin D ($1,25\,(OH)_2\,D_3$): acts mainly to stimulate intestinal absorption of Ca^{2+} and PO_4^{3-}.
- Calcitonin.

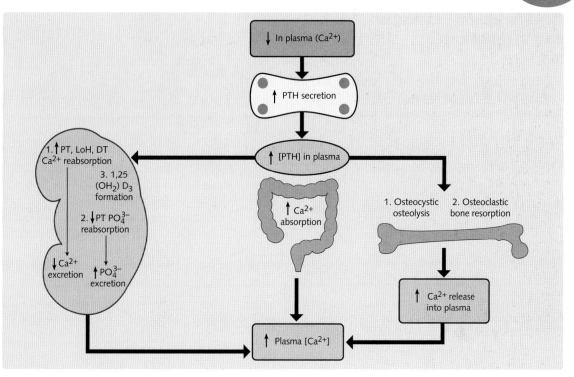

Fig. 8.31 Mechanisms of calcium and phosphate homeostasis.

Parathyroid hormone

This polypeptide controls directly or indirectly the gastrointestinal tract, kidneys and bone. Its secretion by the parathyroid gland is controlled by the surrounding ECF $[Ca^{2+}]$, primarily stimulated by low $[Ca^{2+}]$. High $[PO_4^{3-}]$ also stimulates PTH secretion. Finally, vitamin D alters the sensitivity of the parathyroid gland to Ca^{2+}, and exerts slow, inhibitory effects.

In summary, PTH has four effects on Ca^{2+} homoeostasis (see Fig. 8.3):

1. ↑ Movement of Ca^{2+} from bone into plasma by stimulating:
   - rapid Ca^{2+} retrieval produced by osteocytic osteolysis
   - osteoclast resorption.
2. Stimulate activation of vitamin D, which then promotes intestinal absorption of Ca^{2+}.
3. ↑ Ca^{2+} reabsorption in the PT, LoH and DT through apical membrane Ca^{2+} channels.
4. ↓ PO_4^{3-} reabsorption in the PT.

Although more PO_4^{3-} is absorbed intestinally, in addition to calcium, by the action of vitamin D, it is excreted due to the inhibition by PTH of reabsorption in the PT.

Vitamin D

Vitamin D refers to a closely related sterol family that can be derived from provitamins that form part of the dietary intake or produced from ultraviolet light on the skin. These must undergo a series of metabolic steps in the liver and kidney before having any effect. The main function of vitamin D is to ↑ serum Ca^{2+} and PO_4^{3-} by:

- Enhancing active reabsorption of Ca^{2+} and PO_4^{3-} in the intestine by stimulating the synthesis of proteins involved in their transport.
- ↓ excretion by enhancing movement across the basolateral membrane.
- ↑ Ca^{2+} release from the bone.

In vitamin D deficiency, ↓ Ca^{2+} absorption means that less calcium is available for bone remodelling. In children, osteoid (bone matrix) fails to be mineralized properly, resulting in rickets.

Calcitonin

The parafollicular cells in the thyroid gland produce the peptide calcitonin, which ↓ ECF $[Ca^{2+}]$ by ↓ Ca^{2+} release from bone.

Regulation of potassium and magnesium

Potassium

The vast majority lies intracellularly where it is the main cation. There are approximately 150 g K^+ (3–4 mol) in the body, the distribution being:

- 98% in ICF: normal cell concentration ~150 mmol/L.
- 2% in ECF: normal plasma concentration 3.5–5.0 mmol/L.

The ratio of intracellular:extracellular K^+ is crucial when determining membrane excitability. Hence $[K^+]$ homoeostasis is essential to life.

Renal handling of potassium

After free filtration, 80% is reabsorbed by the PT. This reabsorption is:

- Passive via the paracellular route.
- By diffusion down a concentration gradient, aided by H_2O reabsorption (solvent drag).

Some recycling occurs in the LoH; the descending limb secretes some K^+ and the ascending limb reabsorbs. The thick ascending LoH reabsorbs 10–15% K^+, 50% of which is via the paracellular route (as for Na^+) and 50% transcellular via the apical $Na^+/K^+/2Cl^-$ transporter and Na^+/K^+-ATPase and K^+-channel in the basolateral membrane. Hence, reabsorption here is dependent on that of Na^+.

Less than 10% of the filtered K^+ reaches the distal nephron. In the early DT, there is minimal change in tubular fluid $[K^+]$ due to leakage back similar in magnitude to that reabsorbed. P cells in the late DT and CD secrete K^+ passively into the lumen after active transport via the Na^+/K^+-ATPase pump from the interstitium. Secretion is increased by:

- ↑Intracellular $[K^+]$.
- ↑Tubular flow rate.
- ↑Na^+ delivery from more proximal nephron segments.

The medullary CDs reabsorb K^+, although the amounts will be small if large amounts were secreted proximally to this portion of the nephron.

Regulation of potassium secretion

The net effect of K^+ handling by the distal nephron is adjusted to meet the body's requirements:

- Low K^+ diet: net reabsorption.
- High or normal K^+ diet: net secretion.

K secretion is affected by:

- ↑Plasma $[K^+]$:
  - affects Na^+/K^+-ATPase in the P cells of the cortical CD, which increase in activity (increased uptake from interstitium) with ↑peritubular capillary $[K^+]$
  - stimulates aldosterone production.
- Aldosterone: activates apical K^+ channels (ROMK) in the P cells, thereby promoting K^+ secretion and hence excretion.
- Amount of Na^+ reaching distal nephron: ↑Na^+ delivery to cortical CD results in greater Na^+ reabsorption and more K^+ secreted.
- Tubular flow rate: a high rate would maintain the steep diffusion gradient favouring secretion.
- ADH: enhances Na^+ permeability of epithelial cells, making the lumen more negative, which favours K^+ secretion.

Disturbances of potassium

Hypokalaemia   Either K^+ depletion or intracellular shift is the underlying mechanism for this condition. *Causes include:*

- Reduced K^+ intake.
- Gastrointestinal (GI) losses: vomiting, diarrhoea, bowel ileus or obstruction.
- Drugs: excess insulin, steroids, carbenoxolone.
- Diuretic abuse.
- Metabolic alkalosis.
- Hyperaldosteronism: can be primary or secondary, resulting in ECF depletion, organ failure of heart, kidney or liver.
- Renal tubular acidosis.

Generally, no clinical features occur with $[K^+] > 2$–2.5 mmol/L.

The following events can occur in hypokalaemia:

- Decreased membrane excitability: i.e. they are hyperpolarized. This presents as fatigue, muscle weakness and cramps that progress upwards from the lower extremities. In prolonged, severe cases, death occurs by respiratory muscle paralysis.
- Altered conversion of glucose to glycogen by the liver: this normally requires K^+ and might manifest as an abnormal glucose tolerance test.
- Cardiac arrhythmias, including asystole: with hypokalaemia, it takes longer for cardiac muscles to repolarize. Characteristic ECG changes include ST depression, flattened or inverted T waves and U waves.

- Metabolic alkalosis from decreased K^+ delivery to the distal nephron: normally, Na^+ reabsorption in this portion occurs by exchange with either K^+ or H^+. If $[K^+]$ is low then exchange will occur preferentially with H^+, which is then excreted, i.e. H^+ is lost.
- Impaired response to ADH: with thirst and polyuria because the patient is unable to concentrate his or her urine.

Treatment should always involve identification and treatment of the underlying cause, with care to avoid rebound hyperkalaemia. Potassium salts might need to be administered, either orally or intravenously, with ECG monitoring.

Hyperkalaemia   This is due to either extracellular shift of K^+ or reduced renal excretion. Causes include:

- Increased intake of K^+.
- Cell death or hypoxia: GI bleed, catabolism, rhabdomyolysis.
- Metabolic acidosis: DKA, lactic acidosis.
- Insulin deficiency: Addison's disease, diabetes.
- Drugs causing release of intracellular K^+: beta-blockers, excessive digoxin.
- Drugs impairing renal K^+ excretion:
    - impair K^+ secretion: ACE inhibitors, NSAIDs, heparin
    - render nephron insensitive to aldosterone: spironolactone, amiloride.
- Decreased plasma [aldosterone].
- Renal failure.

Clinical features are generally not seen with concentrations <7 mmol/L. The following events happen:

- Increased excitability and decreased threshold for depolarization: symptoms range from slight paraesthesia to severe muscle weakness with loss of tendon jerks. If the threshold is lowered so that it is less than the resting potential, repolarization is prevented, with flaccid paralysis.
- Cardiac arrhythmias: might be the first presentation of hyperkalaemia and are likely to occur with values >7 mmol/L. Characteristic ECG changes include prolonged PR interval, broad QRS complexes and tall-tented T waves.

Treatment includes rectification of the underlying cause whilst correcting the K^+ levels with ECG monitoring. It includes:

- Intravenous insulin with dextrose (to drive K^+ intracellularly).

- Intravenous calcium gluconate (to protect against membrane hyperexcitability).
- Calcium resonium: an exchange resin that binds K^+ in exchange for Ca^{2+} (promoting K^+ excretion).

If the above measures fail, haemodialysis or haemofiltration can be considered.

Magnesium

The second most important cation within the cells of the body. It has the following roles:

- Mitochondrial energy production: regulated by Mg^{2+} ATP.
- Protein synthesis.
- Cell membrane K^+ and Ca^{2+} channel regulation.
- Dietary intake is normally 300 mg, of which only half is absorbed. Plasma concentration is 0.7–1.0 mmol/L. Body Mg^{2+} totals 28 g, which is widely distributed:
    - bone: stores >50%
    - ICF: approximately 45%
    - ECF: <1%.

Hypomagnesaemia

Mg^{2+} derangements are usually accompanied by K^+, Ca^{2+} or acid–base disturbances. Causes include:

- Reduced intake: especially protein–energy malnutrition.
- GI tract loss: including vomiting and diarrhoea.
- Drugs affecting the kidney: loop diuretics and gentamicin.
- Other metabolic conditions: acute pancreatitis, hyperparathyroidism, primary or secondary hyperaldosteronism.
- Renal disease: renal tubular acidosis, Gitelman's syndrome.

Clinical features:

- Neuromuscular: including tremor.
- Behavioural: including agitation.
- Neuropsychiatric symptoms: including confusion.
- Cardiac arrhythmias: especially torsade de pointes.

Renal handling

Generally, the kidney maintains homoeostasis by excreting the same amount of Mg^{2+} as is absorbed. Approximately 75% is filtered (25% bound to plasma proteins), most of which is ionized, although a small amount is complexed with anions (including citrate, and PO_4^{3-}).

- PT: passive paracellular reabsorption of ~15% filtered load, permeability of epithelial cells is lower for Mg^{2+} than either Na^+ or Ca^{2+}.
- Thick ascending limb of LoH: is the main site for absorption; $\approx 65\%$ of Mg^{2+} is reabsorbed here. Reabsorption is mainly paracellular driven by the lumen positive pd. Additionally, reabsorption is aided by Na^+/Mg^{2+} antiport and basolateral Mg^{2+}-ATPase.
- DT: reabsorbs 5–10%.

Regulation of magnesium reabsorption

A T_m-dependent mechanism applies, with the T_m value = normal filtered loads, thus maintaining plasma values. Intrinsic regulation is by the cells of the thick ascending limb of LoH: If cells encounter tubular fluid with low $[Mg^{2+}]$ (secondary to $\downarrow$ filtered loads), they transport more Mg^{2+}.

PTH acts on the thick ascending limb to increase reabsorption and therefore decrease excretion.

Regulation of erythropoiesis

Cells of the inner cortex and peritubular interstitium of the kidney produce > 80% of the body's erythropoietin (EPO). EPO synthesis, which is mediated by prostaglandins, occurs in response to a reduction in PO_2 in the kidneys as a result of hypoxia – most importantly, hypoxic anaemia and renal ischaemia.

EPO acts on erythrocyte stem cells of the bone marrow, stimulating increased production of red blood cells (RBC) and thus improving the blood's oxygen-carrying capacity. EPO has a half-life of 5 hours. Its secretion can be high or low, depending on the type of renal disease:

- High rate of secretion, resulting in polycythaemia (high RBC), is a feature in polycystic kidney disease and renal cell carcinoma.
- Inappropriately low secretion is seen in chronic renal failure. In addition, 90% of people who have had their kidneys removed or are in end-stage renal disease and are on haemodialysis develop anaemia.

The decrease in EPO production in chronic renal failure cannot be compensated for by the rest of the body and intravenous or subcutaneous recombinant EPO is administered to rescue the anaemia, providing that is the sole reason for the anaemia. This involves a regimen of 30–50 U/kg three times a week, to achieve a haemoglobin count of 10–12 g/dL.

Failure with EPO treatment can occur, most commonly with iron-deficient anaemia, which requires oral or IV ferrous sulphate. In addition, active bleeding or malignancy (anaemia of chronic disease) can cause EPO treatment 'apparently' not to work.

EPO treatment-related complications are usually precipitated by rapid increases in haemoglobin concentration and haematocrit. These include hypertension, thrombotic events and fitting.

SELF-ASSESSMENT

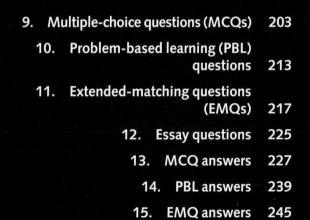

Indicate whether each answer is true or false.

Chapter 1 Introduction to physiology

1. Fluids
a. The body comprises 50–60% fluids in lean adults.
b. Total body fluid is two-thirds intracellular and one-third extracellular.
c. Extracellular fluid comprises around 80% plasma and 20% interstitial fluid.
d. K^+ and Cl^- are predominantly intracellular ions.
e. Interstitial fluid is predominantly intracellular.

2. Organelles 1
a. All normal cells contain a nucleus.
b. Mitosis refers to cell division during the reproduction cycle.
c. 50s ribosomes are smaller than 30s ribosomes.
d. Some ribosomes contain DNA.
e. Agranular endoplasmic reticulum (ER) synthesizes fatty acids.

3. Organelles 2
a. The Golgi apparatus is usually situated distal to the nucleus.
b. Mitochondria are single-membraned, ovoid structures.
c. All the mitochondrial membrane is granular.
d. Lysosomes and peroxisomes have single membranes.
e. The microfilament tubulin is found in muscle cells.

4. Energy production by the cell
a. ATP is a nucleotide containing the base adenine, a pentose sugar and two phosphate molecules.
b. Each covalent phosphate bond broken liberates about 30.6 kJ/mol.
c. The majority (95%) of ATP is formed in the mitochondrial membrane.
d. Oxidative phosphorylation produces more ATP than the Krebs cycle.
e. Forty ATP molecules are formed from one mole of glucose.

5. Genetics 1
a. Genes are made from RNA.
b. DNA is a double helix comprising phosphoric acid, a deoxyribose sugar and four nitrogenous bases.
c. Adenine and guanine are purines.
d. Bases connect to each other via covalent bonds.
e. Guanine binds to thymine and adenine binds to cytosine.

6. Genetics 2
a. Every three successive bases of DNA code for one amino acid.
b. The four possible bases can code for a possible 64 amino acids.
c. Some amino acids are represented more than once by one codon.
d. In RNA, deoxyribose is replaced with mannose.
e. Messenger RNA (mRNA) transfers amino acids to the ribosomes to manufacture proteins.

7. Protein synthesis and cell movement
a. mRNA is formed during transcription.
b. Splicing of mRNA removes the extrons and combines the introns.
c. Once formed, the mRNA remains in the nucleus.
d. tRNA bearing amino acids binds to the complementary codon on the mRNA.
e. Cilia are composed of nine single and two double microtubules.

8. Cell membranes
a. Fat-soluble substance can readily pass through the lipid bilayer.
b. Integral proteins are transmembranous.
c. The glycocalyx is positively charged.
d. Tight junctions are protein tunnels that form between cells.
e. Spot desmosomes are filamentous adhesions between nearby cells, which serve as mechanical reinforcements.

9. Transport
a. Diffusion can be defined as the movement of molecules from an area of low concentration to higher concentration.
b. Simple diffusion requires a carrier protein.
c. A hypertonic solution will cause cells to crenate.
d. The Na^+/K^+-ATPase is energy dependent.
e. Symport is a cotransport system in which two substances move in the same direction by means of a common carrier.

Chapter 2 Physiology of the blood and body fluids

10. Body fluids 1

a. Plasma constitutes 75% of extracellular fluid (ECF).
b. Osmolality is the number of osmoles of solute per unit weight of solvent.
c. Proteins are non-diffusible but attract cations into their compartment.
d. NaCl is the greatest contributor to ECF osmolality.
e. Intracellular fluid (ICF) has a higher osmotic pressure than ECF.

11. Body fluids 2

a. Systemic capillary pressure is maintained at 20 mmHg higher than that of ISF.
b. Normal intake of water is 1–2 L a day.
c. Normal ECF volume is 15 L.
d. Minimal daily water loss is about 1200 mL.
e. Radioactive water can be used to measure total body volume.

12. Blood

a. There are approximately 10 L of blood in the average adult male.
b. Plasma constituents make plasma an important buffer.
c. Neutrophils are involved in antigen–antibody reactions.
d. Leukocytes form 5% of the cellular elements of the blood.
e. Platelets are derived from macrophages.

13. O_2 transport in blood

a. Dissolved O_2 is sufficient to meet the body's basal metabolic requirement (BMR).
b. Haemoglobin binds O_2 when its iron atom is in the ferric state (Fe^{3+}).
c. Above a PaO_2 of 70 mmHg, large changes in PaO_2 have little effect on haemoglobin saturation.
d. Acidosis results in a lower O_2 saturation.
e. The dissociation curve for HbF is positioned left to that for HbA (adult haemoglobin).

14. CO_2 and CO transport in blood

a. 90% arterial CO_2 is transported as bicarbonate (HCO_3^-).
b. The formation of carbamino compounds occurs less readily with reduced haemoglobin (Hb).
c. PCO_2 has a greater variability than PO_2.
d. CO shifts the oxyhaemoglobin curve to the right.
e. CO prevents O_2 binding to free binding sites on haemoglobin molecules with which it has bound.

15. Haemostasis

a. Platelet activation is the first step in the formation of a platelet plug.
b. The intrinsic clotting pathway involves activation of factor VII.
c. The intrinsic pathway can be measured by accelerated thromboplastin time (APTT).
d. Venous thrombi have a larger platelet composition than arterial thrombi.
e. Factor XIII is involved in clot retraction.

Chapter 3 Physiology of the nervous system

16. Overview of the nervous system

a. Afferent receptors include receptors for vision and hearing.
b. The CNS consists of the nerve fibres in your fingers.
c. 'Efferent' refers to motor fibres.
d. The autonomic system consists of the parasympathetic, sympathetic and mesenteric divisions.
e. The autonomic system has a two-neuron chain connected by a synapse between the CNS and the effector organ.

17. Neuronal excitation and inhibition

a. The resting potential difference in neurons is approximately −90 mV.
b. The Nernst equation is similar to the Goldman–Hodgkin–Katz equation but takes into consideration other ions.
c. Na^+/K^+-ATPase moves three Na^+ ions out in exchange for every two K^+ ions transported.
d. Action potentials are self-propagating.
e. The relative refractory period of an action potential is where no further action potential is possible.

18. Ionic mechanism of action potentials

a. Hodgkin and Huxley demonstrated the action potential using the axons of giant squids.
b. Depolarization occurs when Na^+ channels open and the membrane potential tends towards 0.
c. Depolarization is associated with the downstroke on the action potential.
d. The entry of K^+ causes the membrane potential to go towards the resting levels.
e. After potentials are due to the Na^+ and Ca^{2+} channels not returning to their previous states.

19. Myelinated and unmyelinated fibres

a. Myelin comprises proteins and lipids, which form a sheath around all nerves.
b. There are roughly twice as many myelinated fibres as unmyelinated.
c. Myelin is produced by Schwann cells.
d. The node of Ranvier is an axonal myelinated area that increases the speed of nerve conductance.
e. The thicker the axon, the slower the speed of conductance.

20. Synaptic transmission

a. Chemical synapses have direct channels (usually gap junctions) that conduct electricity from one cell to another.
b. The entry of Na⁺ into the presynaptic membrane is the direct cause of the neurotransmitter-containing vesicles migrating towards the synaptic cleft.
c. Excitatory postsynaptic potentials (EPSP) allow Na⁺ to pass through, causing excitation and propagation of the action potential down the postsynaptic neuron.
d. Spatial summation is the activation of many terminals in close proximity to each other to increase the chance of a neuron firing.
e. The effects of acetylcholine on synaptic transmission are restricted to excitation.

21. Sensation

a. Sensation can be defined as the conscious or subconscious awareness of external/internal stimuli.
b. Somatic sensation provides information from sensory receptors in the skin and mucous membranes.
c. Encapsulated nerve endings are naked dendrites.
d. Merkel discs are found in the basal layer of the skin.
e. Pacinian corpuscles are stretch receptors.

22. Pain

a. Fast pain occurs 1–2 seconds after the stimulus has been applied.
b. Slow pain travels along small diameter, unmyelinated C fibres.
c. The periaqueductal grey is a pain-regulating area in the medulla.
d. Exogenous opiate-like peptides are derived from three large molecules: proenkephalin, propiomelanocortin and prodynorphin.
e. The metabolites of morphine can be more potent than morphine itself.

23. Brainstem

a. Most cranial nerves originate from the brainstem.
b. The white matter in the medulla oblongata contains sensory tracts only.
c. Nuclei for cranial nerves V–VIII are present in the pons.
d. The inferior colliculi are concerned with vision.
e. Damage to the reticular formation can result in prolonged coma.

24. Autonomic nervous system

a. The autonomic nervous system is under voluntary control.
b. The sympathetic fibres originate in the spinal cord between cord segments T1 and L2. They pass into the sympathetic chain and then to the relevant tissues.
c. The parasympathetic fibres contain both pre- and postganglionic neurons.
d. The parasympathetic fibres innervating the adrenal medulla increase the release of norepinephrine (noradrenaline).

e. As well as being in the intestinal wall, the Meissner's and Auerbach's plexi are present in the pancreas.

25. Vision

a. The neural retinal layer comprises five cell levels.
b. Rods produce colour vision.
c. Cones have three types of photopigment, which absorb blue, green and yellow–orange light.
d. Isomerization is a change in conformation of retinol from the *trans* to the *cis* form.
e. In the dark, the main ion involved in vision is Na⁺.

26. Optics

a. Myopia is when images of far objects do not fall on the retina.
b. Relaxation of the ciliary muscle relaxes the lens to aid accommodation.
c. The fovea is the area of poorest visual acuity.
d. The retinal fibres terminate in six discrete layers of the lateral geniculate nucleus, with the ipsilateral eye fibres in layers 2, 3 and 5 and contralateral in 1, 4 and 6.
e. Ocular dominance columns enable perception of the depth of an object.

27. Hearing

a. Sound is measured in decibels (dB).
b. The stapes is attached to the inner surface of the eardrum and to the incus.
c. The eustachian tube connects the middle ear and the oropharynx and allows pressure in the middle ear to equal atmospheric pressure.
d. Endolymph has a similar content to intracellular fluid, whereas perilymph resembles extracellular fluid.
e. Eight tonotopic maps are present in the auditory cortex.

28. Olfaction

a. There are approximately 10–100 million olfactory receptors.
b. Olfaction detection propagates along cranial nerve II.
c. Damaged olfactory receptor cells cannot regenerate.
d. The lateral olfactory area is in the frontal lobe.
e. Anosmia is a condition where someone has lost the sense of smell.

29. Taste

a. As well as being present on the tongue, taste buds appear on the epiglottis.
b. The gustatory receptor cells project numerous microvilli.
c. Saltiness is involved with a release of Ca²⁺.
d. Sweetness operates via a second-messenger system.
e. The conscious perception of taste is given by the frontal lobe.

Chapter 4 Physiology of the musculoskeletal system

30. Introduction

a. Bone plays a role in haemopoiesis.
b. Skeletal muscle is non-striated.
c. Cardiac muscle can be voluntary controlled.
d. Smooth muscle is non-striated.
e. Cartilaginous joints are united by hyaline or fibro-cartilage.

31. Muscle microstructure

a. Fasciculi are covered by perimysium.
b. Thick filaments are mainly actin.
c. The A band contains myosin filaments.
d. Sarcoplasm is the muscle fibre matrix where the muscle fibrils are suspended.
e. Sarcoplasmic reticulum is the endoplasmic reticulum in the muscle fibres.

32. Neuromuscular junction (NMJ)

a. Myelinated fibres arising from the posterior horn of the spinal cord innervate the skeletal muscle fibres.
b. Acetylcholine (ACh) is the main neurotransmitter at the NMJ.
c. Each muscle fibre has one neuromuscular junction.
d. The terminal boutons contain non-voltage-gated Ca^{2+} channels.
e. The postsynaptic membrane contains nicotinic ACh receptors.

33. Secretion of acetylcholine (ACh)

a. The entry of Ca^{2+} into the presynaptic membrane attracts the ACh vesicles to the neural membrane.
b. The end-plate potential is due to the entry of Na^+.
c. Two molecules of ACh are required to bind to the ACh receptor to activate it.
d. Monoamine oxidase breaks down ACh to acetic acid and choline.
e. The action potential travels through the T tubules to get deep within the muscle fibres.

34. Muscle contraction

a. Myosin is composed of six heavy polypeptide chains.
b. The myosin head has an ATPase enzyme attached.
c. The actin filament comprises actin, tropomyosin and troponin.
d. Tropomyosin covers the myosin-binding sites on the troponin chain.
e. Troponin binds to Ca^{2+} to initiate the contraction process.

35. Mechanism of contraction

a. The binding of Ca^{2+} to troponin C causes a conformational change in the troponin complex.
b. Myosin heads from the cross-bridges attach to the actin active site.
c. ATP is hydrolysed by the ATPase on the myosin head to AMP to cause the head to tilt and drag the actin filament with it.
d. Relaxation occurs as Ca^{2+} is actively pumped back into the sarcoplasmic reticulum.
e. Phosphocreatine is a source of energy for muscle contraction.

36. Motor units 1

a. Each skeletal muscle fibre is innervated by a single motor neuron.
b. A motor unit comprises a motor neuron and all the muscle fibres it innervates.
c. The finer the control required for motor function, the higher the numbers of muscle fibres per motor neuron.
d. In isometric contraction, the muscle shortens.
e. In eccentric contraction, the muscle lengthens in a controlled manner.

37. Types of muscle fibre

a. The speed of muscle contraction can be related to the proportion of fast and slow muscle fibres in each muscle.
b. Fast muscle is redder in colour than slow muscle because it contains a lot of myoglobin.
c. Type I fibres have a higher mitochondrial density than type IIB.
d. Type II fibres are more resistant to fatigue than type I.
e. Postural muscles have a high myoglobin content.

38. Central control of movement

a. Central control of movement is achieved by three key regions in the brain: motor cortex, basal ganglia and cerebellum.
b. The primary motor cortex is involved in the planning and ongoing control of voluntary movements requiring the coordination of several muscles.
c. The supplementary cortex is involved in programming motor sequences and bimanual coordination.
d. The basal ganglia regulate complex/skilled patterns of motor activity in association with the cerebrospinal system.
e. Muscle spindles are spindle-shaped organs lying parallel to the skeletal muscle fibres.

39. Posture

a. If one's centre of gravity was outside the base of support, one would lose balance.
b. The visual system is important in maintaining posture.
c. The crossed extensor reflex occurs when one leg is extended and lifted off the ground: the opposite leg flexes more strongly to support the shifted weight of the body.

d. The semicircular canals detect angular acceleration during rotation of the head along three perpendicular axes.

e. Vestibular nuclei are present on the floor of the fourth ventricle and medulla and receive input from cranial nerve IX.

40. Bone

a. Bone matrix comprises 75% water and 25% collagen fibres.

b. Osteoblasts break down the bone matrix.

c. Ossification is the conversion of fibrous tissue or cartilage into bone.

d. Parathyroid hormone decreases plasma Ca^{2+} concentration.

e. Red bone marrow actively forms red blood cells; yellow bone is full of fat and thus is inactive.

Chapter 5 Physiology of the cardiovascular system

41. Introduction

a. The average heart rate is the same in both adults and the newborn.

b. The cardiovascular system is involved in temperature regulation.

c. All arteries carry blood away from the heart.

d. The systemic and pulmonary circulations are often referred to as open circulations.

e. Dextrocardia is a condition whereby the heart forms on the right side of the chest.

42. Cardiac anatomy 1

In relation to the heart the:

a. Diaphragm sits inferiorly.

b. Oesophagus is anterior.

c. Great vessels and bronchi are superior.

d. Thymus is posterior.

e. Right phrenic nerve sits medially.

43. Cardiac anatomy 2

a. The heart consists of three layers: pericardium, myocardium and endocardium.

b. Parietal pericardium is attached to the outer surface of the heart.

c. The myocardium consists of myocytes.

d. The middle layer of the endocardium consists of endothelial cells.

e. The outermost endothelial layer contains Purkinje fibres.

44. Cardiac anatomy 3

a. The left ventricular wall is thicker than the right ventricular wall.

b. The tricuspid valve is situated between the right atrium and right ventricle.

c. The bicuspid valve has two leaflets/flaps.

d. The semilunar valves are located between the left ventricle and the aorta and between the right ventricle and the pulmonary artery.

e. Rheumatic fever can cause damage to the heart valves.

45. Cellular physiology of the heart

Cardiac myocytes:

a. Are 50–100 micrometres in length and 10–20 micrometres in diameter.

b. Usually have a single nucleus.

c. Have no mitochondria.

d. Contain actin and myosin filaments.

e. Are also present in skeletal muscle.

46. Conduction system of the heart

a. The sinoatrial (SA) node is located in the right atrium near the entrance of the superior vena cava.

b. The upstroke in the SA node potential is due to the influx of K^+.

c. Immediately after atrioventricular (AV) node conduction, the action potential reaches the Purkinje fibres.

d. The resting membrane of the heart is approximately 90 mV.

e. Repolarization occurs as voltage-gated K^+ channels close.

47. Cardiac cycle

a. Diastole is where the chambers are relaxed and passive filling occurs.

b. Atrial systole delivers 20–30 mL blood to the relaxed ventricles.

c. Ventricular systole lasts for 1 second.

d. Ventricular contraction is responsible for the 'v' wave of the atrial pressure curve.

e. The end-systolic volume is the volume remaining in the ventricles post ventricular systole.

48. Cardiac output

a. Cardiac output = stroke volume × heart rate.

b. Stroke volume is the volume of blood ejected in one ventricular contraction.

c. Stroke volume is regulated by the preload, contractility and afterload of the heart.

d. Mean arterial blood pressure = total peripheral resistance × cardiac output.

e. Chronotropes increase the force of contractility of the heart.

49. Organization of the vessels

a. Exchange vessels usually consist of thin-walled capillaries.

b. Elastic arteries are not involved in the regulation of blood flow.

c. Arterioles regulate blood flow to capillaries.

d. Capillary density is inversely proportional to the metabolic demands of the tissue.

e. Compared to arteries, the intima and media of veins are a lot thinner.

50. Vascular smooth muscle

a. Smooth muscle is involuntarily activated.
b. Each cell is approximately 30–200 micrometres long.
c. Contraction is due to the influx of Cl^-.
d. Relaxation of smooth muscle is due to the decrease in the number of open ion channels.
e. The sliding filament theory of muscle contraction applies to smooth muscle.

51. Regulation of blood flow

a. Blood flow is proportional to vessel resistance/pressure difference.
b. Blood travels from an area of low to high pressure.
c. Resistance to blood flow is dependent on the lumen diameter, blood viscosity and vessel length.
d. Poiseuille's law states that resistance is proportional to diameter.
e. The shorter the vessel length the lower the resistance to blood flow.

52. Haemodynamics of the venous system

a. Venules and veins are thin-walled, distensible capacitance vessels, which act as a reservoir of blood for cardiac filling.
b. The pressure difference from the venules to the right ventricle is 15–18 mmHg.
c. When standing after a hot bath, the dizziness is due to the venoconstriction of the legs.
d. Venous valves prevent blood returning to the heart.
e. During inspiration blood flow increases to the heart.

53. Capillary dynamics and fluid movements

a. Most capillaries are a double layer of endothelial cells.
b. Hydrostatic pressure in the capillaries serves to drive the fluid out.
c. Oncotic pressure in the capillaries is determined by proteins such as albumin.
d. Simple diffusion through the endothelial cell membrane applies to simple molecules like O_2 and CO_2.
e. Lymph capillaries are blind-ending, bulbous tubes with a monolayer of endothelial cells located in between the cells.

54. Control of blood vessels

a. High temperature causes vasodilation in skin arterioles and veins.
b. Increased PO_2, PCO_2 and K^+ in the plasma cause vasodilation in blood vessels.
c. Histamine-mediated arteriolar vasodilation occurs via the H_2 receptor.
d. Nitric oxide (NO) is an endothelium-dependent contraction factor (EDCF).
e. Norephinephrine (noradrenaline) is used clinically to increase blood pressure by vasoconstricting the peripheral arteries.

55. Cardiovascular receptors and central control

a. The cardiovascular centre is primarily in the medulla oblongata.
b. The defence area in the hypothalamus produces the baroreceptor reflex.
c. Baroreceptor reflexes travel through the vagus and hypoglossal nerves.
d. Carotid sinus massage is used to reduce the heart rate in patients with supraventricular tachycardia.
e. Chemoreceptors are located in nerve terminals sensitive to changes in O_2, CO_2 and H^+.

56. Regulation of circulation in individual tissues

a. The cerebral circulation has a high capillary density.
b. Myoglobin is higher in tonic muscles than in phasic muscles.
c. During hypovolemic shock, the release of antidiuretic hormone is inhibited.
d. The brain has the highest oxygen consumption in humans.
e. The intestines are supplied by the superior and inferior mesenteric arteries.

Chapter 6 Physiology of the respiratory system

57. Ventilation

a. Alveolar ventilation is less than minute ventilation.
b. Unperfused alveoli do not contribute to the physiological dead space.
c. Ventilation is better in the upper zones of the lung because the alveoli are larger.
d. Both helium dilution and body plethysmography may be used to measure functional residual capacity.
e. Functional residual capacity (FRC) is increased in obstructive lung disease.

58. Intrapleural pressure

a. Alveolar volume is crucial to production of a pressure gradient, permitting gas exchange.
b. Intrapleural pressure (P_{PL}) is usually negative relative to alveolar pressure.
c. Inspiratory muscles contraction results in increased transmural pressure.
d. P_{PL} is always negative.
e. Lower regions of the lung have less negative intrapleural pressures.

59. Muscles of respiration

a. The diaphragm is innervated by nerves from the T1–T11 anterior rami.
b. The external intercostals slope downwards and forwards.
c. Internal intercostals are vital for expiration.

d. Contraction of quadratus lumborum forces the diaphragm upwards.

e. Diaphragmatic contraction accounts for 70% intrathoracic volume change in normal inspiration.

60. Compliance

a. Low compliance means there will be a large volume change when a small pressure is applied.

b. Compliance varies linearly with lung volume.

c. Lung compliance is increased with emphysema.

d. Lung compliance is decreased with fibrosis.

e. Total compliance of the respiratory system is the sum of compliances of the lungs and chest wall.

61. Surfactant

a. Surfactant consists mainly of phospholipid.

b. Surfactant contributes to alveolar stability.

c. Smaller alveoli contain greater amounts of surfactant than larger alveoli.

d. Closer packing of surfactant molecules during inspiration than expiration accounts for hysteresis.

e. Respiratory distress syndrome (RDS) results from lack of pulmonary surfactant.

62. Dynamics

a. In laminar flow, volume flow rate $\propto$ radius^4.

b. The trachea and bronchi form 48% of total airways resistance.

c. Airway resistance is decreased with increasing lung volume.

d. Forced expiration results in increased airways resistance.

e. Work of breathing is increased with bronchodilation, which increases volume change.

63. Gaseous exchange in the lungs

a. CO_2 diffusion equilibrium is higher than that of O_2 because of its greater solubility.

b. CO is a diffusion-limited substance.

c. Dalton's law relates partial pressure of a gas to its solubility in a liquid.

d. At rest, capillary transit time is surplus to the time taken for P_aO_2 to equilibrate with P_AO_2.

e. CO_2 is normally diffusion limited.

64. Pulmonary blood flow and pulmonary vascular resistance (PVR)

a. Pulmonary arteries can accommodate two-thirds of right ventricular stroke volume.

b. Pulmonary blood vessels have thinner walls with less vascular smooth muscle than systemic blood vessels.

c. Pulmonary vascular resistance (PVR) is higher in lower regions of the lung.

d. During inspiration, transmural pressure applied to alveolar vessels increases.

e. Extra alveolar vessels distend by radial traction.

65. Pulmonary blood flow and water balance

a. There is continuous blood flow in zone 2 of the lung.

b. $PaO_2 < 70\%$ decreases pulmonary blood flow.

c. Total pulmonary blood flow may be measured using the indicator dilution technique.

d. Surfactant helps to maintain pulmonary water balance.

e. Pulmonary oedema may interfere with gas exchange.

66. Ventilation–perfusion relationships (V/Q)

a. $V/Q = 0$ for dead space.

b. Individuals with shunts have chronically raised $PaCO_2$.

c. Inspired $PO_2 = 100\,mmHg$.

d. Lower V/Q values result in higher P_ACO_2.

e. Higher V/Q values result in higher P_AO_2.

67. Neural central control of ventilation 1

a. The sensory portions of the vagus and glossopharyngeal nerve terminate in the dorsal respiratory group (DRG).

b. Pontine respiratory centres are believed to be most important in generating breathing.

c. Ventral respiratory group contains expiratory neurons.

d. Impulses from the DRG increase lung volume.

e. The diaphragm receives all of its afferent information from the DRG.

68. Neural central control of ventilation 2

a. The apneustic centre inhibits the inspiratory neurons.

b. The pneumotaxic centre increases respiratory rate.

c. Characteristics of apneustic breathing include short expiratory gasps.

d. Inspiratory and expiratory impulses travel separately.

e. Internal intercostals are effectors concerned with expiration.

69. Central chemoreceptors

a. Central chemoreceptors form part of the dorsal respiratory group (DRG).

b. Central chemoreceptors provide 80% of ventilatory drive.

c. Central chemoreceptors are most sensitive to acidosis.

d. The protein content of cerebrospinal fluid is lower than that of blood.

e. Central chemoreceptors do not respond to hypoxia.

70. Peripheral chemoreceptors

a. Peripheral chemoreceptors are most sensitive to hypoxia.

b. Peripheral chemoreceptors monitor arterial values only.

c. Aortic bodies have the most significant respiratory effect.

d. Peripheral chemoreceptors are stimulated by anaemia.

e. Type 2 cells of carotid bodies are believed to monitor PaO_2.

71. Lung receptors and other reflexes

a. Receptors in the lungs send afferents via cranial nerve X.

b. J-receptors respond to change in transmural pressure.

c. Hering Breuer reflexes are associated with lung inflation and deflation.

d. Rapidly adapting pulmonary stretch fibres are located in the interstitium.

e. Increase in blood pressure causes bronchodilation.

72. Responses of the respiratory system

a. The ventilatory response to CO_2 is more sensitive with increased work of breathing.

b. Hypoxia normally significantly contributes to ventilatory drive.

c. Hypercapnia interacts with hypoxia.

d. Increase in minute ventilation in response to exercise is achieved solely through respiratory rate.

e. During aerobic exercise, $PaCO_2$ increases.

73. Altitude

a. PaO_2 is decreased at high altitudes.

b. Hyperventilation is the most important compensatory mechanism to altitude.

c. Polycythaemia raises PaO_2 to near normal levels.

d. At very high altitudes, increased 2,3-DPG shifts the dissociation curve to the right.

e. Hyperventilation occurs as an early response (24–72 hours).

74. Diving

a. A linear relationship exists between depth and pressure.

b. For a gas, increasing depth decreases volume and partial pressure.

c. With scuba diving there is no increase in the work of breathing.

d. Functional residual capacity is reduced when immersed up to the neck in water.

e. Arterial gas embolus is a dangerous complication of decompression sickness.

Chapter 7 Physiology of the gastrointestinal system

75. Upper gastrointestinal tract 1

a. Hypothalamic ventromedial stimulation and serotonin promote feeding.

b. Blood nutrient concentration provides short-term control of appetite.

c. During mastication the masseters and lateral pterygoids act to close the jaw.

d. Secondary ionic modification of saliva occurs in the acinus.

e. Salivary proteins form an important oral defence mechanism.

76. Upper gastrointestinal tract 2

a. Vocal cords are pulled apart during the pharyngeal phase of swallowing.

b. Food is propelled posteriorly towards the pharynx with the mouth closed.

c. Secondary peristalsis occurs in response to incomplete emptying into the stomach.

d. Chemoreceptor trigger zones (CTZs) are stimulated by raised intracranial pressure.

e. During vomiting the larynx and hyoid bone are lowered.

77. Motor functions of the stomach

a. The stomach undergoes receptive relaxation in response to the entrance of food.

b. Rostral to caudal food propagation is permitted by synchronous contractions at different levels.

c. Retropulsion during antral peristalsis by pyloric contraction forms chyme.

d. The pylorus remains partially contracted during stomach emptying.

e. The enterogastric reflex inhibits stomach emptying.

78. Gastric secretions

a. Autodigestion is prevented by a thick bicarbonate-rich layer of mucus.

b. Decreasing pH of gastric contents leads to an increase in acid secretion.

c. The rate of formation of hydrochloric acid is directly related to prostaglandins.

d. Distension of the stomach increases pepsin secretion from the chief cells.

e. Anger and hostility decrease the cephalic phase of gastric secretion.

79. Liver and biliary tract

a. The majority of bile is recycled back to the liver from the small intestine.

b. Micelle formation by bile salts permits the transport of hydrophilic substances.

c. Conjugated bilirubin is converted to urobilinogen by intestinal bacteria.

d. High concentrations of bile acid in the portal vein stimulate bile acid secretion.

e. Between digestive periods, bile flows into the gall bladder.

80. The pancreas

a. Cl^- efflux is crucial to alkaline secretion.

b. Zymogen granules help to prevent pancreatic autodigestion.

c. Duct cells secrete enzymes whereas acinar cells primarily secrete alkaline fluid.

d. Sympathetic stimulation increases pancreatic stimulation.

e. Alkaline fluid secretions occur predominantly with the gastric phase.

81. Digestion and circulation of the small intestine

a. The jejunum is the main site of digestion.

b. Most of the small intestine is supplied by the superior mesenteric artery.

c. Increased motility decreases mucosal blood supply.

d. O_2 can diffuse directly from arteriole to venule.

e. Villus arteriolar nutrient concentration is higher than that of venules.

82. Intestinal digestion and absorption

a. Most products of lipid digestion enter the lymphatic system, not the portal vein.
b. The intestinal brush border facilitates monosaccharide release.
c. Absorption of intestinal di- and tripeptides is via Na^+-dependent active transport.
d. Intrinsic factor–vitamin-B_{12} complexes are taken up into enterocytes by endocytosis.
e. Vitamin D decreases transcellular Ca^{2+} absorption.

83. Colonic bacteria and transport

a. Low peristalsis maintains colonic microflora at twice the number of the small intestine.
b. Aerobic bacteria are particularly important for coagulation.
c. Although H_2O absorption is more efficient in the colon, its diffusion is slower.
d. Glucocorticoids promote colonic H_2O absorption.
e. Sympathetic nerve stimulation protects the colonic mucosa from mechanical damage by the faeces.

84. Colonic motility and defecation

a. Parasympathetic stimulation is responsible for the colocolonic reflex.
b. Haustration results in mainly propulsion of faeces.
c. Acetylcholine initiates taenia coli contraction.
d. Defecation occurs with relaxation of the internal sphincter by faeces entering the rectum.
e. Parasympathetic autonomic reflexes inhibit defecation.

Chapter 8 Physiology of the kidneys and urinary tract

85. Kidney basic structure and function

a. The kidneys receive approximately a quarter of resting cardiac output.
b. Most nephrons lie in cortex and have short loops of Henle.
c. Transition between the thin and thick segments of the ascending limb of the loop of Henle occurs at the junction between the outer medulla and the inner cortex.
d. A deep vertical fissure near the centre of the medial concave border is the point of exit and entry.
e. The active vitamin D metabolite is produced in the kidney through its enzyme, 1-α-hydroxylase.

86. The glomerular filter

a. The glomerular filter has three layers, including a positively charged basement membrane.
b. Gaps in the epithelial podocyte layer are known as slit pores.
c. Mesangial cells aid glomerular filtration with their phagocytic properties.
d. The filter discriminates against molecular size of proteins but not their charge.
e. Glomerular filtration rate is proportional to net Starling forces favouring filtration.

87. Control of glomerular filtration rate (GFR)

a. Autoregulation maintains GFR by adjusting mainly afferent arteriolar resistance.
b. Dilation of either afferent or efferent arteriole increases RBF.
c. Increased NaCl load delivered to the macula densa leads to an increase in GFR.
d. Macula densa cells release vasoconstrictor substances.
e. Mesangial cell contraction increases glomerular capillary pressure by dilating the afferent arteriole.

88. Renal blood flow (RBF) and glomerular filtration rate (GFR)

a. Use of PAH clearance to measure renal plasma flow (RPF) is independent of plasma [PAH].
b. Inulin clearance is independent of changes in tubular flow caused by altered H_2O reabsorption.
c. Creatine clearance gives an index of GFR.
d. Adult GFR values are reached within the first 12 months of life.
e. Dietary protein intake affects both RBF and GFR.

89. Transport mechanisms

a. The 2Na/3K-ATPase in the basolateral membrane of proximal tubule cells is an example of a primary active transport system.
b. $Na^+/K^+/2Cl^-$ is an example of a secondary active transport antiport system.
c. Ion channels outnumber active transporters and permit faster transport.
d. Na^+ transport occurs via both paracellular and transcellular routes.
e. Transport processes of the following are Na^+ dependent: glucose, HCO_3^- and PO_4^{3-}.

90. Sodium chloride transport in the proximal tubule

a. The proximal tubule is responsible for approximately 70% of total Na^+ reabsorption.
b. Most secondary active transport of nutritional substances occurs in the pars convoluta.
c. Most Cl^- reabsorption happens in the pars recta.
d. Secondary to its principle reabsorptive function, fluid leaves the PT more dilute.
e. [HCO_3^-] is relatively low in the later portions of the PT.

91. Transport of glucose and other solutes in the proximal tubule (PT)

a. Glycosuria usually occurs if the plasma glucose level > 10 mmol/L.
b. ↓GFR causes plasma PO_4^{3-} to rise.
c. Antidiuretic hormone increases urea reabsorption in the inner medullary collecting ducts.
d. HCO_3^- exits proximal tubule cells by cotransport with Na^+ across the basolateral membrane.
e. The largest contribution to urinary K^+ excretion is that which escapes reabsorption by the proximal tubule and loop of Henle.

92. The role of the loop of Henle (LoH)

a. The descending limb is very permeable to H_2O whereas the ascending limb is H_2O impermeable.
b. Overall, the LoH reabsorbs more H_2O than NaCl.
c. Tubular fluid has the highest concentration of urea at the tip of the hairpin.
d. Tubular fluid leaves the LoH hyperosmolar to plasma.
e. The countercurrent multiplier results in a hyperosmolar medullary interstitium.

93. The loop of Henle (LoH) and the countercurrent multiplier

a. The thin ascending limb of the loop of Henle contributes little to the mechanisms by which the kidney elaborates concentrated urine.
b. Simple diffusion of H_2O and NaCl through the descending limb epithelia occurs.
c. Cells of the thick ascending limb contain numerous mitochondria for active transport.
d. Thick ascending limb cells are impermeable to H_2O.
e. The vasa recta's low blood flow is important in maintaining a hyperosmolar medulla.

94. Regulation of urea and urine concentration

a. 70% filtered urea is absorbed in the proximal tubule.
b. The extremely high [urea] in the collecting ducts results mainly from secretion by the loop of Henle.
c. Antidiuretic hormone increases medullary [urea].
d. Urea excretion is the main contributor to the obligatory volume of urine.
e. Aldosterone-induced increases in collecting duct Na^+ reabsorption result in large volumes of dilute urine.

95. Body fluid osmolality

a. A minimum change to plasma osmolality of 10 mosmol/kg H_2O stimulates regulatory mechanisms.
b. Increased plasma osmolality decreases the rate of antidiuretic hormone (ADH) secretion.
c. A large drop in plasma volume will preferentially stimulate baroreceptors.
d. ADH release is increased with alcohol.
e. Hyperosmolar urine means that the osmotic clearance is less than urine flow.

96. Body fluid volume and sodium

a. Na^+ content affects effective circulating volume (ECV).
b. Renin is released by decreased afferent arteriolar wall tension.
c. Sympathetic drive from the renal nerves has little effect on Na^+ reabsorption in the proximal tubule.
d. PGI_2 (prostacyclin) is a vasodilator and blocks renin release.
e. Dopamine reduces tubular Na^+ transport partly by decreasing Na^+/H^+ antiport activity.

97. Regulation of body fluid pH

a. Normal arterial pH is 7.3–7.5.
b. Buffers can destroy or create de novo H^+, thereby minimizing pH change.
c. pH is directly proportional to $[H^+]$.
d. HCO_3^- is reabsorbed directly in the kidney by a T_m-dependent mechanism.
e. NH_4^+ excretion is increased in acidosis.

98. Acid–base disorders

a. Correction and compensation show resolution of pH.
b. Respiratory acidosis is corrected by increased H^+ excretion and HCO_3^- reabsorption.
c. Muscular dystrophies can cause respiratory alkalosis.
d. Aspirin overdose produces a metabolic acidosis after initial respiratory alkalosis.
e. Decreased dietary intake causing K^+ depletion in tubular fluid causes metabolic alkalosis.

99. Regulation of calcium (Ca^{2+}) and phosphate (PO_4^{3-})

a. Ionized Ca^{2+} is the only biologically active form of calcium.
b. The pars convoluta is the site for 65% reabsorption of Ca^{2+}.
c. Thiazide diuretics inhibit Ca^{2+} reabsorption.
d. In the proximal tubule, alkaline PO_4^{3-} is converted into its acid form.
e. Renal excretion of PO_4^{3-} is reduced by the following: parathyroid hormone, calcitonin and glucagon.

100. Regulation of magnesium (Mg^{2+}) and potassium (K^+)

a. All K^+ reabsorption is dependent on that of Na^+.
b. K^+ excretion is promoted by aldosterone.
c. Mg^{2+} is the second most important extracellular cation.
d. Mg^{2+} derangements are usually accompanied by acid–base disturbances.
e. Most Mg^{2+} reabsorption takes place in the proximal tubule.

1. Sian Nide decided to work a late shift in the plastics factory where she was a machine operator. Unfortunately, there was an electrical fault in the factory and a massive fire ensued. Sian was rescued after half an hour and taken to the nearest teaching hospital with inhalational injuries. Her blood pressure was low and she was tachycardic. She had vomited a number of times and appeared confused. Her breathing was quite laboured and an arterial blood gas showed that she was hypoxic and had a metabolic acidosis. Fortunately, one of the consultants had been an army doctor and recognised this as cyanide poisoning.

   a. What is the pathology of cyanide poisoning?
   b. What are the antidotes?
   c. How would one manage known cyanide poisoning?

2. W. Imp cannot stand the sight of his own blood. One day, he accidentally cuts himself on some glass and inevitably starts to bleed. He quickly covers his arm as he feels faint, and decides to visit his nearby A&E, in the hope that they may be able to save him. When he is seen by the nurse who uncovers his arm for a better look, W. glances over and is horrified to see half of his lower arm covered in blood (albeit it being dry and a very thin smeared layer). He starts to panic and hyperventilate. The nurse tells him that he does not need any stitches as although the wound was long, it was quite shallow and had already stopped bleeding.

   a. Describe the volume, components and osmolality of the vascular compartment.
   b. Briefly describe the mechanisms involved in haemostasis.
   c. What would happen to W. Imp's O_2 carriage by haemoglobin during hyperventilation; relate this to the O_2 dissociation curve.

3. Mr B.K. Pane is 65 years old and suffers from lower back pain secondary to osteoarthritis. He also occasionally suffers from peptic ulcers. He has been on paracetamol as required for the last 2 years, which has controlled the pain. However, in the last few weeks this has been insufficient and he goes to the GP to seek advice on further pain management.

   a. How is pain regulated?
   b. How do opioids work?
   c. What is the analgesia ladder?
   d. What should the patient be put on next?

4. Mr M.T. Newron is a 55-year-old builder who presented to his GP with a 2-month history of weakness in his hands, especially when gripping a hammer, and lower limbs, and also with difficulty swallowing. His wife accompanied him and added that she found it quite difficult to understand his speech and queried whether her husband had had a stroke. The GP examined him and found no sensory or cognitive loss. The ocular muscles were intact. The GP thought it might be motor neuron disease (MND) and sent Mr Newron for electromyography (EMG) studies.

   a. What is the aetiology and pathology of MND?
   b. What are the features of MND?
   c. What would electromyography typically show?

5. The staff at Ms Rest N. Tremmor's nursing home have noticed that her hands tremble when she is sitting down and that she walks with an unusual gait, which they describe as shuffling. They have commented that her face looks expressionless, which is a sharp contrast considering she was fairly lively before. Her children, who live a long way away, say that her handwriting in the letters they receive is unusually incomprehensible. The nursing home staff call the GP to see if there is anything medically wrong.

   a. Which condition does the patient most likely suffer from?
   b. Describe the structure of the basal ganglia.
   c. Describe the functions of the basal ganglia.
   d. Which cells and neurotransmitter are depleted in the above condition?
   e. What are the names given to the characteristic gait and handwriting?
   f. Outline briefly the drug treatment for the above condition.

6. Ms Mya-Tina Graves has been feeling fairly fatigued lately. She complains of having double vision even though she wears glasses. She has also noticed that she cannot lift her arm to get things from the cupboards and brushing her hair has been a nightmare. Her husband has also commented that she sometimes seems expressionless.

   a. What medical condition does this lady most likely suffer from?
   b. What is the aetiology?
   c. What are the clinical features?
   d. Describe how ACh is released at the neuromuscular junction.
   e. What is the medical and surgical management for such a patient?

7. Mrs Heema Ridge is 35 years old and has no past medical history when she is involved in a road traffic accident. She sustains a fractured pelvis. There is no sign of a head injury. Her observations show a blood pressure of 80/55 mmHg and pulse of 135 beats/min.

   a. What are the determinants of blood pressure?
   b. How will her body try to maintain blood pressure via the renin–angiotensin–aldosterone (RAA) system?
   c. Briefly describe ventricular pressure and volume changes during the cardiac cycle.

8. Mr A. Rhythmia is 80 years old when he is found on the floor in his home. A neurological examination showed left-sided weakness and downgoing plantars. On listening to his heart, the medical student hears an irregularly irregular rhythm. The ECG shows an irregular heart rate of 95 beats/min with absent P waves and normal QRS complexes.

   a. Describe how one beat of the heart is conducted.
   b. What is the connection between the irregularly irregular rhythm and the neurological deficit?
   c. In terms of medication, what will this patient be given to minimize future complications of his heart rhythm?

9. Baby Brett Less was born prematurely at 30 weeks' gestation with respiratory distress secondary to lack of pulmonary surfactant. As a result, she has developed alveolar oedema, which is interfering with gas exchange between alveolus and capillary by increasing the diffusion distance.

   a. What is surfactant and what roles does it have within the lung?
   b. Describe two other ways in which gas diffusion can be limited based on Fick's law of diffusion.
   c. Describe three factors favouring pulmonary oedema formation.

10. Mr Phil M. is an elderly smoker with chronic obstructive pulmonary disease (COPD). He goes to see his GP because he has been feeling unwell with symptoms that include a productive cough and shortness of breath over the last week. On examination, he agrees to let the medical student who is sitting in listen to his chest. When auscultating the left lower lung region, the student yelps with excitement and confirms what she has heard by placing her stethoscope over the patient's throat, commenting on her findings of bronchial breathing.

    a. Describe the airflow that normally occurs in larger airways.
    b. What receptors in the lung respond to cigarette smoke and what reaction do they elicit?
    c. How would lung measurements differ from normal with COPD?

11. Miss Divya T is pregnant. She is referred to Accident and Emergency by her GP suffering with shortness of breath after returning from New Zealand via a direct flight. She is admitted following tests that include an arterial blood gas showing hypoxia, with a provisional diagnosis of pulmonary embolus. The medical team explain that they believe that she has a clot in her lung preventing blood from going to part of it.

    a. What is the V/Q ratio for the lung units affected by the clot, and how does this affect the alveolar gas pressures?
    b. Briefly, describe the receptors that respond to hypoxia.
    c. Describe the two mechanisms by which increased hydrostatic pressure decreases pulmonary vascular resistance.

12. On his way to work, E.T. Tupp cannot resist the lure of glistening doughnuts in the bakery across the road. Despite a full English breakfast, he manages to wolf down seven doughnuts. Shortly afterwards, he experiences some abdominal discomfort but is otherwise well.

    a. Define appetite and briefly discuss alimentary feedback.
    b. What salivary constituents have a role in oral defence?
    c. How is food stored in the stomach; describe the reflex facilitating this.

13. P. Yuke is a constant worrier who also suffers from a duodenal ulcer. Following a particularly stressful day, he feels that he ought to quench his misery with a bottle of whisky, after which he begins to experience stomach cramps. He then proceeds to vomit, noticing in his drunken stupor that the semi-liquid slush bears little resemblance to his last meal.

    a. How is chyme entering the duodenum formed?
    b. What is bile composed of?
    c. Briefly describe when the phases of pancreatic secretion occur in relation to a meal.
    d. What causes vomiting?

14. Carla Wrecktall is an avid fan of fad diets. Having been unsuccessful with her last fifteen, she decides to modify a low-carbohydrate diet to her whims – namely no fruit and few vegetables. As she also eschews drinking plenty of water, Carla consequently becomes constipated, although she notices some slime in the 30 minutes it takes her to defecate.

    a. Describe water absorption in the colon.
    b. How is mucus secretion promoted in the large intestine?
    c. How does mass movement contribute to the urge to defecate?

15. Mrs Rena L. Failure is an elderly hypertensive woman who is referred to hospital by her GP with abnormal blood tests that indicate decreased renal function. Abnormal blood results include the following:

Haemoglobin 8.6g/100 mL($\downarrow$); creatinine 837 µmol/L($\uparrow$); calcium 1.70 mmol/L($\downarrow$) and phosphate 2.41 mmol/L($\uparrow$)

a. Relate her anaemia to decreased kidney function.
b. What is the significance of serum creatinine and what variables should be taken into account when looking at this measurement?
c. Explain her calcium and phosphate concentrations in relation to her kidneys.

16. Mrs X. S. Sugar is a longstanding diabetic who has never been able to give up her sweet tooth. At a recent diabetic clinic (the first of which she has decided to attend), a urine dipstick test showed glucose in the urine (glycosuria). Additionally, she was told by the specialist that her glomerular filtration rate (GFR) was lower than expected for her age.

a. How is glucose normally reabsorbed in the kidney?
b. In what situations would you expect to see glycosuria?

c. Define GFR, including normal values and describe the conditions that apply to the measurement of a substance used in its calculation.
d. Describe the structure of the glomerular filter that produces an ultrafiltrate.

17. S. Tench is a medical student who, on returning from his elective in a deprived village in Africa, experienced torrential diarrhoea. Feeling rather washed out, he pays his local A&E department a visit. Here, among other things an arterial blood gas is performed:

pH 7.29($\downarrow$); HCO_3–8 mmol/L($\downarrow$);$PaCO_2$40 mmHg; PaO_2 100 mmHg

a. What is the normal pH range, and what formula is used in its calculation?
b. What is the acid–base disorder in the scenario? Give two additional causes that would cause this disturbance.
c. Describe two ways the body will try to rectify this imbalance.

1. **The cell 1**
    A. ATP
    B. DNA
    C. Myosin
    D. Actin
    E. Tropomyosin
    F. Water
    G. Meiosis
    H. Na+
    I. Nucleus
    J. Mitochondria
    K. Endoplasmic reticulum
    L. Glycolysis
    M. Oxidative phosphorylation
    N. Krebs cycle
    O. Lysozymes
    N. K+

    Instruction: Match one of the above to each of the statements below:

    1. This ion is predominantly intracellular
    2. The majority component of cells
    3. Contains cell DNA
    4. Describes cell division for reproduction
    5. Double-layered elongated ovoid structures
    6. The main subunit of microfilaments
    7. Nucleotide containing the adenosine, ribose and phosphate
    8. Where the Krebs cycle takes place
    9. This process forms two molecules of ATP from one molecule of glucose
    10. This process forms 34 molecules of ATP from one molecule of glucose

2. **The cell 2**
    A. rRNA
    B. Amoeboid
    C. Adenine
    D. Integral
    E. Thymine
    F. Phospholipid
    G. mRNA
    H. Facilitated
    I. Glucose
    J. Cytosine
    K. Deoxyribose
    L. Guanine
    M. tRNA
    N. Cilia
    O. Ribose
    P. Osmosis

    Instruction: Match one of the above to each of the statements below:

    1. DNA contains which type of sugar?
    2. In DNA, guanine base pairs with…

3. RNA contains which type of sugar?
4. In RNA, uracil replaces which base?
5. This type of RNA transfers amino acids to the ribosomes
6. This type of cell movement moves in a whip-like fashion
7. Has both a hydrophilic and hydrophobic component
8. This type of diffusion requires a carrier protein
9. These proteins traverse the entire lipid bilayer
10. Diffusion of water across a semi-permeable membrane down a concentration gradient

3. **Blood & Body Fluids 1**
    A. Insulin
    B. Monocytes
    C. Osmolality
    D. ~1500 mL
    E. β-globulins
    F. Eosinophils
    G. Sodium thiosulphate
    H. Albumin
    I. Basophils
    J. Fat cell number
    K. γ-globulins
    L. ^{51}Cr tagged red blood cells & haematocrit
    M. Osmolarity
    N. ~1000 mL
    O. Fibrinogen
    P. ^2H_2O
    Q. Neutrophils

    Instruction: Match one of the above to each of the statements below:

    1. Number of osmoles per unit volume
    2. Accounts for the difference in total body water per Kg body weight between males and females
    3. Used in estimations of plasma volume
    4. Used in estimations of extracellular fluid volume
    5. Volume of non renal fluid loss at rest per day
    6. Constituent of plasma that defends against viruses and bacteria
    7. Major contributor to plasma oncotic pressure
    8. Largest % of white blood cells
    9. Largest % of agranular white blood cells
    10. Combat histamine in allergic reactions

4. **Blood & Body Fluids 2**
    A. 0.3 mL/L
    B. adult
    C. ↑ PO_2
    D. Factor XII
    E. Vasodilatation
    F. Sickle cell disease
    G. 90%
    H. Sluggish blood flow
    I. 3 mL/L

J. carbon monoxide inhalation
K. Formation is associated with turbulent blood flow
L. 6 mL/L
M. Fetal
N. ↑PCO_2
Q. 75%
R. Platelet and fibrin components are large
S. Factor V

Instruction: Match one of the above to each of the statements below:

1. Cause of blue palor of the extremities
2. Dissolved O_2 in plasma at 100 mmHg PaO_2
3. Haemoglobin with 2 α & 2 γ chains
4. O_2 dissociation curve is displaced to the left
5. Facilitates unloading of O_2 at tissues
6. Facilitates unloading of CO_2 at the lungs
7. Carriage of CO_2 as HCO_3^- in arterial blood
8. Arterial to venous difference in dissolved CO_2 in the pulmonary circulation
9. Factor NOT activated by thrombin in the clotting cascade
10. Characteristic of arterial thrombi

5. Nervous system 1

A. Central nervous system
B. Electrical
C. Soma
D. Dopamine
E. Nissl bodies
F. Chemical
G. Norepinephrine
H. Terminal boutons
I. Acetylcholine
J. Peripheral nervous system
K. Antidiuretic hormone
L. Nitric oxide
M. Glutamate

Instruction: Match one of the above to each of the statements below:

1. The somatic system is part of which division of the nervous system?
2. The spinal cord is part of which division of the nervous system?
3. These swellings form the presynaptic synapse
4. This contains the neuronal organelles
5. These synapses have direct channels
6. This type of transmission involves the spread of action potentials over the presynaptic terminal
7. A reduction in this is associated with Alzheimer's disease
8. The neurotransmitter involved in hearing at the sterocilia
9. This neurotransmitter stimulates NMDA receptors
10. This neurotransmitter is reduced in Parkinson's disease

6. Nervous system 2

A. 2
B. −90

C. 0
D. −70
E. 30
F. 3
G. 20
H. Na^+
I. K^+
J. Meissner corpuscles
K. Pacinian corpuscles
L. Merkel discs
M. Myelinated A fibres
N. Ruffini corpuscles
O. Free nerve endings

Instruction: Match one of the above to each of the statements below:

1. The Nernst equation predicts the resting membrane potential to be this value in mV
2. The measured resting potential in mV
3. The number of potassium ions exchanged for sodium at the Na^+/K^+-ATPase
4. The value of the overshoot in mV
5. This ion enters the neuron during depolarization
6. The gates for this ion close in repolarization
7. This ion enters the neuron during repolarization
8. Pressure detectors
9. Mechanoreceptors
10. Detects fine touch; is found at the fingertips

7. Nervous system 3

A. Rods
B. Suprachiasmatic nucleus
C. Proencephalin
D. Aspirin
E. Substantia nigra
F. Lateral geniculate nucleus
G. Semi circular canals
H. Cones
I. Reticular formation
J. Organ of Corti
K. Eustachian tube
L. Retina
M. Morphine
N. Cochlea

Instruction: Match one of the above to each of the statements below:

1. Endogenous opioid
2. Exogenous opioid
3. Loosely arranged network of white and grey matter in the brainstem
4. Releases dopamine
5. Its two layers are neural and pigmented
6. These cells produce colour vision
7. These contain rhodopsin
8. Converts mechanical vibrations into nerve impulses
9. This structure is divided into three channels: scala vestibuli, scala tympani and scala media
10. Involved in maintaining balance

8. Musculoskeletal 1

A. Na+
B. Synovial
C. Ball & socket
D. Sarcoplasmic reticulum
E. K+
F. Fibrous
G. Skeletal
H. Ca2+
I. Tropomyosin C
J. Cl−
K. Smooth
L. ATPase
M. Actin
N. Cartilaginous
O. Myosin

Instruction: Match one of the above to each of the statements below:

1. Another name for striated muscle
2. Another name for non-striated muscle
3. The most common type of joint
4. Thin filament is composed of…
5. This ion attracts the acetylcholine vesicles to the neural membrane
6. This ion continues the action potential in the postsynaptic neuron
7. Composed of six polypeptide chains forming a double helix
8. The binding of calcium with this displaces the tropomyosin
9. During the power stroke the head is attached to this filament
10. The main ion in muscular contraction

9. Musculoskeletal 2

A. Caudate
B. Type I
C. Twitching
D. Parathormone
E. Primary motor cortex
F. Tetanization
G. Calcitonin
H. Osteoclast
I. Supplementary cortex
J. Isometric
K. Substantia nigra
L. Putamen
M. Premotor area
N. Type 2B
O. Isotonic
P. Osteoblast

Instruction: Match one of the above to each of the statements below:

1. This type of contraction involves muscle shortening
2. Sequential twitches at low frequency
3. These muscle types have small motor neurons
4. Generally this muscle type is predominantly involved in short-term anaerobic work

5. This muscle type has a low mitochondrial density
6. Lies in the frontal lobe
7. Part of the basal ganglia involved in regulating eye movements
8. Neurons from here release dopamine
9. These cells are involved in bone resorption
10. An increase in this hormone increases osteoclast activity

10. General cardiovascular 1

A. 0.02
B. 5.25
C. Semilunar
D. 50–100
E. 0.1
F. 72
G. −90
H. 0.06
I. 0.0256
J. Bicuspid
K. −70
L. 0.3
M. 25
N. 30–200
O. 0.0625

Instruction: Match one of the above to each of the statements below:

1. These valves are located between the left ventricle and the aorta
2. The resting membrane of the cardiac cell in mV
3. The length of time of ventricular systole in seconds
4. The length of time of atrial systole in seconds
5. The volume of blood ejected at rest by one ventricle in litres/minute
6. The normal range for the length of vascular smooth muscle fibres, in micrometers
7. The average number of heart beats per minute in a healthy adult
8. The normal range for the average length of a myocyte in micrometres
9. Resistance of an artery with a diameter of 5 mm, in units
10. The duration of isovolumetric contraction in seconds

11. General cardiovascular 2

A. Left bundle branch
B. Endothelial
C. Pulmonary
D. Bronchial
E. Hydrostatic
F. Sinoatrial node
G. Pericardium
H. Purkinje fibres
I. Gravity
J. Bundle of His
K. Atrioventricular valve
L. Myocardium
M. Transmural
N. Endocardium
O. Oncotic

Instruction: Match one of the above to each of the statements below:

1. The outermost layer of the heart
2. This layer of the heart is continuous with the endothelial lining of the large blood vessels
3. This vessel carries deoxygenated blood away from the heart
4. Capillaries contain a single layer of these cells
5. The node juxtaposed between the two atria
6. This carries the action potential from the atria to the ventricles
7. These nerves, when excited, result in ventricular contraction
8. This pressure keeps fluid within vessels
9. This pressure pushes fluid out of vessels
10. This pressure decreases along the length of the capillary

12. General cardiovascular 3

A. Atrial natriuretic peptide
B. Antidiuretic hormone
C. Dietary fats
D. Angiotensin-converting enzyme
E. α-receptors
F. Lactate
G. Epinephrine
H. Angiotensinogen
I. Serotonin
J. Renin
K. β-receptors
L. Trapezius
M. Aldosterone
N. Norepinephrine

Instruction: Match one of the above to each of the statements below:

1. An increase in the concentration of this can cause vasodilation of peripheral vessels
2. Responsible for transporting fat-soluble vitamins in the blood
3. Is a vasoconstrictor released from platelets
4. Is the primary secretion from the adrenal medulla
5. Is secreted by the adrenal cortex
6. Is produced in the hypothalamus
7. Has high affinity for alpha receptors
8. Agonists at these receptors have chronotropic and ionotropic effects of the heart
9. Is produced in the kidney
10. Is produced by the atria

13. Respiratory system 1

A. Pulmonary fibrosis
B. 3–4 L
C. Visceral pleura
D. 3×10^9
E. Non respiratory bronchioles
F. Parietal pleura
G. Dead space volume
H. Respiratory bronchioles

I. Scalene
J. Trachea
K. 3×10^8
L. T1 – T11
M. Diaphragm
N. C3 – C5
O. Internal intercostal
P. COPD
Q. 2–3 L

Instruction: Match one of the above to each of the statements below:

1. Layer covering the thoracic cavity
2. Number of alveoli in human lung
3. Respiratory tree represented by generation number 0
4. In health volume that best describes functional residual capacity
5. Inspiratory muscle normally involved in quiet breathing
6. Nerves supplying the external intercostals arise from spinal cord segments
7. Respiratory tree represented by generation numbers 17–19
8. Condition in which residual volume would be expected to increase
9. In health volume that best describes physiological dead space
10. Part of the respiratory tree with the largest resistance to air flow

14. Respiratory system 2

A. Emphysema
B. – 6–7
C. Type I alveolar cells
D. $\Delta P/\Delta V$ where P = pressure, V=volume
E. Asthma
F. Increases physiological dead space
G. 0.2
H. $\Delta V/\Delta P$
I. Silicosis
J. Reduces fluid exudation from pulmonary capillaries
K. β_1-adrenoceptors
L. Leukotrienes
M. 0
N. Type II alveolar cells
O. Increase $[CO_2]$ in conducting airways
P. β_2-adrenoceptors

Instruction: Match one of the above to each of the statements below:

1. Functional residual capacity (FRC) decreases
2. Relative to atmospheric pressure, alveolar pressure (cm H_2O) at tidal volume
3. Formula for calculation of lung compliance
4. Cells producing dipalmitoyl phosphatidodylcholine
5. Arithmetic difference between compliances of chest wall and lung at FRC
6. Decreases airway resistance
7. Relative to atmospheric pressure, intrapleural pressure (cm H_2O) at tidal volume
8. Lung compliance increases

9. An effect of surfactant
10. Receptors responsible for bronchodilation

15. Respiratory system 3

A. Barbiturates
B. ~120 mmHg
C. PCO_2 / PO_2: 40 mmHg / 100 mmHg
D. ↑Renal HCO_3^- excretion
E. Tractus solitarius
F. Hypercapnia
G. PCO_2 / PO_2: 45 mmHg / 70 mmHg
H. Type 1
I. ↑pH
J. Hypoxia
K. HCO_3^-
L. Apneustic
M. Ambiguus
N. H^+
O. ~90 mmHg
P. Pneumotaxic
Q. Type 2

Instruction: Match one of the above to each of the statements below:

1. PO_2 of inspired air reaching alveoli in person at altitude 4000 metres
2. Major stimulus to the ventilatory response of central chemoreceptors
3. Nucleus primarily associated with dorsal respiratory group neurons (DRG)
4. To what is the blood-brain barrier relatively impermeable
5. Chemosensitive cells in the carotid body
6. An early compensatory response to ascent to high altitude
7. Increases the ventilatory response to $PaCO_2$
8. Pontine centre that stimulates inspiration
9. Major stimulus to the ventilatory response of peripheral chemoreceptors
10. Arterial blood gases during moderate aerobic exercise

16. Respiratory system 4

A. $Pv > P_A$
B. Unchanged
C. Difference in diffusion coefficients
D. Partial pressure gradients are similar
E. Increases
F. ↑ diffusion distance
G. Acetylcholine
H. ↑ thickness diffusion pathway
I. Decreases
J. Hypoxia
K. $P_A > Pv$
L. Nitric oxide (NO)
M. ↓ surface area

Instruction: Match one of the above to each of the statements below:

1. Reason for reduced gaseous exchange in emphysema
2. Transmural pressure in alveolar vessels during normal inspiration

3. Reason for reduced gaseous exchange in left ventricular failure
4. Ventilation:perfusion ratio (V/Q) from apex to base of lung
5. Total pulmonary vascular resistance during deep inspiration
6. Reason for similarity between rates of diffusion of CO_2 & O_2 across alveolar membranes
7. Differences between alveolar (P_A) and venous (Pv) pressures in Zone 2 of the lung
8. Decreases pulmonary blood flow
9. Alveolar PO_2 (P_AO_2) when V/Q decreases
10. Pulmonary vascular resistance in extra-alveolar vessels during inspiration

17. Gastrointestinal system 1

A. Superior mesenteric artery
B. Digastric
C. Pancreatic duct
D. Duodenum
E. Ventromedial nucleus
F. Ileum & jejenum
G. Thorax
H. Serotonin
I. Cystic duct
J. Inferior mesenteric artery
K. Stomach
L. Ventrolateral nucleus
M. γ-aminobutyric acid
N. Masseters
O. Colon
P. Acetylcholine

Instruction: Match one of the above to each of the statements below:

1. Feeding centre in the brain
2. Neurotransmitter involved in the inhibition of feeding
3. Region of gastrointestinal tract where % fluid reabsorption is greatest
4. Where *Pseudomonas* spp is found
5. Neurotransmitter responsible for the secretion of salivary amylase
6. Muscles involved in jaw opening during mastication include
7. Where an increase in pressure occurs during vomiting
8. Blood vessels supplying duodenum and ileum
9. Peristaltic waves occur at ~3–4/min
10. Duct into which gall bladder empties concentrated bile

18. Gastrointestinal system 2

A. Na^+/H^+ antiport
B. Acetylcholine
C. H^+-ATPase
D. Gastrin
E. Somatostatin
F. Norepinephrine
G. Ca^{2+}
H. Antral
I. Histamine

J. cAMP
K. Prostaglandins
L. H$^+$/K$^+$-ATPase
M. Secretin
N. Body
O. Cholecystokinin(CCK)

Instruction: Match one of the above to each of the statements below:

1. Inhibits gastric acid secretion from parietal cells
2. Increases gastric mucus secretion
3. Protein carrier responsible for proton secretion in oxyntic cells
4. Stimulates acid production via cAMP as second messenger
5. Neurotransmitter that increases gastric motility
6. Hormone that stimulates flow of alkaline pancreatic fluid
7. Hormone that increases secretion of bile
8. Stimulates gastrin secretion in the cephalic phase
9. Causes contraction of the gallbladder
10. Principle region of gastric mucosa responsible for gastrin secretion

19. Gastrointestinal system 3

A. Galactose
B. Pancreatic duct
C. Vitamin B$_{12}$
D. Aminopeptidase
E. Conjugated bilirubin
F. K$^+$
G. Stomach
H. Fructose
I. Colon
J. Vitamin E
K. Enterokinase
L. Fe^{2+}
M. Chenodeoxycholic acid
N. Triglyceride
O. Na$^+$

Instruction: Match one of the above to each of the statements below:

1. Absorption in the terminal ileum depends on secretion from parietal cells
2. Recycled via the enterohepatic circulation
3. Absorption is dependent on secretions from the gallbladder
4. Gastrointestinal site where the CFTR regulator for Cl$^-$ secretion is found
5. Converted in the stomach to a form readily absorbed in the duodenum
6. Major component of chylomicrons
7. Cation absorbed in the ileum and secreted in the colon
8. Is absorbed by facilitated transport utilising Na$^+$-independent carrier proteins
9. Intestinal brush border enzyme responsible for activation of pancreatic enzymes
10. Where H$_2$O absorption depends on an existing osmotic gradient?

20. Kidneys & the urinary tract 1

A. Glomerular
B. Interlobar
C. Para-aminohippuric acid
D. Renal blood flow
E. Peritubular
F. Descending limb of the loop of Henle
G. Radial
H. Vasa recta
I. Glomerular filtration rate
J. Proximal convoluted tubule
K. Creatinine
L. Pars recta
M. Effective renal plasma flow
N. Distal convoluted tubule
O. Inulin

Instruction: Match one of the above to each of the statements below:

1. Can be measured using the Fick principle
2. Blood vessels that give rise to the afferent arterioles
3. Consists of cuboid epithelia cells with few microvilli on the apical surface
4. Capillaries in Bowman's capsule
5. The straight portion of the proximal tubule
6. Contains specialised cells that monitor NaCl load delivered to the tubule
7. Clearance of this solute is independent of its plasma concentration
8. Hydrostatic pressure relatively constant along the length of this capillary
9. Consists of simple squamous epithelia cells with few mitochondria
10. Clearance of this solute gives an accurate estimate of GFR

21. Kidneys & the urinary tract 2

A. Mg^{2+}
B. HCO$_3$$^-$
C. Creatinine
D. K$^+$
E. Glucose
F. CO$_2$
G. Aspirin (salicylate)
H. Na$^+$
I. NH$_4$$^+$
J. Ca^{2+}
K. H$_2$O
L. Urea

Instruction: Match one of the above to each of the statements below:

1. Paracellular reabsorption in the proximal tubules of >80% filtered load
2. Transcellular reabsorption of almost 100% of the filtered load
3. Is secreted by the thin ascending limb of the loop of Henle (LoH)
4. Solute transported across proximal tubule apical membrane by a protein symporter
5. Reabsorption from the proximal tubules <20% of the filtered load

6. Crosses apical membranes of renal tubular epithe-lial cells by simple diffusion
7. Is passively reabsorbed by the thin ascending limb of the LoH
8. Reabsorption from the proximal tubules ~50% of the filtered load
9. Can be secreted across proximal tubule epithelium in significant amounts
10. Reabsorption from the thick ascending limb of LoH is ~60% of the filtered load

22. Kidneys & the urinary tract 3

A. Dopamine
B. Medullary collecting duct
C. Atrial natriuretic peptide
D. Distal convoluted tubule
E. Thick ascending limb of loop of Henle
F. Angiotensin I
G. Vasa recta
H. Aldosterone
I. Thin descending limb of LoH
J. Renin
K. Pars recta
L. Proximal convoluted tubule
L. Antidiuretic hormone
M. Thin ascending limb of loop of Henle
N. Angiotensin II
O. Cortical collecting duct

Instruction: Match one of the above to each of the statements below:

1. Na$^+$ reabsorption occurs by thiazide-sensitive protein symporter
2. Increases H$_2$O permeability of collecting ducts
3. Fluid leaving this part of the nephron is always hypo-osmotic to plasma
4. Increases Na$^+$ reabsorption in cortical collecting ducts
5. Cl$^-$ reabsorption occurs by furosemide-sensitive protein carrier
6. Reduces Na reabsorption in inner medullary collecting ducts

7. Is secreted from juxtaglomerular cells in the afferent arteriole
8. Acts a countercurrent exchanger
9. Nephron site where amiloride reduces K$^+$ secretion
10. Increases Na$^+$ reabsorption in the proximal tubules

23. Kidneys & urinary tract 4

A. Increases plasma volume by 6.7%
B. Decreases urine osmolality
C. Decreases body weight/extracellular fluid volume
D. Decreases plasma osmolality
E. Increases titratable acid excretion
F. 1% increase in total body fluid volume
G. Increases plasma PaCO$_2$ and urine pH but decreas-es plasma pH
H. Decreases antidiuretic hormone secretion
I. Increases mean arterial pressure
J. Increases K$^+$ excretion
K. Increases extracellular fluid volume by 2.5%
L. Decreases aldosterone secretion
M. Increases hydrostatic pressure in glomerular capillaries

Instruction: Match one of the above to each of the statements below:

1. Inhibition of carbonic anhydrase activity
2. Increased dietary protein intake
3. Urinary excretion in compensated respiratory acidosis
4. Chronic decrease in dietary salt intake
5. Ingestion of 1 L isosmotic saline solution by 70 Kg healthy volunteer
6. Voluntary hyperventilation
7. Administration of a furosemide to a H$_2$O-loaded subject
8. Excessive sweating
9. Increased vasa recta blood flow
10. Ingestion of NH$_4$Cl

1. Describe the structure and function of the mammalian cell.

2. Explain how one molecule of glucose is converted to ATP. Also explain how other substrates are used to make ATP.

3. Explain how an amino acid is manufactured in the cell.

4. Describe movement between the various fluid compartments within the body and with the external environment.

5. Describe the factors affecting oxygen transport by the blood and effects on gas exchange.

6. Explain how pain is sensed in the body and the different methods to control it.

7. Describe the events from when light enters the eye to an image being formed.

8. Describe the action potential and how it leads to muscle contraction

9. Describe the different proprioceptors and how their structure affects function.

10. Describe the anatomy and physiology of the conducting system of the heart.

11. Discuss how the structure of blood vessels facilitates their function.

12. Outline how blood vessels are regulated.

13. Compare and contrast the variation of blood flow and ventilation in the lung, commenting on the consequences for gas exchange.

14. Describe the effects of exercise on ventilation.

15. Outline central control of breathing and involvement of chemoreceptors.

16. Compare and contrast salivary gland secretion with that of alkaline pancreatic fluid.

17. Describe the intestinal flora and discuss their importance.

18. Outline acid–base balance by the kidneys.

19. Describe features of the loop of Henle that enable it to perform its function.

20. Discuss the roles of the renal hormones.

Chapter 1 Introduction to physiology

1. Fluids
a. True
b. True
c. False. Extracellular fluid comprises around 80% interstitial and 20% plasma
d. False. Just K^+ is the predominant intracellular ion
e. False. Interstitial fluid is extracellular

2. Organelles 1
a. False. Mature red blood cells do not
b. False. That is meiosis
c. False. 30s ribosomes are smaller than 50s ribosomes
d. False. Ribosomes contain RNA only
e. True

3. Organelles 2
a. False. The Golgi apparatus is usually situated proximal to the nucleus
b. False. They are double-membraned, ovoid structures
c. False. All the mitochondrial membrane is smooth
d. True
e. False. They are found in organelles

4. Energy production in the cell
a. False. ATP has three phosphate molecules
b. True
c. False. The majority (95%) of ATP is formed in the mitochondrial matrix
d. True
e. False. 38 ATP molecules are formed from one mole of glucose

5. Genetics 1
a. False. They are made from DNA
b. True
c. True
d. False. They connect via hydrogen bonds
e. False. Guanine binds to cytosine and adenine binds with thymine

6. Genetics 2
a. True
b. True
c. True
d. False. In RNA, deoxyribose is replaced with ribose

e. False. mRNA transfers the genetic code from the nucleus to the cytoplasm

7. Protein synthesis and cell movement
a. True
b. False. Splicing of mRNA removes the introns and combines the extrons
c. False. It moves out of the nucleus
d. True
e. False. Cilia are composed of nine double and two single microtubules

8. Cell membranes
a. True
b. True
c. False. The glycocalyx is negatively charged so it can repel other negative objects
d. False. Tight junctions form impermeable bond between cells
e. True

9. Transport
a. False. Diffusion is the movement of molecules from an area of high concentration to lower concentration
b. False. Simple diffusion of a molecule passes though the membrane opening
c. True. A hypotonic solution will cause cells to swell
d. True
e. True

Chapter 2 Physiology of the blood and body fluids

10. Body fluids 1
a. False. ECF is 75% interstitial fluid (ISF) and 25% plasma
b. True. Normal plasma values are 280–295 mosmol/kg H_2O- Osmolarity, is measured per unit volume
c. True. Proteins exhibit the Donnan effect. The concentration of diffusible ions would still be equal
d. True. Na^+ and Cl^- are the principal ECF cation and anion and account for 80% ECF osmolality
e. False. Normally they are equal with continual exchange of fluid

11. Body fluids 2

a. False. Capillary pressure declines along the length of a systemic capillary

b. False. It is 2–3 L: 2 L from food and drink and 400 mL from oxidative metabolism

c. True. Normal ECF volume is 20% body weight

d. True. This must be balanced by intake. It may be excessive with some diarrhoeas

e. True. Radioactive water – tritium (^3H_2O) or deuterium (^2H_2O) – are used by the dilutional principle

12. Blood

a. False. It is 8% total body weight or 5 L in the average adult

b. True. Plasma contains proteins, which are important buffers, and HCO_3^-, which buffers protons

c. False. It is lymphocytes that generate specific immune responses. Neutrophils are phagocytic

d. False. They comprise <1% of cellular elements

e. False. They are fragments derived from megakaryocytes in the bone marrow

13. O_2 transport in blood

a. False. Dissolved O_2 delivery is 15 mL/min – not sufficient for a basal metabolic rate of 250 mL/min

b. False. The iron atom needs to be in the ferrous state (Fe^{2+}) to bind with O_2

c. True. This value corresponds to the flatter part of O_2 dissociation curve

d. True. pH <7.4 reduces haemoglobin (Hb) affinity for O_2, so there is easier dissociation at any PO_2 (right shift)

e. True. HbF has a higher affinity and will be 20–50% more saturated than HbA at low PO_2 levels

14. CO_2 and CO transport in blood

a. True. The corresponding value for venous blood is 60%

b. False. It occurs more readily with reduced Hb as it is less acidic

c. True. The CO_2 dissociation curve shows there is no saturation

d. False. It causes a left shift, $\downarrow O_2$ unloading at tissues

e. True. It also has an affinity for Hb 250 times that of O_2

15. Haemostasis

a. False. Activation follows adhesion of platelets to the damaged blood vessel and collagen

b. False. Factor VII is acted upon by tissue factor, forming part of the extrinsic pathway

c. True.

d. False. Arterial thrombi have a large platelets, unlike venous thrombi

e. True. It cross-links fibrin fibres, permitting further compression of the clot

---

Chapter 3 Physiology of the nervous system

16. Overview of the nervous system

a. True

b. False. The CNS consists of the brain and spinal cord

c. True

d. False. The autonomic system consists of the parasympathetic, sympathetic and enteric divisions

e. True

17. Neuronal excitation and inhibition

a. False. It is –70 mV

b. False. The Goldman–Hodgkin–Katz equation takes into consideration other ions

c. True

d. True

e. False. The relative refractory period of an AP is when a further AP is possible but a larger than normal stimulus is required

18. Ionic mechanism of action potentials

a. True

b. True

c. False. Depolarization is associated with the upstroke on the action potential

d. True

e. False. They are due to the Na^+ and K^+ channels not returning to their previous states

19. Myelinated and unmyelinated fibres

a. False. Myelin forms a sheath around some nerves

b. True

c. True

d. False. The node of Ranvier is unmyelinated

e. False. The thicker the axon the faster the speed of conductance

20. Synaptic transmission

a. False. Chemical synapses are based on neurotransmitters

b. False. It is Ca^2, not Na^+

c. True

d. True

e. False. ACh has both inhibitory and excitatory effects

21. Sensation

  a. True

  b. True

  c. False. Encapsulated nerve endings are surrounded by a connective tissue substance

  d. True

  e. False. Pacinian corpuscles are pressure receptors

22. Pain

  a. False. Fast pain occurs 0.1 seconds after the stimulus has been applied

  b. True

  c. False. The periaqueductal grey is a pain-regulating area in the midbrain and pons

  d. False. They are endogenous opiate-like peptides

  e. True. The liver metabolizes morphine to morphine-6-glucuronide, which is more potent than morphine itself

23. Brainstem

  a. True

  b. False. The white matter in the medulla oblongata contains both sensory and motor tracts

  c. True.

  d. False. The inferior colliculi are concerned with hearing

  e. True. The reticular activating system is involved in consciousness

24. Autonomic nervous system

  a. False. The autonomic nervous system is under involuntary control

  b. True

  c. True

  d. False. It is the sympathetic fibres innervating the adrenal medulla that increase the release of norepinephrine (noradrenaline)

  e. True

25. Vision

  a. True. The five cell layers are: photoreceptive, bipolar, ganglion and two synaptic cell layers

  b. False. Rods help vision in the dark; they have no role in colour vision

  c. True

  d. False. Isomerization is a change in conformation of retinol from the *cis* to the *trans* form

  e. True. Na^+ entry via Na^+-gated channels causes depolarisation

26. Optics

  a. True. Hence short-sightedness

  b. False. It is the contraction of the ciliary muscle that relax the lens to aid accommodation

  c. False. The fovea is the area of greatest visual acuity

  d. True

  e. True

27. Hearing

  a. True

  b. False. The malleus is attached to the inner surface of the eardrum and the incus

  c. True

  d. True

  e. False. Six tonotopic maps are present in the auditory cortex

28. Olfaction

  a. True

  b. False. Along cranial nerve I (olfactory)

  c. False. Basal stem cells are present which form new olfactory receptor cells

  d. False. The lateral olfactory area is in the temporal lobe F

  e. True

29. Taste

  a. True. Taste buds are also present on the soft palate and the pharynx

  b. False. The gustatory receptor cells project only a single microvillus

  c. True

  d. True

  e. False. The conscious perception of taste is given by the parietal lobe

Chapter 4 Physiology of the musculoskeletal system

30. Introduction

  a. True

  b. False. It is striated muscle

  c. False. It is involuntarily controlled

  d. True

  e. True

31. Muscle microstructure

  a. False. Fasciculi are covered by epimysium

  b. False. Thick filaments are predominantly myosin

  c. True

  d. True

  e. True

32. Neuromuscular junction (NMJ)

  a. False. Myelinated fibres arise from the anterior horn

  b. True

  c. True

d. False. They are voltage gated

e. True

33. Secretion of acetylcholine (ACh)

a. True

b. True

c. True

d. False. Acetylcholinesterase is responsible; monoamine oxidase is responsible for oxidation of neurotransmitters such as norepinephrine and serotonin

e. True

34. Muscle contraction

a. False. Myosin is composed of two heavy and four light chains

b. True. The ATPase cleaves ATP for contraction

c. True

d. False. Tropomyosin covers the actin

e. True

35. Mechanism of contraction

a. True

b. True

c. False. ATP is hydrolysed to ADP by the ATPase on the myosin

d. True

e. True

36. Motor units 1

a. True

b. True

c. False. The finer the control required for motor function, the fewer the numbers of muscle fibres per motor neuron

d. False. In isometric contraction the muscle neither lengthens nor shortens during contraction

e. True

37. Types of muscle fibre

a. True

b. False. Fast muscle has less myoglobin and is whiter in colour than slow muscle

c. True

d. False. Type II fibres are less resistant to fatigue than type I

e. True

38. Central control of movement

a. True

b. True

c. True

d. False. It is the corticospinal system

e. True

39. Posture

a. True

b. True

c. False. The crossed extensor reflex is when one leg flexes and lifts off the ground the opposite leg extends more strongly to support the shifted weight of the body

d. True

e. False. Vestibular nuclei receive input from the vestibulocochlear nerve (CN VIII)

40. Bone

a. False. Bone matrix comprises 25% water, 25% collagen fibres, 50% crystallized mineral salts

b. False. Osteoblasts are bone-forming cells

c. True

d. False. Parathyroid hormone serves to increase Ca^{2+} in the blood

e. True

Chapter 5 Physiology of the cardiovascular system

41. Introduction

a. False. It is higher in newborns

b. True. It controls peripheral blood flow to regulate heat exchange

c. True. The pulmonary artery also carries blood away from the heart

d. False. They are closed, otherwise blood would come out uncontrollably

e. True

42. Cardiac anatomy 1

a. True

b. False. The oesophagus is posterior to the heart

c. True

d. False. The thymus is superior

e. False. The right phrenic nerve is lateral to the heart

43. Cardiac anatomy 2

a. True

b. False. The visceral pericardium is attached to the outer surface of the heart

c. True

d. False. The middle layer of the endocardium consists of connective tissue

e. True. It also contains veins and nerves

44. Cardiac anatomy 3

a. True. It pumps the blood further-hypertrophy

b. True

c. True. 'Bi' means 'two'

d. True

e. True.   It is an inflammatory disease usually caused by a streptococcal infection. An immune response is triggered that also attacks the valves

45. Cellular physiology of the heart
a. True
b. True.   Some might have two nuclei
c. False. They have a dense mitochondrial network as they are very energy dependent
d. True
e. False.   They are only present in cardiac tissue

46. Conduction system of the heart
a. True
b. False. It is due to the influx of Na^{2+}
c. False. The conduction goes to the bundle of His before it reaches the Purkinje fibres
d. False. The resting membrane of the heart is approximately $-90\,mV$ *not* $+90\,mV$
e. False. Repolarization occurs as voltage-gated K^+ channels open to increase efflux of K^+

47. Cardiac cycle
a. True
b. True
c. False. It lasts for 0.3 s
d. False. It is responsible for the 'c' wave; the 'v' wave is a consequence of passive filling of the atria
e. True

48. Cardiac output
a. True
b. True
c. True
d. True
e. False.   Chronotropes increase the rate; inotropes increase the force of contractility of the heart

49. Organization of the vessels
a. True.   To allow exchange of fluids
b. False.   Elastic arteries are usually involved in the regulation of flow (e.g. Windkessel effect in the aorta and other elastic arteries)
c. True.   By vasodilatation and vasoconstriction
d. False. It is proportional to the metabolic demands of the tissue
e. True

50. Vascular smooth muscle
a. True
b. True
c. False. It is due to the influx of Ca^{2+}
d. True.   Due to the decrease in the number of Ca^{2+} channels
e. True.   This causes contraction and relaxation of the smooth muscle

51. Regulation of blood flow
a. False. $$Flow = \frac{pressure\ difference}{vessel\ resistance}$$
b. False.   Blood travels from an area of high to low pressure
c. True
d. False. Resistance is proportional to $length/radius^4$
e. True

52. Haemodynamics of the venous system
a. True
b. True.   This helps drive the blood to the right atrium
c. False. Vasodilation occurs so less blood goes to the brain and you feel dizzy
d. False. They are important in venous return
e. True.   During inspiration the diaphragm flattens, causing negative intrathoracic pressure and an increase in intra-abdominal pressure. This increases blood flow to the heart

53. Capillary dynamics and transport of solutes
a. False.   They are a single layer
b. True
c. True
d. True
e. True.   This is the definition of lymph capillaries

54. Control of blood vessels
a. True.   This regulates temperature
b. False.   Decreased PO_2, increased PCO_2 and K^+ causes vasodilation
c. False.   It is via the H_1 receptor. The H_2 receptor causes venodilatation and arterial vasoconstriction
d. False.   Nitric oxide (NO) is an endothelium-dependent relaxing factor (EDRF)
e. True

55. Cardiovascular receptors and central control
a. True
b. False.   It is the depressor area in the hypothalamus that produces the baroreceptor reflex
c. False.   Baroreceptor reflexes travel through the vagus and glossopharyngeal nerves
d. True
e. True

56. Regulation of circulation in individual tissues
a. True.   As the brain has a high oxygen demand
b. True
c. False.   Antidiuretic hormone is increased to conserve water
d. False.   It has the third highest after the heart and kidney
e. True

Chapter 6 Physiology of the respiratory system

57. Ventilation

a. True. Last portion of inhaled air will occupy the anatomical dead space

b. False. Physiological dead space = Anatomical dead space + Alveolar dead space (unperfused)

c. False. Ventilation is better in the lower regions where the smaller alveoli distend better

d. True. The former measures only ventilated lung volume, whereas the latter includes trapped air

e. True. Obstructive disease leads to air trapping, increasing the lung volume (FRC) after normal expiration

58. Intrapleural pressure

a. True. Alveolar volume may be used in the equation pertaining to Boyle's law: pressure α 1/volume

b. True. Lung and chest wall pull in opposite directions → subatmospheric pressure in between (P_{PL})

c. True. Inspiration causes chest expansion, lowering intrapleural pressure

d. False. It is positive in forced expiration

e. True. P_{PL} is subject to gravity so increases (becomes less negative) as you go down

59. Muscles of respiration

a. False. The T1–T11 anterior rami supply the external intercostal muscles; the phrenic nerve (C3–C5) supplies the diaphragm

b. True. Contraction therefore elevates the lower ribs and pulls them forwards (pump handle action)

c. False. Expiration is usually passive

d. False. It aids expiration by fixing the ribcage downwards

e. True. Usual contraction of 1–2 cm increases the length of the thorax

60. Compliance

a. False. Compliance = Change in volume/change in pressure, so the converse will be true

b. False. It is lower when lung volume < residual volume (RV) and also at high volumes, because alveoli are already stretched

c. True. Interalveolar septa and constituents responsible for elastic properties of the lung are destroyed

d. True. The lung is stiffer with increased elastic recoil, providing further resistance to expansion

e. False. Although the compliances are in series, they must be added as reciprocals to give the inverse of Total compliance

61. Surfactant

a. True. Notably, it contains dipalmitoyl phosphatidodylcholine (DPPD)

b. True. It is important, and so is mechanical interdependence

c. False. The same amount is spread thicker in smaller alveoli, compensating for radial differences

d. False. Closer packing occurs in expiration, lowering surface tension and increasing compliance

e. True. Respiratory distress syndrome (RDS) is related to prematurity as surfactant is normally produced from 34 weeks' gestation

62. Dynamics

a. True. Poiseuille's law applies, so doubling the radius increase flow rate 16-fold

b. True. They form 80% lower respiratory tract resistance, which itself is 60% total airways resistance

c. True. Radial traction supports bronchi and smaller airways with increased lung volumes

d. True. Positive PPL → Negative transmural pressure → Airway constriction and resistance to airflow

e. False. Work of breathing is reduced by decreasing dynamic resistance

63. Gaseous exchange in the lungs

a. False. They have similar rates as slower CO_2 release compensates for its higher solubility

b. True. CO strongly combines with haemoglobin, causing a slow rise in partial pressure

c. False. Dalton's law: total pressure in a gas mixture is equal to the sum of the partial pressures of the gases in the mixture

d. True. Equilibrium between PaO_2 and P_AO_2 takes 0.25 s: one-third of the transit time

e. False. It is only diffusion limited with abnormal gas exchange membranes in some diseases

64. Pulmonary blood flow and pulmonary vascular resistance (PVR)

a. True. It is enabled by high compliance and great distensibility

b. True. This suits the low pressure system of the pulmonary circulation

c. False. Greater hydrostatic pressures in this region result in recruitment and distension, ↓PVR

d. False. Higher alveolar pressure outside compresses the vessel and ↑resistance

e. True. These run through lung parenchyma or vessels located at junctions of alveolar septa

65. Pulmonary blood flow and water balance

a. False. Postcapillary venules open and close with systole and diastole, respectively

b. True. Adjacent precapillary muscle contraction occurs in response

c. True. Average blood flow based on cardiac output = amount of dye/[dye] × time

d. True. It lowers surface tension that would otherwise favour transudation into alveoli

e. True. If the peribronchial and perivascular spaces become engorged, diffusion distance is increased

66. Ventilation–perfusion relationships (V/Q)

a. False. V/Q is ∞ for dead space since Q = 0, there is no perfusion

b. False. $PaCO_2$ is usually slightly low as initial hypercapnia stimulates ventilation

c. False. Inspired PO_2 is 150 mmHg, higher than alveolar air

d. True. Q > V: more perfusion results in greater CO_2 release into alveoli from blood

e. True. There is less blood to take away O_2 from the alveoli

67. Neural central control of ventilation 1

a. True. Integration of sensory information may occur before influencing breathing control

b. False. Medullary respiratory centres, dorsal respiratory group (DRG) or pre-Botzinger cells are thought to be responsible

c. True. There are both inspiratory and expiratory neurons in all three nuclei

d. True. Lung volume increases steadily with increasing signals

e. False. Nucleus retroambiguus type 2 inspiratory neurons (VRG) also project to the diaphragm

68. Neural central control of ventilation 2

a. False. It prolongs the inspiratory ramp of the dorsal respiratory group (DRG), lengthening inspiration

b. True. This occurs secondary to the effect of shortening inspiration

c. True. Prolonged inspiration associated with lung hyperinflation is another feature

d. True. Separate inspiratory and expiratory fibres exist in spinal cord white matter

e. True. They receives afferents from the nucleus retroambiguus's expiratory neurons

69. Central chemoreceptors

a. False. They are located bilaterally in the medulla, distinct from the ventral respiratory group (VRG) and dorsal respiratory group (DRG)

b. True. This occurs in response to elevated [H⁺] in surrounding ECF

c. False. Increased $PaCO_2$ has the greatest effect through changing ECF [H⁺]

d. True. Consequently CSF buffers pH changes less well so that pH varies largely

e. True. Hypoxia does not alter [H⁺] and has no effect

70. Peripheral chemoreceptors

a. True. Response is greatest when PaO_2 drops below 50–60 mmHg

b. True. Arteriovenous difference is minimal despite high metabolic rate due to huge blood flow rate

c. False. In humans, carotid bodies are of greatest respiratory significance

d. False. It responds only to PaO_2, from which it derives it O_2 supply, not carriage by haemoglobin

e. False. Type 1 (glomus) cells detect and respond to hypoxia; type 2 cells are supportive

71. Lung receptors and other reflexes

a. True. The vagus nerve transmits information from all three types of lung receptor

b. False. J receptors are sensitive to oedema and vascular congestion

c. True. Lung inflation and deflation respectively inhibit and stimulate the inspiratory muscles

d. False. They are located at branch points of the tracheobronchial tree

e. True. In addition, arterial baroreceptors mediate hypoventilation and bradycardia in response

72. Responses of the respiratory system

a. False. Along with age and narcotics, it decreases CO_2 ventilatory response

b. False. Responses occur with much lower PaO_2 than normal values

c. True. Ventilatory stimulus is greater when they occur together than the sum of their separate effects

d. False. Increased respiratory rate and tidal volume account for the increased minute ventilation

e. False. $PaCO_2$ stays normal or slightly low with aerobic exercise but rises with anaerobic metabolism

73. Altitude

a. True. Composition of air is unchanged but barometric pressure decreases

b. True. It minimizes $PaCO_2$ thereby maximizing P_AO_2 according to the alveolar gas equation

c. False. PaO_2 is still low, however, raised haemoglobin return total arterial O_2 to near normal

d. False. Respiratory alkalosis shifts the curve to the left, so there is sufficient O_2 loading at the lungs

e. False. This is an immediate response that persists

74. Diving

a. True. Pressure increases by 760 mmHg (1 atm) with every 10 m increase in depth

b. False. Partial pressure is increased with depth; PO_2 doubles at 10 m

c. False.   Higher gas densities increase the work of breathing needed to overcome ↑airways resistance

d. True.   Positive pressure of water opposes chest elastic recoil, effectively halving FRC

e. True.   Passage through the CNS circulation risks paralysis or brain damage

---

Chapter 7 Physiology of the gastrointestinal system

75. Upper gastrointestinal tract 1

a. False.   They both inhibit feeding

b. False.   This is achieved through mechanical feedback from the alimentary tract during feeding

c. True.   Jaw-closing muscles also include the temporal muscles

d. False.   After acinar secretion, saliva is modified along the duct via active transport mechanisms

e. True.   Proteins provide a protective tooth covering, known as an acquired pedicle

76. Upper gastrointestinal tract 2

a. False.   Vocal cords are pulled together, closing the trachea

b. True.   The mouth is closed during the voluntary phase of swallowing

c. True.   Oesophageal distension (also by reflux), stimulates secondary peristalsis

d. False.   Chemoreceptor trigger zone (CTZ) stimulants include drugs, upper GI tract lesions and prolonged vestibular nuclei stimulation

e. False.   Both are raised so as to extend the upper oesophageal sphincter

77. Motor functions of the stomach

a. True.   Decreased muscle tone as part of the vas-ovagal reflex facilitates food storage

b. False.   Peristaltic waves are staggered so that distal areas have a lag phase

c. True.   Reflux of antral contents allows the breakdown and churning of food with secretions

d. True.   The passage of small amounts of chyme occurs while reflux of bilious contents is prevented

e. True.   Presence of chyme in the duodenum has negative effects on gastric emptying

78. Gastric secretions

a. True.   Alkaline mucus coats the entire mucosal surface, protecting against gastric acid

b. False.   Decreasing pH of the contents in the antral region leads to decreased gastrin secretion, hence acid secretion

c. False.   This is true for histamine, whereas prostaglandins inhibit secretion of hydrochloric acid

d. False.   Distension stimulates the secretion of pepsinogen, the inactive precursor of pepsin

e. False.   This is true for fear and depression, whereas anger and hostility have the opposite effects

79. Liver and biliary tract

a. True.   This is known as the enterohepatic circulation, which occurs via the portal vein

b. False.   They enable lipid transport in solution, thus assisting with absorption

c. True.   In addition some free bilirubin is formed in this way

d. False.   Low portal vein concentrations of bile acid stimulate secretion but inhibit their synthesis

e. True.   Sphincter of Oddi constriction between meals results in bile diversion towards the cystic duct

80. The pancreas

a. True.   Cystic fibrosis transmembrane regulator (CFTR)-facilitates Cl^- exchange with intracellular HCO_3^-

b. True.   The acidic pH of zymogen lies outside the optimal range for proteolytic enzymes

c. False.   The opposite is true

d. False.   Sympathetic gland arteriolar vasoconstriction decreases secretion

e. False.   An increase in predominantly enzyme secretion (5–10%) occurs with the gastric phase

81. Digestion and circulation of the small intestine

a. False.   Main digestion takes place by enzymatic activity within the duodenum

b. True.   The celiac trunk and inferior mesenteric artery provide lesser contributions

c. True.   The portion of blood received by the muscular layers increases with motility levels

d. True.   Close proximity of vessels enable direct diffusion, bypassing the villus capillary

e. False.   Net diffusion means that nutrient-rich venous blood leaves the intestine

82. Intestinal digestion and absoprtion

a. True.   Only short-chain fatty acids and glycerol pass into the portal system directly

b. True.   Local oligo- and disaccharidases act on starch breakdown products

c. True.   Active transport not necessarily involving Na^+ accounts for amino acid uptake

d. False.   Ileal receptors bind to IF–vitamin-B_{12} complexes with uptake only of vitamin B_{12}

e. False.   The opposite occurs by increased synthesis of basolateral membrane Ca^{2+}-ATPase and apical membrane Ca^{2+}-binding proteins

83. Colonic bacteria and transport

a. False.   Colonic bacteria is 10^3 that of the small intestine

b. False.   Anaerobic bacterial production of vitamin K is vital for coagulation

c. True.   Solvent drag accounts for 90% of water absorption: smaller pores resist passive diffusion

d. True.   Glucocorticoids increase solvent drag by increasing ATPases

e. False.   Sympathetic stimulation inhibits secretion of protective mucus

84. Colonic motility and defecation

a. False.   Sympathetic stimulation produces reflex relaxation of bowel adjacent to a contracting portion

b. False.   Non-propulsive circular muscle contraction occurs, facilitating mixing of chyme

c. True.   Sequential haustration can also be initiated by the release of substance P

d. False.   Relaxation of the external sphincter, which is under voluntary control, permits defecation

e. False.   The intrinsic reflex is augmented by the para-sympathetic system by increasing distal motility

Chapter 8   Physiology of the kidneys and urinary tract

85. Kidney basic structure and function

a. True.   The kidneys receive approximately 1100 mL/min of blood

b. True.   80–85% of the nephrons are cortical; the remainder are juxtamedullary

c. False.   Transition occurs between the inner and outer medulla

d. True.   The ureter, blood vessels and nerves enter or leave the kidney via the fissure, or hilus

e. True.   This enzyme hydroxylates the liver metabolite ($25(OH) D_3$) at position 1

86. The glomerular filter

a. False.   All three layers carry strong negative charges, which discourage the passage of plasma proteins

b. True.   Slit pores are formed from gaps between pedicles/processes at the ends of podocyte cells

c. True.   Escaped capillary macromolecules are phagocytosed, keeping the basement membrane clear

d. False.   Molecular size, charge and shape are important

e. True.   Glomerular filtration rate is dependent on glomerular capillary and Bowman's capsular oncotic pressure (π) and hydrostatic pressure (P) (Starling forces)

87. Control of glomerular filtration rate (GFR)

a. True.   Autoregulation of RBF and GFR occurs over pressure range mean arterial 90–180 mmHg

b. True.   Dilation of each arteriole alone will increase RBF; there are opposite effects on P_{CAP}; efferent dilation $\downarrow P_{CAP}$, afferent dilation $\uparrow P_{CAP}$

c. False.   T-G feedback operates to correct changes in filtered load (GFR). $\uparrow$load $\rightarrow \downarrow$ GFR to protect against Na loss

d. True.   Afferent arteriole constriction will correct for an increase in GFR

e. False.   Mesangial cell contraction will reduce glomerular capillary surface area $\rightarrow \downarrow$ GFR

88. Renal blood flow (RBF) and glomerular filtration rate (GFR)

a. False.   PAH is filtered & secreted (Tm transport system). C_{PAH} = RPF depends on complete removal from plasma. At low plasma [PAH], renal venous [PAH] ~0; at high concentration renal venous [PAH] >>0

b. True.   Altered H_2O reabsorption alters [inulin] but not amount of inulin excreted; U_{inulin} V/P is constant

c. False.   Creatinine clearance is used

d. True

e. True.   RBF, GFR and glomerular capillary pressure are increased

89. Transport mechanisms

a. True.   Many other reabsorptive mechanisims depend on this transporter

b. False. $Na^+/K^+/2Cl^-$ is a cotransporter (symport)

c. False.   Ion channels do permit rapid transport, but are relatively few in number

d. True.   In addition to paracellular routes, transcellular transport occurs via ion channels, symports and antiport

e. True.   Amino acids, lactate and Cl^- reabsorption are also dependent on that of Na^+

90. NaCl transport in the proximal tubule

a. True

b. True. Na^+ is symported with metabolic substances, and antiported in early parts of the proximal tubule

c. True.   [Cl^-] builds up along the PT; reabsorption down its concentration gradient occurs in the pars recta

d. False.   Tight coupling between reabsorption of Na^+ and H_2O keeps fluid osmolality unchanged

e. True. HCO_3^- reabsorption in the early PT $\rightarrow$ osmolality and neutrality is maintained with increasing [Cl^-]

91. Transport of glucose and other solutes in the proximal tubule (PT)

a. True. Plasma glucose > 10 mmol will exceed the threshold or T_m of some nephrons

b. True. Usually < 20% of filtered PO_4^3 is excreted and importantly buffers urinary H^+

c. True. Reabsorption of urea is linked to controlling water balance

d. True HCO_3^- and Na^+ are transported across the basolateral membrane on a $3HCO_3^-$, $1Na^+$ symport

e. False. Most K^+ excreted is derived from K^+ secretion in late distal tubule and cortical collecting duct

92. The role of the loop of Henle (LoH)

a. True. The descending limb reabsorbs H_2O whereas the ascending limb reabsorbs NaCl

b. False. More NaCl is reabsorbed than water: 25% versus 15% of the filtered loads

c. False. Highest [urea] will be in late medullary collecting duct fluid – due to the differential permeability of the cortical collecting duct to H_2O and urea – urea moves out of the medullary CD down this gradient.

d. False. More solute than H_2O is removed by the LoH, so it leaves hypo-osmolar: 100 mmol/L

e. True. The ascending limb extrudes solutes, creating and maintaining a hyperosmolar medulla

93. The loop of Henle (LoH) and the countercurrent multiplier

a. False. Ascending thin limbs of LoH are important since passive Na^+ (&Cl^-) reabsorption occur: this is the single osmotic effect that is multiplied

b. True. H_2O diffuses from lumen to interstitium, while Na^+ and Cl^- diffuse in the opposite direction

c. True. Apical $Na^+/K^+/2Cl^-$ symporters rely on the basolateral Na^+/K^+-ATPase

d. True. Generation of osmotic pressure difference between interstitium and tubule depends on this impermeability

e. True. Medullary osmotic gradient is not dissipated; solute in interstitium >> solute entering blood

94. Regulation of urea and urine concentration

a. False. ~50% is reabsorbed in the PT

b. False. Secretion in the loop occurs but the high concentration relates to H_2O reabsorption in later nephron segments

c. True. Antidiuretic hormone increases the permeability of the medullary collecting ducts (CD) to urea, which then diffuses into the medullary interstitium

d. True. Urea is the major urinary solute; it will be important in determining obligatory loss

e. False. Aldosterone increase Na^+ reabsorption in the cortical collecting duct, H_2O is freed and will be reabsorbed; urine flow will be neither large nor dilute

95. Body fluid osmolality

a. False. Regulatory mechanisms will be stimulated by changes as little as 3 mosmol/kg H_2O

b. False. There is a higher rate of antidiuretic hormone (ADH) secretion, conserving H_2O and lowering plasma osmolality

c. True. Maintaining perfusion pressures by volume is a greater priority, hence plasma becomes hypo-osmolar

d. False. Alcohol decreases ADH release, distorting osmoregulation

e. False. Clearance is greater than urine flow since a volume of osmotically free H_2O is reabsorbed

96. Body fluid volume and sodium

a. True. Na^+ content will affect ECF and ECV because it is the principal extracellular cation

b. True. This results from decreased systemic BP and renal perfusion pressure secondary to low ECV

c. False. Nerve endings containing catecholamines act directly on tubule cells

d. False. It is an important vasodilator, but responds to reduced ECV by mediating renin release

e. True. In addition, it inhibits Na^+/K^+-ATPase

97. Regulation of body fluid pH

a. False. Normal pH range is 7.35–7.45, equivalent to 45 nmol/L > [H^+] > 35 nmol/L

b. False. They can only combine with or release H^+

c. False. pH is inversely proportional to [H^+]

d. False. HCO_3^- conversion products only are reabsorbed, which then reform intracellularly

e. True. There is ↑ glutamine production with renal NH_4^+ synthesis and excretion, conserving and generating HCO_3^-

98. Acid–base disorders

a. True. However, unlike correction, compensation still has derangements of PCO_2 and/or HCO_3^-

b. False. Correction is achieved through the system at fault, since in this case there is compensation

c. False. Work of breathing is affected, with decreased CO_2 elimination and metabolic acidosis

d. True. There is initial hyperventilation, then metabolic acidosis by addition of acid

e. True. Na^+ reabsorption will occur preferentially by exchange with H^+ rather than K^+ in the collecting duct

99. Regulation of calcium (Ca^{2+}) and phosphate (PO_4^{3-})

a. True.   It constitutes <50% of plasma Ca^{2+}, which makes up 1% of total body Ca^{2+}

b. True.   Transport is largely passive and paracellular

c. False. Loop diuretics decrease reabsorption, unlike thiazides, which enhance Na^+/Ca^{2+} exchange

d. True. H^+ secretion in the proximal tubule has this effect

e. False. All these factors promote renal PO_4^{3-} excretion unlike insulin, for which the statement is true

100. Regulation of magnesium (Mg^{2+}) and potassium (K^+)

a. False. This is true for the loop of Henle (LoH) but 80% reabsorption occurs in the proximal tubule (PT), passively aided by H_2O

b. True.   Aldosterone activates apical K^+ channels, promoting its secretion and subsequent excretion

c. False. It is the second most important intracellular cation

d. True.   They are also accompanied by disturbances in K^+ or Ca^{2+}

e. False. PT reabsorbs 15%, whereas the thick ascending limb of LoH is the main site for reabsorption

1a. Cyanide reversibly inhibits cellular oxidizing enzymes, which contain ferric iron such as cytochrome oxidase. This prevents the transfer of electrons in the electron transfer chain, which results in hypoxia.

1b. The antidotes are the following:
- Dicobalt edentate: is the treatment of choice for definitive cyanide poisoning.
- Thiosulphate: this works by causing an osmotic diuresis to help rid the cyanide.
- Nitrites: these convert a proportion of the body's haemoglobin to methaemoglobin, which contains ferric as opposed to the ferrous form of iron.
- Oxygen: this has a synergistic antidotal action when used in combination with sodium thiosulphate and sodium nitrite.

1c. If the diagnosis is certain:
- 100% oxygen.
- Dicobalt edentate.
- If no response give sodium nitrite followed by sodium thiosulphate.
- Cardiorespiratory support.

2a. The average adult contains 5 L of blood (8% of total body weight), of which 55% is plasma and 45% cellular elements. Plasma is a watery ground substance containing dissolved solutes and proteins in suspension. These proteins cause the osmolality of the plasma to be higher than those of interstitial fluid and intracellular fluid by 1 mosmol/kgH$_2$O. Cellular elements include erythrocytes, leukocytes and platelets.

2b. Haemostasis involves three mechanisms:
- Vasoconstriction: limits blood loss.
- Formation of a platelet plug: involves platelet adhesion, activation and aggregation.
- Coagulation: involves the clotting cascade with intrinsic and extrinsic pathways.

2c. Haemoglobin is already 97–100% saturated with O$_2$ at normal arterial values (100 mmHg) with little elevation by hyperventilation. This is because increasing the PO$_2$ over 70 mmHg by large amounts only increases saturation by a little, which corresponds to the plateau or flat portion of the graph. There is a risk of alkalosis if he continues to hyperventilate with the effect of causing a leftwards shift of the curve, permitting easier O$_2$ uptake and binding at any PO$_2$.

3a. Pain is regulated in two centres:
- Peripheral regulation: activity in the low-threshold mechanoreceptors can inhibit the spinothalamic nerve impulses, e.g. rubbing the area of pain.
- Central regulation: the central nervous system has its own pain control areas:
  - periaqueductal grey area of the midbrain and pons
  - raphe magnus nucleus: located in the lower region of the pons and superior medulla
  - the reticular formation regions in the dorsal horn; nucleus reticularis paragigantocellularis and locus coeruleus.
  - Stimulation of opiate and 5-HT receptors in these areas can inhibit the pain pathway and cause analgesia.

3b. Opioids mimic endogenous opioid peptides and act on μ- and γ-morphine receptors in the spinal cord, brainstem and hypothalamus to block pain signals.

3c. The World Health Organization (WHO) has produced an analgesic ladder to be used as a guide for prescribing analgesics. Patients who do not experience pain relief on one step of the analgesic ladder should progress to the next step. Oral analgesic drugs are usually the first-line treatment for treating pain. The choice of analgesic should be based on the severity of the pain rather than the stage of the patient's disease. Analgesics should be taken regularly and the dose gradually increased, as necessary:
- Step 1: non-opioid analgesic, e.g. paracetamol, NSAIDs. Adjuvant drugs to enhance analgesic efficacy, treat concurrent symptoms that exacerbate pain, and provide independent analgesic activity for specific types of pain may be used at any step (e.g. NSAIDs).
- Step 2: if the pain is not controlled then a mild opioid such as codeine should be added. Combinations of opiods with paracetemol include cocodamol, coproxamol.
- Step 3: this is used when higher doses of opioid are necessary. An example is morphine. The dose of the stronger opioid can be titrated upwards, according to the patient's pain, as there is no ceiling dose for morphine.

3d. He could be put on a NSAID such as ibuprofen. However, because of the fact that he has had gastric problems in the past, and because NSAIDs can exacerbate stomach ulcers, it would be more appropriate to give him something like codydramol.

4a. The aetiology of MND is unknown, but about 7% of cases show a family link. The pathology involves thinning of the anterior roots of the spinal cord, loss of neurons in anterior horns, cranial nerve nuclei, and the motor cortex.

4b. The presentation of MND depends on the particular pathways involved: features usually develop gradually. The initial complaint can be a weakening hand grip. Parts of the body affected include:
- Hands: wasting, weakness, fasciculation.
- Bulbar symptoms: dysarthria and dysphagia.
- Lower limb: weakness and spasticity.
- Breathing difficulty.

4c. Electromyography (EMG) shows a typical pattern of severe chronic denervation. These manifest themselves on the EMG as spontaneous fibrillation potentials, reduced number of spikes on activity due to the reduced number of motor neurons and an increased duration and amplitude of action potentials of the remaining units.

5a. Parkinson's disease. This is characterized by an insidious onset with slowing of emotional and voluntary movement, muscular rigidity, postural abnormality and tremor

5b. This comprises five structures on either side of the brain: caudate nucleus, putamen, globus pallidus, subthalamic nucleus and substantia nigra.

5c. The basal ganglia have the following functions:
- Regulate complex/skilled patterns of motor activity in association with the corticospinal system, e.g. writing numbers sequentially, suturing a wound during surgery.
- Cognitive control of motor pattern sequences, e.g. seeing and recognizing that something is imminently dangerous and running away from it.
- Control of the timing and magnitude of movements, e.g. writing the word 'physiology' slowly or quickly. Or even writing it in large and small type. For both examples the proportion of the letters will be the same.
- Saccadic eye movements via connections to the frontal eye fields. Saccades are rapid intermittent eye movement, such as that which occurs when the eyes fix on one point after another in the visual field. The purpose of this type of movement is to move the eyes as quickly as possible, so that the point of interest will be centred on the fovea.

5d. Substantia nigra. Neurons from here release dopamine which helps regulate subconscious activity in the muscles. An absence of these cells leads to Parkinson's disease.

5e. The gait is known as festinating because the patient takes very small and rapid steps for a short distance. The writing is called micrographia. The writing starts off as normal but then the size of the letters decreases.

5f. Drug treatment focuses on increasing the production of dopamine in the brain, slowing down its destruction, provision of additional neurotransmitter, and/or using agonists to dopamine receptors.

Levodopa, amino acid precursor to dopamine, crosses the blood-brain barrier (b-bb), (dopamine does not). Transfer across b-bb is increased when levodopa is combined with carbidopa or benserazide (reduce the peripheral conversion to dopamine).

Dopamine agonists (e.g. apomorphine, bromocryptine, cabergoline, pramipexole, ropinirole,) can be used alone or as adjunct therapy to carbidopa-levodopa therapy

Monoamine oxidase inhibitors (e.g. selegiline) and *catechol-O-methyltransferase inhibitors* slow the destruction of naturally occurring dopamine and that produced from carbidopa-levodopa therapy.

Co-enzyme Q10 – involved in the mitochondrial electron transfer system to produce energy – concentration is low in patients with the disease; over-the-counter supplements may slow the progression of early stage of the disease.

Anticholinergics – to control the tremor, were the main treatment before carbidopa-levodopa therapy but the benefits can be offset by the side effects.

6a. Myasthenia gravis.

6b. Myasthenia gravis is an autoimmune disease in which antibodies result in a loss of muscle acetylcholine (ACh) receptors (AChRs).

6c. The clinical features of myasthenia gravis are:
- External ocular muscles: are usually the first muscles affected; the patient complains of diplopia.
- Limb weakness: characteristically, worsened by exercise. The shoulder girdle is commonly affected with the patient finding it difficulty to raise their arms above their head or brushing their hair.
- Bulbar: loss of facial expression; the patient appears unable to smile.
- Respiratory: shortness of breath.

6d. (1) Nerve impulse reaches the neuromuscular junction; (2) voltage-gated Ca^{2+} channels open and calcium diffuses into the terminal bouton; (3) the Ca^{2+} attracts the acetylcholine (ACh) vesicles to the neural membrane; (4) the vesicles fuse with the neural membrane and empty their ACh into the synaptic cleft; (5) when two of the ACh molecules bind to the ACh receptor (AChR) it opens; (6) Na^+ enters the channel and creates a potential change at the muscle membrane, called the end-plate potential (EPP); (7) the EPP initiates an action potential at the muscle membrane and causes muscle contraction; (8) the ACh in the synaptic space continues to activate the AChR, so when muscular contraction is no longer required the ACh must be removed.

6e. Oral anticholinesterase medication, e.g. pyridostigmine or neostigmine; thymectomy required if there is a thymoma because of the risk of local infiltration; immunosuppression with corticosteroids ± cytotoxic agents is also highly effective in inducing remission of disease.

7a. **Cardiac output (CO) = Stroke volume × Heart rate**
   - Stroke volume (SV) = the volume of blood ejected in one ventricular contraction; heart rate (HR) = number of ventricular contractions in one minute; total peripheral resistance (TPR) = resistance to blood flow in the circulatory system. Therefore:

   Mean arterial blood pressure (MABP) = CO × TPR

7b. A fall in BP causes the release of renin from the juxtaglomerular cells of the kidney. Renin converts liver α_2 globulin, angiotensinogen to angiotensin I. In the lungs, angiotensin-converting enzyme (ACE) converts angiotensin I to angiotensin II, which: (1) stimulates the adrenal cortex to secrete aldosterone which increases salt and water retention in the kidneys; (2) acts on the vascular smooth muscle and causes vasoconstriction; (3) increases norepinephrine (noradrenaline) release; and (4) increases cardiac contractility.

7c. The cardiac cycle can be divided into three stages:
   1. Atrial and ventricular filling (diastole): all the chambers are relaxed and there is passive filling of both the atria and ventricles.
   2. Atrial systole (0.1 s): these contract and add about 25 mL to the relaxed ventricles. The volume in the ventricle at this point is known as the end diastolic volume (EDV). The ventricles are relaxed at this stage.
   3. Ventricular systole (0.3 s): both the ventricles contract. The pressure increases inside the ventricle and shuts the atrioventricular valves. For about 0.06 s both the semilunar (SL) and AV valves are shut; this is known as isovolumetric contraction – cardiac muscle fibres are contracting but not yet shortening.

   When the ventricular pressure exceeds aortic and pulmonary pressure, both the SL valves open (ventricular ejection) and blood is expelled into the aorta and pulmonary artery. The amount of blood ejected per ventricle is called the stroke volume (70 mL). The volume remaining in the ventricles post ventricular systole is called the end-systolic volume.

8a. Spontaneous depolarization in the sinoatrial (SA) node results in a pacemaker potential. The SA node is located in the right atrium near the entrance of the superior vena cava. When this reaches threshold it causes an action potential (AP) in the myocytes. The AP propagates to both atria via the gap junctions. This causes the atria to contract in synchrony and pump blood into the ventricles.

The AP travels along the cardiac muscle and come to the atrioventricular (AV) node, which is in the septum sandwiched by the two atria. The AP then enters the bundle of His, which carries it from the atria to the ventricles. It then enters the left and right bundle branches, which extend through the interventricular septum and carry the AP at high velocity toward the apex of the heart. The AP eventually reaches the Purkinje fibres, a network of fine fibres extending through the ventricular walls. These cause the ventricles to contract, pushing blood in the right ventricle into the pulmonary artery and blood from the left ventricle into the aorta via the semilunar valves.

8b. **The irregularly irregular pulse is classically associated with atrial fibrillation. This is where there is abnormal electrical activity between the SA and AV nodes, indictaing erratic and irregular contraction of myocytes not coordinated atrial contraction. Synchrony between atria and ventricular contraction is lacking. One of the main complications with this is that clots in the arterial tree can become dislodged and cause occlusion in the cerebral circulation. This would cause an embolic stroke.**

8c. Warfarin affects both the intrinsic and extrinsic clotting pathways. It is a vitamin K antagonist; blocking vitamin K-dependent γ carboxylation of glutamate residues, resulting in the production of modified factors VII, IX, X and prothrombin (γ carboxylation confers essential Ca^{2+} binding properties on these factors). Patients are put onto warfarin to minimise the risk of clots forming.

9a. Surfactant is mainly phospholipids, notably dipalmitoyl phosphatidodylcholine (DPPD) and proteins whose properties, through decreasing surface tension, are important to the prevention of pulmonary oedema, alveolar stability and hysteresis.

9b. A decrease in surface area of the membrane (e.g. emphysema) or decrease in partial pressure difference of gas between alveoli and capillary may decrease the rate of diffusion. Decreased solubility and increased molecular weight of the substance would also have the same effect.

9c. Other factors favouring pulmonary oedema besides those discussed below include decreased capillary oncotic pressure or interstitial fluid hydrostatic pressure:
   - Increased capillary hydrostatic pressure that may result from excess intravenous fluid administration or outflow obstruction favours transudation.
   - Decreased reflection coefficient (membrane impermeability to solute particles) such as occurs with septic shock, causes the capillary endothelium to become more leaky, allowing solutes out, which osmotically drag water with them.

10a. Turbulent airflow occurs in the larger airways. This is characterized by breakdown in streamlines and eddy currents. It may be noisy and erosive and is more likely with high flow rates, high gas density and in wide-bore vessels.

10b. Cigarette smoke stimulates irritant receptors in the upper and lower airways, which send afferents via the vagus nerve. Responses include bronchoconstriction mediated by the vagus nerve and cough.

10c. Airway obstruction leads to trapping of air behind closed airways. Hence, there is increase in residual volume (RV), functional residual capacity (FRC) and total lung capacity (TLC). FEV_1 and FVC would be below 80% of predicted values for age, gender and height. FEV_1 : FVC ratio < 80% indicating an obstructive lung disease.

11a. V/Q = ∞ since Q = 0; no gas exchange will occur between these alveoli and their capillary:
- P_AO_2 is higher than normal because there is no blood to take away O_2.
- P_ACO_2 is lower than normal because there is no blood releasing CO_2.

11b. Peripheral chemoreceptors: most importantly, carotid body type 1 (glomus) cells are sensitive to hypoxia. Their response increases significantly when PaO_2 falls below 50–60 mmHg at normal $PaCO_2$.

11c. Increased hydrostatic pressure decreases pulmonary vascular resistance by two mechanisms:
1. Recruitment: conductance of blood through previously unused capillaries
2. Distension: dilatation of individual capillary segments with increase in radius.

12a. Appetite reflects the desire to eat certain foods and can be independent of hunger. Alimentary feedback, limiting oral intake, can occur through:
- Oral activity.
- Stomach/duodenal distension.
- Chemical content of food stimulating release of gastrointestinal hormones and humoral factors.

12b. Salivary oral defence is achieved by:
- Immunoglobulins and bactericidal substances.
- Proteins: which form an acquired pedicle.
- High $[HCO_3^-]$: protects against caries.
- Ca^{2+} and PO_4^{3-}: which promote mineralization.

12c. Decreased muscle tone of the fundus and upper stomach walls (receptive relaxation) caused by food entering the lower oesophageal sphincter (receptive reflex mediated by the vagus nerve) permits storage by the fundus. Approximately 1–1.5 L of food may be stored in this area which distends for up to 1 hour.

13a. Staggered peristaltic waves progressively move solid food towards the antrum, which has a thicker muscle layer enabling a much higher pressure and thus grounding of small boluses. Pyloric contraction during antral peristalsis causes retropulsion (reflux to more proximal stomach regions) resulting in churning of food with secretions and the breakdown of solid foods.

13b. Bile is an aqueous solution composed of bile salts, electrolytes, bile pigments, phospholipids and cholesterol.

13c. Pancreatic phases of secretion are:
- Cephalic: with anticipation of food.
- Gastric: with initial distension of food entering the stomach.
- Intestinal: with the entrance of gastric chyme into the duodenum.

13d. Distension, irritation or excitation of upper small intestine and stomach may stimulate vomiting centres which receive input from higher centres and chemoreceptor trigger zones.

14a. The colon is a tight epithelium with lateral intercellular spaces into which Na^+ is actively pumped, generating a region of high osmolality. This osmotic gradient (known as the standing osmotic gradient) favours H_2O movement from the lumen through the cell into the intercellular spaces; hydrostatic pressure increases and isosmotic fluid moves from intercellular space into capillary plasma

14b. Parasympathetic stimulation and either distension or mechanical irritation of the colon favour mucus secretion.

14c. Mass movement refers to the intense contraction between the transverse and sigmoid portions, propelling faeces towards the anus; faeces filling the rectum initiate reflex relaxation of the internal anal sphincter and reflex constriction of the external sphincter causing the urge to defecate.

15a. The haemopoietic hormone, erythropoietin is produced by interstitial cells in the cortex and outer medulla in response to decreasing PaO_2 (hypoxia). Loss of renal function leads to reduced erythropoietin release, reduced stimulation of haematopoietic stem cells in bone marrow and reduced RBCs (reduced [haemoglobin]).

15b. Plasma creatinine (normal range 44–133 μmol/L) is derived from muscle protein creatine. Rate of creatinine production at rest will be constant. Since creatinine is freely filtered, is not reabsorbed by the tubules and is not secreted in significant amounts, its excretion (derived primarily from filtration) will equal rate of production. If GFR decreases there is a transient decrease in creatinine excretion, plasma creatinine increases and excretion returns to the original value. Thus raised plasma [creatinine] indicates that GFR is reduced. Factors to be considered include: muscle mass; large muscle mass, large plasma [creatinine]: binding to plasma protein, (normally ~5%); changes in binding lead to changes in amount filtered: amount secreted, the greater the amount secreted the greater contribution to excretion and the less reliable is creatinine clearance as an assessment of GFR.

15c. Reduced GFR of renal failure leads to reduced PO_4^{3-} excretion. Continued PO_4^{3-} absorption from the GI tract leads to increased plasma $[PO_4^{3-}]$. This increased plasma $[PO_4^{3-}]$ complexes with Ca^{2+} and is deposited in bone thereby reducing plasma $[Ca^{2+}]$. Increased plasma $[PO_4^{3-}]$ reduces production of calcitriol thereby reducing Ca^{2+} absorption by GI tract and adding to the lower plasma Ca^{2+}. PTH is released in response to low plasma Ca^{2+} but in severe renal failure may not be able to increase PO_4^{3-} excretion to lower plasma concentration sufficiently to prevent deposition of Ca^{2+} in bone and non skeletal sites (blood vessels, heart – metastatic calcification).

16a. Virtually all filtered glucose is reabsorbed by the proximal tubules by Tm-dependent mechanisms. Apical transport is via a secondary active Na^+/glucose (SGLT1, SGLT2) and via facilitated diffusion across the basolateral membrane (GLUT1, GLUT2).

16b. Glycosuria will be seen in any condition in which the filtered load of glucose exceeds the Tm for the reabsorptive mechanisms (~2 mmol/min) through either plasma glucose reaching the threshold value (10 mmol/L: diabetes mellitus; Type I & II) at normal GFR or when GFR is elevated and plasma glucose is raised but <threshold values (gestational diabetes)

16c. GFR is the volume of fluid that is separated from plasma proteins and the cellular elements of blood in unit time.
Normal values are 85–125 mL/min for women & 95–140 mL/min for men. GFR declines with age

Measurement of GFR requires measurement of the clearance of a substance that is freely filtered (not bound to plasma proteins) and passes through the tubule without modification (not reabsorbed, not secreted, and not metabolised by the kidney).

Amount filtered = Amount excreted
$GFR \times P_A = U_A \times V$ where P & U are, respectively, plasma and urine concentration of substance A, and V is urine flow rate.
$C_A (GFR) = \dfrac{U_A \times V}{P_A}$ mL/min where C_A is the clearance of substance A

Clearance of a substance can be defined as the volume of plasma that needs to be filtered to account for the amount of solute appearing in the urine in unit time. Inulin, a polysaccharide of fructose, fulfils the criteria listed above; its clearance gives a reliable measurement of GFR.

16d. Filtering membrane is composed of three distinct parts – the capillary endothelial cells, basement membrane, epithelia cells of Bowman's capsule Capillary endothelial cells have *pores (fenestrations)* of diameter 60–70 nm which serve to limit passage of the cellular elements of blood. The basement membrane has of three layers (*lamina rara interna, lamina densa, lamina rara externa*), consists of proteoglycans with strong –ve charge,

The epithelium consists of highly specialised cells, *podoctes* which are attached to the basement membrane by foot-like processes known as *pedicels*. These are separated by filtration slits which helps to determine the size of molecule that can traverse the membrane. Also, as with the case of the basement membrane, charge of the molecule is also important - the podocytes and pedicals are negatively charged.

17a. Normal pH range 7.38 – 7.42
Calculated using the Hendersen-Hasselbalch equation

$$pH = pK + \log_{10} \frac{[HCO_3^-]}{(S \times PaCO_2)}$$

pK = association constant for the equation:

$$CO_2 + H_2O \leftrightarrow H_2CO_3 \leftrightarrow H^+ + HCO_3^-$$

HCO_3^- plasma $[HCO_3^-]$
$PaCO_2$ partial pressure of CO_2 in arterial plasma
S   solubility coefficient for CO_2 in plasma

17b. pH is very acid,
$PaCO_2$, mid normal range (37–43 mmHg); PaO_2, at the top of the normal range
HCO_3^- –8 mmol/L This is referred to as the anion gap calculated as $([Na^+] + [K^+]) – ([Cl^-] + [HCO_3^-])$. In some cases $[K^+]$ is ignored.

It can be used to classify the type of metabolic acidosis detected. High value > –11 mmol/L reflects loss of HCO_3^- through buffering that has not been replaced by Cl^- but by another unmeasured anion (depending on the acid causing the acidosis)

Normal range –3 to –11 mmol/L; This is interpreted as loss of HCO_3^- that has been replaced by Cl^- to maintain electrical neutrality.

In this patient the anion gap is within the normal range hence the acid – base disorder is through excessive loss of HCO_3^- in diarrhoea – metabolic acidosis without compensation from reduced $PaCO_2$.

Ketoacidosis: e.g. unncontrolled diabetes mellitus; overproduction of non-volatile acids, e.g. β-hydroxybutyric acid Anion gap > 11 mmol/L

Renal failure: inability to excrete sufficient of the daily metabolic acid production. Anion gap < 11 mmol/L
Lactic acidosis: e.g. shock Anion gap > 11 mmol/L

17c. $\text{pH}\,\alpha\;\;\downarrow\downarrow\downarrow\dfrac{HCO_3^-}{PaCO_2}$

1. Respiratory compensation to lower $PaCO_2$

   $CO_2 + H_2O \leftarrow H_2CO_3 \leftarrow H^+ + HCO_3^-$; this also reduces plasma HCO_3^-

   $\text{pH}\,\alpha\;\;\dfrac{\downarrow\downarrow\downarrow HCO_3^-}{\downarrow\downarrow PaCO_2}$

2. Renal compensation through the excretion of protons and reclamation and production of HCO_3^- returning plasma $[HCO_3^-]$ to normal

   $\text{pH}\,\alpha\;\;\dfrac{\uparrow\uparrow\uparrow HCO_3^-}{\downarrow\downarrow PaCO_2}$

   Final correction needs to be through respiratory changes returning $PaCO_2$ to normal.

1. **The cell 1**
    1. N
    2. F
    3. I
    4. G
    5. J
    6. D
    7. A
    8. J
    9. L
    10. M

2. **The cell 2**
    1. K
    2. J
    3. O
    4. E
    5. M
    6. N
    7. F
    8. H
    9. D
    10. P

3. **Blood & Body Fluids 1**
    1. M
    2. J
    3. L
    4. G
    5. N
    6. K
    7. H
    8. Q
    9. B
    10. F

4. **Blood & Body Fluids 2**
    1. H
    2. I
    3. M
    4. J
    5. N
    6. C
    7. G
    8. I
    9. D
    10. K

5. **Nervous system 1**
    1. A
    2. J
    3. H
    4. C
    5. B
    6. F
    7. I
    8. M
    9. M
    10. D

6. **Nervous system 2**
    1. B
    2. D
    3. A
    4. E
    5. H
    6. H
    7. I
    8. K
    9. L
    10. J

7. **Nervous system 3**
    1. C
    2. M
    3. I
    4. E
    5. L
    6. H
    7. A
    8. J
    9. N
    10. G

8. **Musculoskeletal 1**
    1. G
    2. K
    3. B
    4. M
    5. H
    6. A
    7. O
    8. I
    9. M
    10. H

9. **Musculoskeletal 2**
    1. O
    2. F
    3. B
    4. N
    5. N
    6. E
    7. A
    8. K
    9. H
    10. D

10. General cardiovascular 1
1. C
2. G
3. L
4. E
5. B
6. N
7. F
8. D
9. I
10. H

11. General cardiovascular 2
1. G
2. N
3. C
4. B
5. K
6. J
7. H
8. O
9. E
10. E

12. General cardiovascular 3
1. F
2. C
3. I
4. G
5. M
6. B
7. N
8. K
9. J
10. A

13. Respiratory system 1
1. F
2. K
3. J
4. B
5. M
6. L
7. H
8. P
9. G
10. J

14. Respiratory system 2
1. I
2. M
3. H
4. N
5. M
6. O
7. B
8. A
9. J
10. P

15. Respiratory system 3
1. O
2. N
3. E
4. N
5. Q
6. D
7. J
8. L
9. J
10. C

16. Respiratory system 4
1. M
2. I
3. F
4. I
5. E
6. C
7. K
8. J
9. I
10. I

17. Gastrointestinal system 1
1. L
2. H
3. O
4. O
5. P
6. B
7. G
8. A
9. K
10. I

18. Gastrointestinal system 2
1. E
2. K
3. L
4. I
5. B
6. M
7. M
8. B
9. O
10. H

19. Gastrointestinal system 3
1. C
2. M
3. J
4. B
5. L
6. N
7. F
8. H
9. K
10. I

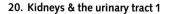

20. Kidneys & the urinary tract 1

1. D
2. G
3. N
4. A
5. L
6. N
7. O
8. A
9. F
10. O

21. Kidneys & the urinary tract 2

1. D
2. E
3. L
4. E
5. A
6. F
7. H
8. L
9. G
10. A

22. Kidneys & the urinary tract 3

1. D
2. L
3. E
4. H
5. E
6. C
7. J
8. G
9. O
10. N

23. Kidneys & urinary tract 4

1. G
2. M
3. E
4. C
5. A
6. J
7. J
8. C
9. B
10. E

A

A band, 57
abdominal muscles, 101
absorption
  intestinal, 152–6
  oral, 139
acclimatization, high altitude, 130–1
accommodation, 49
ACE see angiotensin-converting enzyme
acetyl-CoA, 7–8
acetylcholine (ACh), 39, 58–9
  exocrine pancreas, 150
  gastric function, 143
  vascular control, 86, 91
acetylcholine receptors (AChR), 58, 59
acetylcholinesterase, 58, 59
acid–base balance, 93, 189–95
  input and output, 189
  physiological buffers, 190
  renal control, 178, 190–1
acid–base disturbances, 191–5
  compensation, 192, 193
  correction, 192
acidosis, 191
  chemoreceptors, 89, 124
  Davenport diagram, 192, 193
actin, 57, 59–60, 74
actin filaments, 59–60
action potentials (AP), 35–7
  cardiac myocytes, 75–6
  control of breathing, 122
  muscle cells, 59
  myelinated and unmyelinated fibres, 36–7
  sinoatrial node, 76
  speed of conduction, 37
  synaptic transmission, 38–9, 59
activated partial thromboplastin time (APTT), 30
active transport, 13, 18
  capillary wall, 83
  intestine, 152, 156–7, 158
  primary, 173
  renal tubule, 173, 174
  secondary, 173
acute mountain sickness, 130

adaptation, 41
adenosine diphosphate (ADP), 6, 29
adenosine monophosphate (AMP), 6
adenosine triphosphate see ATP
ADH see antidiuretic hormone
adrenal steroids, 187
adrenaline (epinephrine), 87, 89
adrenocorticotrophic hormone (ACTH), 188
aerobic exercise, 129
afferent arterioles, glomerular, 164, 171, 172
after potentials, 36
afterload, 78
airflow
  into lungs, 98–9, 100
  patterns, 105–6
airways, 94
  conducting, 94
  dynamic compression, 107
airways resistance (P_{AR}), 105–8
  factors affecting, 106–7
  measurement, 107–8
  sites, 106
  transmural pressure and, 107
albumin, 22, 167
aldosterone, 187–8
  gastrointestinal actions, 152, 158
  renal actions, 179, 183, 188, 198
alkali ingestion, 194
alkalosis, 191, 193, 194–5
altitude, high, 109, 129–31
alveolar gas equation, 130–1
alveolar pressure (P_A), 98, 99, 114
alveolar ventilation (V[*]), 96
alveoli
  blood vessels, 113
  gaseous exchange, 108–11
  stability, 104–5
  water balance, 116
amacrine cells, 49
amino acids, 39, 154
  renal handling, 177, 178
  transport, 154
amino aciduria, 177
ammonium (NH_4^+), 191, 192
amoeboid movement, 11
amphotericin, 195

amylase
  gastric, 143
  pancreatic, 149, 153
  salivary, 137, 138, 139, 153
anaemia, 166, 200
anaerobic exercise, 129
anal sphincters, 159–60
analgesia, 43–4
anatomical dead space (V_D), 95, 96
angiotensin I, 187
angiotensin II, 88, 158, 165, 187
angiotensin-converting enzyme (ACE), 88, 95, 187
  inhibitors, 185, 187
anion gap, 194
anorexia, 135
anterior chamber, 46
antidiuretic hormone (ADH; vasopressin), 183, 184–6
  cardiovascular regulation, 87, 88
  cellular actions, 178, 183, 184–5
  deficiency, 186
  drugs and, 185
  fate, 185
  K^+ regulation, 198
  osmoregulation, 186–7
  syndrome of inappropriate secretion (SIADH), 185
antiport, 14, 173
antithrombin III, 32
aortic arch baroreceptors, 88
aortic bodies, 89, 124–5
apneustic centre, 122–3
appetite, 135–7
aquaporins, 176, 184–5
arcuate arteries, 164
arterial blood gases (ABG), 192, 194
  see also $PaCO_2$; PaO_2; pH, arterial blood
arterial blood pressure, 81
arterial thrombi, 30
arteries, 71, 78
  elastic, 78
  muscular, 78
  structure, 72, 78, 79
arterioles, 71, 72, 79
arteriovenous anastomoses (AVA), 90
ascorbic acid, 154